T3-AKD-598

THE TEACHING OF MATHEMATICS from Counting to Calculus

THE TEACHING OF MATHEMATICS from Counting to Calculus

HAROLD P. FAWCETT
The Ohio State University

KENNETH B. CUMMINS
Kent State University

Charles E. Merrill Publishing Co.
A Bell & Howell Company
Columbus, Ohio

Standard Book Number: 675-09512-3

Library of Congress Catalog Card Number: 69-19768

1 2 3 4 5 6 7 8 9 10—73 72 71 70

Printed in the United States of America

PREFACE

This book is the outgrowth of continuing experiences by the authors to serve their students by helping them become involved in the process of creating mathematics new to them. It is designed for use in courses which deal with the teaching of mathematics and for study by teachers in service. For many years the authors of this text have been deeply interested in what is known today as "the discovery approach" to the teaching of mathematics. The fundamental ideas of mathematics are the result of long study and to emphasize the "finished product" at the expense of the *process* which produced it is to stifle the spirit of the subject. Although the rules and principles of mathematics are important, the *real* values of this subject reside in the processes by which these principles were developed — processes which reflect the creative power of man's intellectual genius—and wherever mathematics is taught with little emphasis on these processes we are likely to find a teacher who feasts the memory but starves the reason.

Our sole purpose is to clarify our interpretation of "the discovery approach" and to encourage its use in the mathematics classrooms of tomorrow, but not our intention or purpose to duplicate material found in books on methodology which are in current use. To assist in the achievement of this objective we have included many illustrations and student study guides consistent with this approach which may serve as suggestions to the reader as he plans his own teaching procedures. Indeed, the book includes examples and suggestions by which students can be encouraged to participate actively in developing and organizing their mathematical findings into logical structures in a spirit of exploration and investigation which can make learning truly an exciting experience.

In keeping with the manner of teaching encouraged by the authors, the teacher may wish to supplement the text with study guides to consider more deeply topics which class and teacher interest indicate might be pursued with profit. The readings for students at the close of each chapter are helpful in keeping them in contact with current thinking and experimentation, while portions of the text will be particularly helpful in serving special interests of both students and teachers.

To express appreciation to all those who have helped in one way or another to produce this book through its several years of development in classroom use — the work of part-time student typists and helpers, encouragement by staff members, students and teachers, along with points of view gained from former teachers and associates, and from speakers at professional meetings — is indeed difficult and quite impossible. In the later stages, however, thanks are due to Mrs. Julie Froble and to Miss Carol Govedich of Kent State University for their care in typing and in general clerical assistance.

Too, the encouragement of the staff of Charles E. Merrill Publishing Company and especially the timely help and advice of Mrs. Diana August, Production Editor, have been invaluable.

CONTENTS

1	The Call of Mathematics	1
2	A Survey of Methodology	17
3	Discovery Approaches	47
4	On Teaching Arithmetic	81
5	Teaching Algebra	149
6	Teaching Geometry	201
7	Teaching Advanced Topics in the Secondary School	315
8	The Teaching of Calculus	369

1

The Call of Mathematics

> The call of mathematics is. . . to our physical well-being. . . but it is also to our spiritual well-being. . . it is the call of the soul.
>
> David Eugene Smith[1]

1.1. The Continuing Call

The mathematics teacher has a singularly important role in our modern world. All fields depend upon mathematics as a tool for discovery and expansion. Your students will look to you for guidance concerning the part mathematics plays in our environment and for help in discovering the nature of mathematical tools and structures. They will depend upon you to justify the study of mathematics.

You are already aware that there are two different but interdependent and supporting facets, or faces, of mathematics and that these affect both our everyday living and our thinking and intellectual pursuits in a never-ending way. Of course, we cannot expect our students to realize this immediately. One facet of mathematics uses its method, its algorithms, and its results in our homes, stores, and industries, and as a pattern in our thinking in disciplines other than mathematics. The other facet investigates the disci-

---

[1] David Eugene Smith, *The Poetry of Mathematics and Other Essays* (New York: Scripta Mathematica, 1934), pp. 14–27.

pline of mathematics itself; investigates and scrutinizes its fruits to see if conclusions develop in a valid manner from its beginnings. These two faces invite us to let mathematics serve as well as beckoning us to study its very foundations; one face looks outward, the other, inward. These two facets comprise "the call of mathematics."

It is the challenging responsibility of the teacher to cultivate this call of mathematics in the classroom and to emphasize its two sides, or faces. The *outward* facet furnishes and devises mathematics to serve the practical needs of society and other disciplines. It reaches out to give greater control of our environment through a language and a symbolism by which man has been able to harness the mighty forces of nature. This facet came to the rescue of ancient peoples attempting to relocate boundaries and is now helping modern scientists build pathways to the moon. The *inward* face examines the structure and the method of mathematics. It reaches inward to see if its own house is in order: do the postulates form a consistent system? are the methods of argument acceptable? do proposed conclusions arise validly from the assumptions? This inward-looking face may probe to see if a general principle or a structure can be perceived to solve all problems of a certain type. These two facets can be described further in this manner: one ". . . is the mathematics of Newton, Clerk Maxwell, and Einstein. It is the mathematics responsible for expressing cosmic relations of time and space in a universal code . . ." The other face is " . . . (the) internal world of ideas, ideas created by its own processes, and (it) examines the relationships between them. It is the mathematics of Gauss and George Cantor. It extends conceptual insights and emphasizes structure."[2]

Urgent and constant attention to both aspects is a must for teachers who would carry on a live, effective program in mathematics. The inward face sees if a hypothetical result fits into the logical pattern of the discipline. For example, through the examination of many parallelograms by measurement of their angles, the student may propose that the opposite angles are congruent and hence have a useful, outward-looking property. But when he seeks to see if this result can be *deduced* from previous elements in a deductive system then he is working with the inward-looking face. In algebra, the solving of one problem, such as:

> There are three consecutive integers whose sum is 15. What are the integers?

emphasizes the practical, outward-looking face of mathematics; but if this is all that is done, the teacher leaves his students with an incomplete picture

---

[2] Harold P. Fawcett, "Reflections of a Retiring Mathematics Teacher," *The Mathematics Teacher*, 57 (November, 1964), 450–6.

of mathematics—incomplete because the inward-looking face is ignored. Let him at some time generalize the problem and emphasize the structure essential in the solution of *all* such problems:

> There are three consecutive integers whose sum is S. What are the integers?

The solution of this problem calls for insight into the structure of the process and provides a solution, $N = (S - 3)/3$, for all such problems. In the inward-looking facet it is interesting and meaningful to investigate the role of each part of the problem in the expression for the solution.

In the following exercises, we shall consider further these two aspects of the call of mathematics.

Exercises

1. Consider the problem: There are three consecutive integers such that the sum of four times the first plus twice the second plus the third is 53. What are the numbers?
   Solve this problem. Then reword it so it is more general and rework the generalized problem so the structure of the process is revealed. You might extend its generalization in several directions. In all such cases we are probing into mathematical structure.
2. One general problem to which that in No. 1 can be extended is "The sum of k consecutive integers is S. What are the integers?" Show the solution to this problem. Note the role of each part of the problem in the expression which is the solution.
3. Here is a problem on coins: A certain collection consists of 36 dimes and quarters. The value of the collection is \$4.20. How many quarters and how many dimes are there? Work this problem. Now formulate a new problem and work it so the structure of the solution and process are exhibited.
4. One way to tell if text writers of algebra are emphasizing both aspects of mathematics is to examine lists of exercises in their texts. Examine an algebra text on consecutive numbers and on coin problems to see if there are exercises which bring out both faces. Experience in consulting texts on various items helps develop a professional point of view on curricula and methodology. Summarize your findings.
5. In coordinate geometry one can determine easily the length of the line segment determined by the points $(-1, 2)$ and $(4, 3)$. An emphasis on the inward-looking face of mathematics will result in what general relation?
6. In calculus one looks outward when he computes the derivative of the function $\{x, x^3\}$ to be the function $\{x, 3x^2\}$. The inward-looking face reveals that the derivative of $\{x, x^n\}$ is the function $\{x, nx^{n-1}\}$. You might review this emphasis by deducing this theorem.

7. In the National Council of Teachers of Mathematics Third Yearbook, Chapter 2 is entitled "Mathematics in the Training for Citizenship."[3] Note D. E. Smith's discussion on the "reach of mathematics." How does he emphasize the "outward reach?"

8. In the essay mentioned in No. 7 above are suggested seven reasons for believing that mathematics can help develop good citizenship. Write those and be able to elaborate on them. Consider carefully how they are related to the two faces of mathematics.

9. David Eugene Smith[4] in his essay, "The Call of Mathematics," says that

   > The call of mathematics is, then, to our physical well-being, and this is always recognized, but it is also to our spiritual well-being, and this we must not fail to recognize if our labors are not to be in vain.

   Referring to this second "call" Smith says, ". . . it is the call of the soul, precisely as in the case of music, of painting, and of other fine arts, or of science, or of letters."
   Comment on how Smith's second call might be the "call of the soul." How can this second aim in teaching mathematics be translated into classroom accomplishment? You might read this and other essays in *The Poetry of Mathematics and Other Essays.*

10. As teachers we should use every opportunity to emphasize both aspects of mathematics. Quotations which can be examined and placed at appropriate locations in the classroom help to create a picture of mathematics as ever growing and the subject of thought of some of the best minds. From your reading, begin to assemble a list of quotations on mathematics and on the teaching of mathematics. The yearbooks referred to above will be most helpful in developing such a list.

1.2. Helping Students to Feel This Call

It is not enough to tell students that mathematics plays a vital role in our civilization. Classroom activities must be planned or engineered to help them experience this role. "Bringing mathematics to life" may seem an empty phrase, but the fact remains that the teacher should give careful thought to the contention that bringing mathematics to life in the classroom may be far more important than following any rigid curriculum. Well-considered plans are necessary for the effective teaching of mathematics and

---

[3] National Council of Teachers of Mathematics, *Selected Topics in the Teaching of Mathematics*, Third Yearbook (New York City: Bureau of Publications, Teachers College, Columbia University, 1928), p. 12.

[4] David Eugene Smith, *The Poetry of Mathematics and Other Essays, op. cit.*

they should emphasize its growing importance, both as an instrument in the service of man and as an ordered system of ideas.

Experiencing the call of mathematics is encouraged when students relive some of the great moments in mathematical history as mathematics unfolds by means of their own experimentation and investigation. The presence of the inward aspect of this call is felt as students attempt to organize and discuss critically their own conjectures and deductions; that of the outward call is emphasized when students see mathematics at work in practical situations. To help teachers cultivate in their students both aspects of this call of mathematics is one of the great aims of succeeding chapters of this book.

As one faces the reality of teaching he can do several things to awaken in his students this call of mathematics. Among these are the use of bulletin boards, the use of exhibits and models, the emphasis of mathematical viewpoints and processes in courses usually regarded as nonmathematical, and improved methodology in the classroom. Opportunity in nonmathematical fields may arise as a teacher works in his minor area or as he enriches his mathematics courses with topics which demonstrate its role in other fields. We shall consider a number of these procedures which sensitize the student to the fascination of this powerful science.

Bulletin Boards

Well-arranged bulletin boards can do much to help the student feel the call of mathematics. There is no reason for the walls of a mathematics classroom to remain bare and lifeless while the use of charts, exhibits, pictures, and other items of mathematical interest can help to create a highly desirable atmosphere for exploring mathematic's magic and mystery.

Donavon A. Johnson[5] gives many good suggestions concerning the effective use of bulletin boards. Among these are attractiveness, unity of theme, simplicity and clearness, arrangement, and the use of quotations. Bulletin-board activities pay big dividends since they can incite interest in many ways and help to display the work of other students. Bulletin boards which contain manipulative items might well be considered.

Exercises

1. Plan a display of materials which you can collect or have aleady collected to show the role of mathematics in some subject-matter area or to illustrate the outward-looking face of mathematics in general. Suggest a pattern for

[5] Donovan A. Johnson, "How to Use Your Bulletin Board," *How-To Series*, 1, National Council of Teachers of Mathematics (Washington: NCTM, 1953).

the display and suggest some eye-catching title which somehow includes the observer. Sketch on paper how you wish this to appear. Use the principles outlined by Johnson.

2. Scan a newspaper or a magazine and select several pictures or articles which would be good bulletin-board material to call to the attention of your students the value of the study of mathematics.

Exhibits and Models

Various ways in which mathematics serves humanity can be emphasized through exhibits of concrete items. For example, the problem of determining the dimensions of the open box with maximum volume which can be made from a piece of cardboard of given length and width can be made the subject of a very instructive exhibit, as in Fig. 1.1. One simply mounts boxes

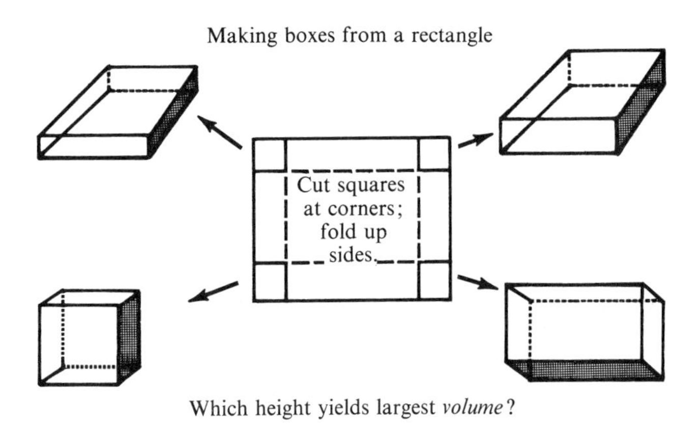

Figure 1.1

which have been made from pieces of cardboard, say 10″ by 12″, by cutting into the corners and folding up the sides. The length of cut x determines the volume:

$$V = x(12 - 2x)(10 - 2x)$$

and there is a value of x for which the volume is a maximum (see Fig. 1.2). Indeed, this problem and exhibits of this nature can become the center of discussion at almost any level of study after one knows how to find the volume of a rectangular solid. The methods of attack may be different, but this, again, shows the development of mathematics itself. The proposal of such questions through the use of models and exhibits not only helps to define problems with tactile and visual clearness, but also emphasizes continually and forcibly how mathematics responds to the needs of society. Too,

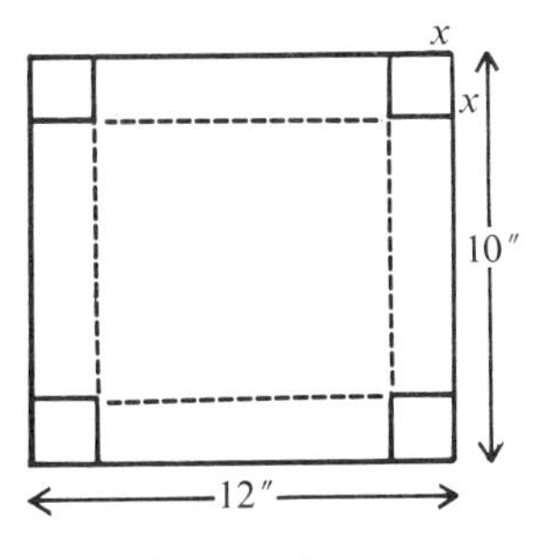

Figure 1.2

the formulation of problems from physical models helps the student to translate physical properties and conditions into mathematical language and back again, and this is a very necessary skill if the student is to appreciate deeply the outward-looking facet of mathematics.

Through models and exhibits entire areas of mathematics can be introduced, as well as the mathematical principles found or used in different disciplines. Introductory books on topology, for example, suggest bulletin-board material for showing interesting surfaces which under certain definitions are surprisingly equivalent. Mathematical expressions of scientific laws discovered in the laboratory and the ever-present mathematical principles found in nature provide rich sources to illustrate how thoroughly mathematics is present in the world. The arrangements of leaves on stems, the spirals in sea shells are interesting ways to show that there seems to be mathematics in the most unthought-of places.

Not to be overlooked are models or exhibits which move and parts of which the student can handle or operate, whether they be of practical or of theoretical interest. Simple working models are parallel rulers, proportional dividers, and the "circle square." More elaborate ones are parabolic and elliptical reflectors, the brachistochrone, models in probability, and others.

Begin now to develop a list of such models and exhibits.

Mathematical Topics in Other Subjects

The teacher who is sufficiently fortunate to handle classes in both mathematics and science has a unique opportunity to show how mathematics becomes useful as a means of summarizing behavior and as a tool in more detailed research. Many students have been won to mathematics in science courses if there is convincing and effective evidence that mathematics is useful, not merely as a computational device, but also in the study of data and in the development of general laws. The physical, biological, and social sciences are especially rich in such possibilities.

If the teacher does not have the privilege of conducting other courses in which he can develop consideration for the outward face of mathematics, it may be necessary for him to make more effort in the mathematics courses themselves to show how mathematics helps to answer questions in other disciplines. This is not difficult. Direct-proportion considerations, for example, may stem from problems on bending, and one may consider at the same time certain topics in strength of materials; the graphing of inequalities may have basic origins in situations which are related to linear programming. Although the use of mathematics in the natural setting of the science is most convincing relative to its amazing power and utility, many a teacher can effectively widen the horizons of his students in the mathematics classroom to help the students see mathematics at work in the world. There is no end to such opportunities in the realm of science and related fields. Also, there is room here for much interdiscipline teacher communication and for professional friendship, with opportunity to develop healthy intellectual attitudes toward the relation of mathematics to other fields. There is a good chance for team teaching and for cooperative planning on projects, papers, and reports.

In *chemistry* and *physics*, for example, one is often interested in relations between two varying quantities. One may seek to learn how the stretch of a spring varies with the load placed upon it, or how the illumination on a screen caused by a light source of given intensity varies with the distance away from the source, or how the viscosity of a liquid varies with temperature. All such studies require mathematics. Mathematics can help to translate raw data into tentative hypotheses and, in doing this, the student can gain much insight into the outward call of mathematics and see mathematics at work in the natural setting of the science. Some illustrations of helping the student to gain this insight will be presented in the chapter on the teaching of algebra.

Opportunities for cooperative work with the *biological sciences* are numerous also. The outreach of mathematics to areas related to biology ranges from the most elementary arithmetic in, let us say, the percent gain of weight of seeds due to moisture absorption, to studies in symmetry, spirals, differential equations expressing growth, statistics, and other biomathematical disciplines.

The teacher who is interested in *speech*, or who has avenues through which the speech teacher is willing to help, has interesting opportunities to further the cause of mathematics. For example, to help the student realize that the time required for a sound to become inaudible can be governed by the materials used in the room emphasizes again the outward face of mathematics. The knowledge of the design and use of sound reflectors and absorbers, and of mathematico-acoustical problems in general, helps the student to feel the significance of mathematics in his world.

The mathematics teacher who can work with the teacher of *history* has a unique opportunity to help emphasize both areas of mathematics. Emphasis on mathematics, including early mathematical instruments and concepts, can do much to help the student perceive how man has groped to develop mathematics to help him answer some of the pressing questions of his day. Also, for the mathematics-interested history teacher there are some "musts" among abaci, numeral systems, and older methods of computation. As the inward-looking face is emphasized, discussion, at a level appropriate for the group, of some of the crises in mathematics (the paradoxes of Zeno, the irrationality of $\sqrt{2}$, the parallel postulate, non-Euclidean geometry, non-Aristotelian logic, and others) can be a fruitful and far-reaching contribution to enlarging the mathematical horizons of all students. We would hope that through such a combination of interests the teacher could bring his students to sense a rise in mathematical thought, to feel that mathematics has changed history, and, of course, to know that history is changing mathematics.

The *social science* teacher who is interested in mathematics may well convince his students that the time is past when social studies is considered devoid of mathematics. Contrary to this once-popular conception, social studies is dependent on mathematics more than ever before for tools with which to work and for a way to give social studies a deductive character. Mathematics is responding too, by developing new concepts and skills. Under guidance of the mathematics-interested teacher, the students might undertake work in index numbers, graphs, percent increase or decrease, and descriptive statistics. An introduction to elementary ideas in linear programming is not out of place in suitable social science subjects; even now this topic is receiving attention among high-school teachers.[6] In geography portions of social studies there is the fascinating problem of constructing flat maps of our nearly spherical earth. The mathematics of a south polar gnomonic projection is not beyond the ability of students who are acquainted with the tangent function of first-year algebra. Through this they can be introduced to the entire field of cartography.[7]

The *language* teacher also has a wealth of material to help emphasize and to keep the student aware of the role of mathematics in the lives of people. Languages often shed light on the mathematical customs of people. We think, at the moment, of such words as *score*, *stock*, and *bankrupt*, which tell something about early financial practices in England, and of the words *minutes*, *seconds*, *primes*, which come from *partes minutae primae* and *partes minutae secundae*, from the Babylonians through Latin. Too, the

---

[6]Donavon Lichtenberg and Marilyn Zweng, "Linear Programming Problems for First-Year Algebra," *The Mathematics Teacher*, 53 (March, 1960).

[7]United States Navy Department, *Aircraft Navigation Manual*, H. O. 216, Hydrographic Office, p. 10.

numeral systems in various languages are different; witness the *eighteen* (ten and eight) in English and *achtzehn* in German as opposed to *duodeviginti* (two from twenty) in Latin. We note the similarity of *novem, neun, nine, neuf, nueve,...* to the word for *new* in the various languages which gives rise, at least, to the suggestion that perhaps at one time many numeral systems were based on *eight* and that "nine" meant using a finger on a "new" person.[8] As we study the history, customs, and famous personalities of a country through its native language we can point out contributions it has made to mathematics: perhaps the kind of abacus, if any, and other mathematical instruments.

The teacher of *art* can easily discuss the mathematics of perspective drawing and the golden-section ratio in architecture and other forms.

Finally, we would urge that it be the conscientiously assumed obligation of every teacher of mathematics to structure mathematics experiences for students in as many ways as possible. Since every high-school student has had at least one year of algebra or general mathematics, in most instances teachers can develop, either directly or in cooperation with other teachers, two or three topics of mathematics which grow out of a subject at hand in such fields as science, social science, or language. These experiences, emerging from the "natural" habitat of the nonmathematics classroom, can show in a most dramatic way what service mathematics provides.

Exercises

1. Suggest two or three topics which could be used to emphasize the outreach of mathematics. Write several sample problems or exercises for each topic. In selecting these topics for a given situation, one should be guided by the amount of mathematics which the student will have experienced.
2. To help the discussion of this section suggest some definite items for teacher-student use: Select one topic mentioned in this section and elaborate on it to show how you could use it to encourage students to feel the call of mathematics. You might wish to look further into the subject of acoustics, into mathematical expressions in psychology, into the matter of index numbers, or into other topics. In doing this you may find it helpful to consult encyclopedias or the references at the end of this chapter.

1.3. Emphasizing by Methodology

Indeed, the wise use of bulletin boards, models, and exhibits helps the student sense the outreach of mathematics. Teachers who show the role of mathematics as they work in other subjects do much to herald the ever-

---

[8]Levi Leonard Conant, *The Number Concept, Its Origin and Development* (New York: MacMillan and Company, 1931), pp. 218–30.

present call of mathematics to students, many of whom have had little more than algebra or some elementary work in geometry.

However, the most effective and meaningful way to help the student sense the significant role of mathematics in intellectual progress is by a carefully planned methodology in the classroom. This approach should show mathematics growing and unfolding as it reaches out to serve others and as it uses its own methods to investigate its foundations and to look forward to new conclusions. By "its own methods" we do not mean any solemn or majestic outgiving of theorems as from a logical engine, but excitement-filled conjectures made by students themselves, to be tested by them as they attempt to deduce them. Mathematics calls for questions about our environment as well as for an investigation of its own growing and expanding discipline.

The teacher is effective who can help the students experience mathematics as a servant and, at the same time, guide them into the romance of discovery and critical examination of some mathematics of their own. The teacher of arithmetic who encourages students to discover an algorithm for adding fractions, having worked out answers otherwise, is stressing the challenge of mathematics. The teacher of analytic geometry who suggests "how good it would be" to have a means, or an algorithm, to tell the area of a triangle in terms of coordinates of the vertices will undoubtedly stimulate the proposal of several such methods. Indeed, student-centered teaching is a most significant key in encouraging student discovery and it is to this phase of teaching that the writers plan to give much attention.

1.4. Summary

The two faces of mathematics can be emphasized by good classroom use of display boards, news clippings, and static and working models.

An effective way to help students sense the role of mathematics in our civilization is to develop several mathematics topics both useful and pertinent to nonmathematics courses which the teacher may find himself conducting. This requires extra reading, research, and classroom planning, but it is interesting and rewarding to both the teacher and the student. Such topics as these help in the students' realization that mathematics *serves*.

Both facets of the reach of mathematics, especially the inward facet may be emphasized in the mathematics classroom as student-centered approaches help students see mathematics growing as a result of their own efforts.

There is a rising feeling that some programs in mathematics are emphasizing the *inward call* with little concern for the *outreach* of mathematics as it provides increased understanding of our environment. A discussion of this controversy appears in the article "On the Mathematics Curriculum of the High School."[9]

The references within the chapter and bibliography at the close of this chapter fall into two categories. The references include those from which quotations are taken and which might be looked into by the student, not only to gain the full context of the quotation, but to explore other thoughts of these writers. The bibliography has been assembled by students who have found articles interesting, pertinent, and applicable to the material discussed in this chapter. These may serve the reader to explore his interests further and, in the university classroom, they may furnish suggestions for student reports, contributions, and continued study. These references emphasize also that much thinking is being done in the teaching of mathematics and that the study of journals provides an opportunity by which one can develop a very effective methodology and understanding.

Bibliography

Other helpful sources for browsing and reading on the call of mathematics may be found in many places. Some are

Adler, Irving, "The Changes Taking Place in Mathematics," *The Mathematics Teacher*, LV (October, 1962), pp. 441-51.

Atkins, R. A., "Mathematics in Elementary Photography," *School Science and Mathematics*, LV (March, 1955), pp. 175-8.

Baumgartner, W. S., "Effective Mathematics in Industry, "*The Mathematics Teacher*, XLIX (May, 1956), pp. 356-9.

Bell, Clifford, "What Every Teacher Should Know about the Uses of Mathematics," *The Mathematics Teacher*, LVI (May, 1963), pp. 302-6.

Burington, Richard S., "Contemporary Applications of Mathematics," *The Mathematics Teacher*, XLIX (May, 1956), pp. 322-9.

Butler, Charles H., and F. Lynwood Wren, *The Teaching of Secondary Mathematics* (New York: McGraw-Hill Book Company, Inc. 1960), Chapter 2.

Cain, R. W. and E. C. Lee, "Analysis of the Relationship Between Science and Mathematics at the Secondary School Level," *School Science and Mathematics*, LXIII (December, 1963), pp. 705-13.

Cairns, Stewart Scott "Mathematics Education and the Scientific Revolution," *The Mathematics Teacher*, LIII (February, 1960), pp. 66-74.

Cairns, Stewart Scott, "Mathematics, Missiles, and Legislation," *The Mathematics Teacher*, LI (November, 1958), pp. 514-20.

---

[9]"On the Mathematics Curriculum of the High School," *The Mathematics Teacher*, 55 (March, 1962), 191-5.

Clifford, Edward L., "An Application of the Law of Sines: How Far Must You Lead a Bird to Shoot in on the Wing," *The Mathematics Teacher*, LIV (May, 1961), pp. 346-50.

Fischer, I.,' 'How Far Is It from Here to There," *The Mathematics Teacher*, LVIII (February, 1963), pp. 123-30.

Grosch, H. R. J., "Science and Mathematics," *The Mathematics Teacher*, L (March, 1957), p. 231.

Guggenbuhl, Laura, "An Unusual Application of a Simple Geometrics Principle," *The Mathematics Teacher*, L (May, 1957), pp. 322-4.

Hammer, Preston, "The Role and Nature of Mathematics," *The Mathematics Teacher*, LVII (December, 1964), pp. 514-21.

Jones, Phillip S., "The History of Mathematics as a Teaching Tool," *The Mathematics Teacher*, L (January, 1957), pp. 59-64.

Kadushin, I., "Mathematics in Present Day Industry," *The Mathematics Teacher*, 35 (October, 1942), pp. 260-4.

Kennedy, E. S., "Al-Biruni on Determining the Meridian," *The Mathematics Teacher*, LVI (December, 1963), pp. 635-7.

Kinney, Lucien B., and C. Richard Purdy, *Teaching Mathematics in the Secondary School* (New York: Rinehart and Company, Inc., 1952), Chapter I.

Kinney, L. B., "Why Teach Mathematics?" *The Mathematics Teacher*, 35 (April, 1942), pp. 169-74.

Kline, Morris, *Mathematics in Western Culture* (New York: Oxford University Press, 1953).

Kline, Morris, *Mathematics and the Physical World* (New York: Thomas Y. Crowell Company, 1959).

Landon, M. W., "Mechanics of Orbiting," *The Mathematics Teacher*, LII (May, 1959), pp. 361-4.

Lankford, F.G., "Mathematics for Citizen and Consumer," *Ohio Schools*, XXVII (October, 1949), p. 160.

Lee, Everett S., "The Engineer in Industry—Seen and Unseen Mathematics," *Mathematics Teacher* (February, 1955), p. 66 ff.

McKinney, W. M., "Measuring the Earth," *The Journal of Geography*, LXIV (November, 1965), pp. 350-6.

Mark, S. J., "How Do Levers Help Us?" *The Instructor*, LXXV (October, 1965), p. 81.

Mathematics Staff of the University of Chicago, "Coloring Maps," *The Mathematics Teacher*, L (December, 1957), pp. 546-50.

Moorman, R. H., "Mathematics and Philosophy," *The Mathematics Teacher*, LI (January, 1958), pp. 28-37.

Mucci, Joseph F., "Probability and the Radioactive Disintegration Process," *The Mathematics Teacher*, LIV (December, 1961), pp. 606-8.

National Council of Teachers of Mathematics, *Mathematics in Modern Life*, Sixth Yearbook (New York City: Bureau of Publications, Teachers College, Columbia University, 1931).

National Council of Teachers of Mathematics, *The Place of Mathematics in Secondary Education*, Fifteenth Yearbook.

Pack, E., "Mathematics and Science," *Education Digest*, XXIV (November, 1958), pp. 34-5.

Powers, L. S., "Correlation of Science and Mathematics in the Junior High School," *School Science and Mathematics*, LIV (October, 1954), pp. 571-3.

Rasmussen, Othom, "Mathematics Used in Courses of Various Departments in a University," *The Mathematics Teacher*, XLVIII (April, 1955), pp. 237-42.

Rees, Mina, "The Impact of the Computer," *The Mathematics Teacher*, LI (March, 1958), pp. 162-8.

Rees, Mina, "The Nature of Mathematics," *The Mathematics Teacher*, LV (October, 1962), pp. 434-40.

Rettaliata, J. T., "Mathematics and Science: Partners in Progress," *School Science and Mathematics*, LXIV (March, 1964), pp. 173-9.

Rosenbaum, R. A., "Mathematics, the Artistic Science," *The Mathematics Teacher*, LV (November, 1962), pp. 530-4.

Rosenberg, Herman, "Great Challenges of Mathematics Education," *The Mathematics Teacher*, LV (May, 1962), pp. 360-8.

Rosenberg, H., "Modern Applications of Exponential and Logarithmic Functions," *School Science and Mathematics*, LX (February, 1963).

Rosenberg, Herman, "Values of Mathematics for the Modern World," *The Mathematics Teacher*, LIII (May, 1960), pp. 353-8.

Rosskopf, M. F., "The Place of Mathematics in General Education," *School Science and Mathematics*, XLIX: 565-570 (October, 1949).

Ruchlis, H., "Basic Concept: The Impossibility of Continuous Growth; Geometric Progressions and Its Relationship to Biology, Nuclear Explosions and Economics," *School Science and Mathematics*, LXV (May, 1965), pp. 416-24.

Runkel, Philip J., "Quantification in the Social Sciences," *The Mathematics Teacher*, LV (January, 1962), pp. 20-33.

Schaaf, William L., "Guided Missiles ... and Mathematical Education." *The Mathematics Teacher*, XLIX (October, 1956), pp. 477-8.

Schaaf, William L., "How Modern Is Modern Mathematics," *The Mathematics Teacher*, LVII (February, 1964), pp. 89-97.

Schaaf, William L., "Just What Is Mathematics?" *The Mathematics Teacher*, XLVIII (April, 1955), pp. 264-6.

Schaaf, William L., "Mathematics and Billiards," *The Mathematics Teacher*, L (May, 1957), pp. 384-5.

Scheid, Francis, "Clock Arithmetic and Nuclear Energy," *The Mathematics Teacher*, LII (December, 1959), pp. 604-7.

Scheid, Francis, "Intuition and Fluid Mechanics," *The Mathematics Teacher*, LIII (April, 1960), pp. 226-34.

Slook, Thomas H., "Designing RAPID, an Analogue Computer," *The Mathematics Teacher*, LVII (March, 1964), pp. 149-51.

Smiley, Charles H. and David Peterson, "No Space Geometry in the Space Age," *The Mathematics Teacher*, LIII (January, 1960), pp. 18-21.

Springer, C. F., "The Meaning of Mathematics," *The Mathematics Teacher*, XLVIII (November, 1955), pp. 453-9.

Stover, Donald W., "Projectiles," *The Mathematics Teacher*, LVII (May, 1964), pp. 317-22.

Stretton, William C., "The Velocity of Escape," *The Mathematics Teacher*, LVI (October, 1963), pp. 400-2.

Swineford, Edwin J., "Ninety Suggestions on the Teaching of Mathematics in the Junior High School," *The Mathematics Teacher*, LIV (March, 1961), pp. 145-8.

Teller, Edward, "The Geometry of Space and Time," *The Mathematics Teacher*, LIV (November, 1961), pp. 505-14.

Thompson, R. A., "Using High School Algebra and Geometry in Doppler Satellite Tracking," *The Mathematics Teacher*, LVIII (April, 1965), pp. 290-4.

Trimble, A. C. "Mathematics in General Education." *The Mathematics Teacher*, L (January, 1957), pp. 2-5.

Varberg, Dale E., "The Development of Modern Statistics, *The Mathematics Teacher*, LVI (April, 1963), pp. 252-7.

Wallin, Don, "An Application of Inequalities," *The Mathematics Teacher*, LIII (February, 1960), pp. 134-5,

Wick, J. W., "Physical Mathematics," *School Science and Mathematics*, LXIII (November, 1963), pp. 619-22.

Wilansky, Albert, "Algebra and Geometry: Mathematics or Science?" *The Mathematics Teacher*, LIV (May, 1961), pp. 339-43.

Wildermuth, Karl P., "Application of the Cosine Function," *The Mathematics Teacher*, LVI (May, 1963), pp. 319-20.

Wilks, S. S., "New Fields and Organizations in the Mathematical Sciences," *The Mathematics Teacher*, LI (January, 1958), pp. 2-9.

Willerding, M. F., "Challenge of Practical Applications," *School Science and Mathematics*, LVII (June, 1957), pp. 437-46.

Wilson, Raymond H., Jr., "The Importance of Mathematics in the Space Age." *The Mathematics Teacher*, LVII (May, 1964), pp. 290-7.

Winthrop, H., "Mathematics in the Social Sciences," *School Science and Mathematics*, LVII (January, 1957), pp. 9-16.

Wren, F. Lynwood, "Mathematics in Focus," *The Mathematics Teacher*, LXVIII (December, 1955), pp. 514-24.

2

A Survey of Methodology

> . . . a teacher is a learning engineer . . . as such (he) must first know the total mathematics he will teach. . . he must also know how the whole structure is put together in the minds of his students.
>
> Clark and Fehr[1]

2.1. Introduction

Current teaching practice spreads over a spectrum ranging from nearly impersonal dictation of information, with lack of concern as to how students learn, to the encouragement of discovery or invention of nearly all of the mathematics by the student. There are some teachers who teach entirely by telling and drilling, resulting in manipulative rote learning. Others help the student to relive in a somewhat dramatic way high points of mathematical history, and thus maximize creativity. There are perhaps only a few teachers at the extremes of this span of classroom methodology, although many classes are strictly teacher centered with little for the student to do but to copy, observe, and imitate. Few teachers use one method to the exclusion of all others. Classroom techniques may vary with the type of student, the time, the teacher, and the matter at hand. One really expresses his philosophy of teaching through his methodology.

---

[1]John R. Clark and Howard F. Fehr, "Learning Theory and Improvement of Instruction—A Balanced Program," *The Learning of Mathematics; Its Theory and Practice*, Twenty-First Yearbook (Washington: National Council of Teachers of Mathematics, 1953), p. 348.

It might be expected that throughout the years there have evolved many highly respected methods of teaching and, indeed, this is the case. It is the purpose of this chapter to classify these existing methods—although it is readily apparent that there might be overlapping—and to point out significant features of each.

Just as mathematics itself is changing, certain methodology is receiving increasing emphasis. Genuine interest is being shown in patterns of teaching which stress the contributions of students through their own mathematical experience, discovery, creativity, and by thinking about mathematics in their own way. Such attitudes in teaching may not be new, but they are presently of fresh interest to many. Gibby suggests that we should turn our thoughts to newer methods of teaching:

> Mathematics has expanded explosively in the last twenty years—but we go on teaching in the way we have always taught—often even the way we were taught ourselves. We have got ourselves in a mess, and we must wake up.[2]

Some of these changes in the emerging stream of methodology are reflected in what follows and in the references and bibliography for additional reading at the close of this chapter. For a discussion of classic methods, however, we follow somewhat closely the organization of Arthur Schultze in *The Teaching of Mathematics in Secondary Schools.*[3] For more and different illustrative examples the reader is referred both to Schultze and to a similar work by J. W. A. Young.[4]

2.2. Synthetic and Analytic Approaches in Teaching

Before we attempt to describe these two approaches in the classroom, let us attack the given geometry problem in two different ways.

Given: Fig. 2.1 with

(A) $\overline{AB} \cong \overline{AH}$

(B) $\overrightarrow{\mathbf{AF}}$ bisecting $\angle HAB$

To show:[5] (F) $\overline{FH} \cong \overline{FB}$

---

[2]W. A. Gibby, "Mathematics in Secondary Schools," *New Approaches to Mathematics Teaching*, edited by F. W. Land (London: MacMillan and Company, Ltd., 1963), p. 107.

[3]Arthur Schultze, *The Teaching of Mathematics in Secondary Schools* (New York: The Macmillan Company, 1914).

[4]J. W. A. Young, *The Teaching of Mathematics in the Elementary and the Secondary School* (New York: Longmans, Green and Company, 1924).

[5]The writers prefer the use of the words "assumed data" rather than the usual "given" in problems for proof or investigation because the items listed

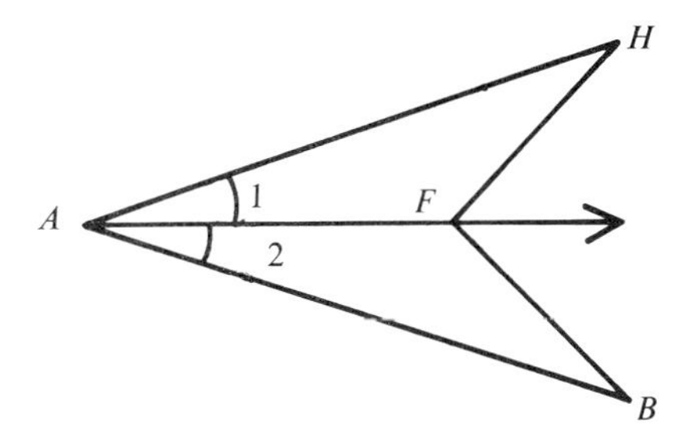

Figure 2.1

Argument:

(A)	$\overline{AB} \cong \overline{AH}$	(Given)
(B)	$\overrightarrow{\mathbf{AF}}$ bisects $\angle HAB$	(Given)
(C)	$m\angle 1 = m\angle 2$	(Definition of bisection)
(D)	$\overline{AF} \cong \overline{AF}$	(Line congruent to itself)
(E)	$\triangle AFH \cong \triangle AFB$	(SAS)
(F)	$\overline{FH} \cong \overline{FB}$	(Corresponding parts of congruent figures) Q. E. D.

Note that the step which is labeled (B) implies immediately that $m\angle 1 = m\angle 2$ and hence, to help describe this approach to the proof, we write (B) $\rightarrow$ (C). Also, (A), (C), and (D) $\rightarrow$ (E) $\rightarrow$ (F). It is apparent that in the arrangement of steps above it was demonstrated precisely how (B) along with (A) led directly to the conclusion (F). The pattern of the proof was

---

are those being assumed arbitrarily by the investigator. Also the use of "conjecture to be tested" is preferred over that of "to show," for the former reflects a certain spirit in the classroom which should be encouraged. As the student examines figures by measurement, or even by intuition, he feels that certain general properties might be true and these seemingly promising conclusions we call conjectures. In the "proof" or "investigation" the student attempts to see if these conjectures can be deduced from the assumed data within the logical framework of the system. If so, the conjecture will become a theorem. This point of view in teaching will be elaborated on much more in succeeding pages, especially in the chapter on the teaching of geometry. Finally, although the figure used is a general case, the student should be encouraged to state carefully made generalizations on what he has proved.

$$(B) \longrightarrow \left.\begin{matrix}(A)\\(C)\\(D)\end{matrix}\right\} \longrightarrow (E) \longrightarrow (F)$$

and in writing this proof in the style above we moved directly from the known or given statements to the unknown. Each step in the proof led directly to a succeeding step which was a logical consequence of a previous one. The steps are so written that by the time one comes to the end of the proof, all the steps have fit perfectly and climax with the conclusion. An argument written in this style is said to be done in *synthetic* style. This is the manner in which most proofs appear and which students study.

Written in synthetic style, a proof has some good and some less desirable features. A proof in synthetic style is concise, elegant, and a masterpiece of mathematical beauty. It is this latter feature which "stirs the soul" as suggested by David Eugene Smith.[6] The good features are overwhelming if the reader understands or has full knowledge of the proof. If he does not, however, it is often not apparent why steps are taken as they are until the very close of the proof when it is seen that "everything ends all right." Proofs learned in synthetic style are difficult to reconstruct if the steps have been forgotten, because the students may not understand why they were taken in the first place.

There is another method of attack which we call the *analytic* approach. We shall illustrate this method with the same problem; letter names of the steps taken will remain the same as those used above (see Fig. 2.2).

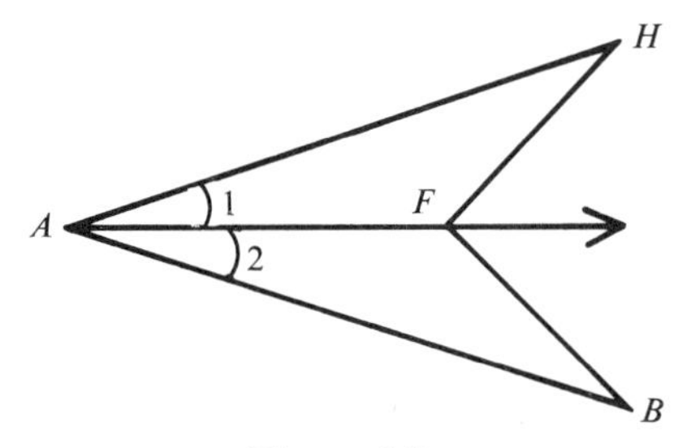

Figure 2.2

Given: (A) $\overline{AB} \cong \overline{AH}$

(B) $\overrightarrow{AF}$ bisecting $\angle HAB$

To show: (F) $\overline{FH} \cong \overline{FB}$

---

[6]David Eugene Smith, *The Poetry of Mathematics And Other Essays* (New York: Scripta Mathematica, 1934), pp. 14–27.

Argument: Of course, if a student could construct an argument in synthetic style to show directly that (A), (B) $\rightarrow$ (F), he would write it immediately. If such is not the case we could start in the manner suggested below though we cannot predict what will be said in the classroom. We begin by asking questions either of ourselves or our class:

1. How can we show that $\overline{FH} \cong \overline{FB}$?

   We expect the answer that $\overline{FH} \cong \overline{FB}$ if they are corresponding parts of congruent figures.

2. In what figures might we look for congruence?

   $\triangle$s AHF and ABF (E)

3. What methods do we have at our command to show that two triangles are congruent?

   ASA, SAS, SSS

4. Do we have ASA?

   We do have $\angle 1 \cong \angle 2$ since AF is an angle bisector and we have (C)

   $$\overline{AH} \cong \overline{AB} \tag{A}$$

   and

   $$\overrightarrow{\mathbf{AF}} = \overrightarrow{\mathbf{AF}} \tag{D}$$

   but these facts do not furnish ASA.

5. Do we have SAS? Yes (!) for $\overline{AH} = \overline{AB}$ (A)

   and

   $$\overline{AF} = \overline{AF} \tag{D}$$

   and

   $$\angle 1 = \angle 2 \tag{C}$$

Note that in our explanation and discussion above we are saying that (F) will be a valid consequence if (E) is a valid consequence: (E) is a valid consequence if (A), (C), and (D) are. But (A), (C), and (D) are either given or are readily shown; hence we see that we can retrace our steps and argue (F) as a valid conclusion.

In the analytic approach we reason that we can write (F) if we can "do" step (E). We can "do" step (E) if we can "do" steps (A), (C), and (D), but, indeed, we now see how steps (A), (C), and (D) can be accepted. Therefore, we can reverse the steps quickly to show how one step implies another and can reconstruct and record the proof in synthetic style.

Proofs and problems in fields other than geometry lend themselves to consideration by analytic methods.

Given: $a > 0, b > 0$

To show: $\sqrt{ab} \leq \frac{a+b}{2}$

Argument: Let us use an analytical approach.

1. For $\sqrt{ab}$ to have meaning, what must be true about a and b? a and b must be both positive or both negative. Indeed they are both positive by hypothesis and hence the condition that $\sqrt{ab}$ have meaning is satisfied.

2. Now $\sqrt{ab} \leq (a+b)/2$ is valid if we can show that $ab \leq [(a+b)/2]^2$ is a valid statement; this is valid if $4ab \leq (a+b)^2$ is valid and this is valid if $4ab \leq a^2 + 2ab + b^2$ is valid; this is valid if $0 \leq a^2 - 2ab + b^2$ is valid. But $a^2 - 2ab + b^2$ equals $(a-b)^2$ which is *always* greater than or equal to 0 for $a > 0, b > 0$. The analytic approach has led us to a beginning statement which we know is always valid and we have recognized all along that we could retrace each step if necessary. Hence we can now start with the statement $0 \leq (a-b)^2$ and, by going backward, we can construct a proof in valid steps and record it in synthetic style.

The analytic approach helps one to *find* steps to be rearranged later to record a proof in synthetic style, whereas the synthetic approach to learning a proof asks the student to attempt to *follow* the work of another. The analytic approach encourages exploration and creativity while the synthetic approach often causes the student to wonder why certain steps were taken. The analytic approach fosters trials by the student with immediate subsequent checking, while the synthetic approach may result in mere memorization of steps. Schultze summarizes the comparison of these two approaches in the classroom by saying that, "Analysis is the method of discovery, synthesis is the method of concise and elegant presentation."[7]

To present proofs solely in synthetic style hides from the student the way they were discovered. The striking fact is that few proofs arise imme-

[7]Schultze, *op. cit.*, p. 36.

To show: $\overline{BD} \cong \overline{AD} \cong \overline{DC}$

Record the analytic approach steps which you make and then write the argument in synthetic style.

4. Given: *ABCD* a quadrilateral with
E, F, G, H midpoints (Fig. 2.6)

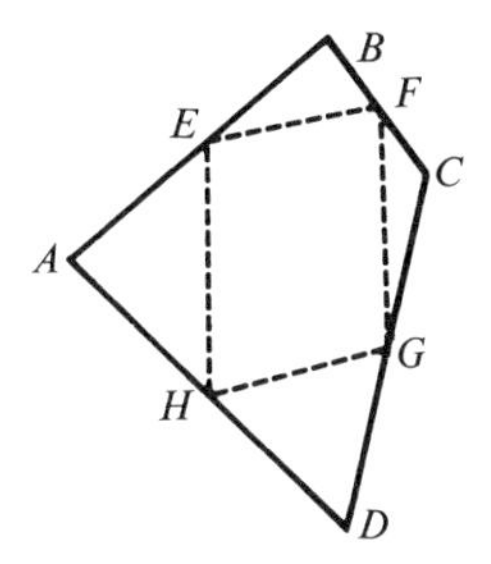

Figure 2.6

To show: *EFGH* forms a parallelogram

Assuming that you have proved the theorem that the line which joins the midpoints of two sides of a triangle is parallel to the third side and has a length equal to one-half that of the third side, record steps which might be made in an analytic approach and then write the proof in synthetic style.

5. If the two opposite sides of a parallelogram are extended by equal lengths, show that the extremities of the extensions are collinear with the intersection of the diagonals of the parallelogram. What questions or steps in analytic approach do you go through? Write the proof in synthetic style.

6. A circle with center at *A* is internally tangent to a circle with center at *B*. Let the point of internal tangency be *C*. Now let any line through *C* intersect the smaller circle at *D* and the larger at *E*. Show that *BE* is parallel to *AD*. Note steps which are taken in an analytic approach; write the proof in synthetic style.

7. Show that the two lines which join the midpoints of the opposite sides of a quadrilateral are concurrent with the line which joins the midpoints of the two diagonals. Record thoughts in an analytic approach; write the proof in synthetic style.

8. One definition for $|a|$ is $\sqrt{a^2}$. Use this definition and axioms on inequalities to show that $|a + b| \leq |a| + |b|$. Record the thoughts you make in an analytic approach and then write a proof in synthetic style.

It should be emphasized that these points of view in teaching are not related to the methods used in synthetic and analytic geometry. The use of analytic methods in geometry means that the results are gained by computation; with such methods two line segments are shown to be congruent by computing their lengths in terms of the coordinates of their endpoints; in synthetic geometry the result might be gained by showing that the two line segments are corresponding parts of congruent figures. Both synthetic and analytic geometries are deductive in character; both have postulates and theorems, although the deductive character of geometry treated by analytic methods is often lost in the details of computational activity.

In this work we shall use the word *approach* when we refer to analytic and synthetic styles in classroom methodology. We shall use the word *method* when we refer to analytic (computational) and synthetic treatments of geometry.

Exercises

1. Deduce the theorem in Exercise 3 by the methods of analytic geometry.
2. Deduce the theorem in Exercise 4 by the methods of analytic geometry.
3. Deduce the theorem in Exercise 7 by the methods of analytic geometry.

2.3. Inductive and Deductive Approaches

Another contrasting pair of approaches prevalent in methodology in the classroom is that of inductive and deductive styles of presentation and experience.

To illustrate differing features of those approaches, let us suppose that a teacher of geometry wishes his students to learn that the sum of the measures of the interior angles in a triangle is 180°. Classroom practice would be different under these two approaches. For each approach, the teacher could do the following:

Inductive Approach	**Deductive Approach**
(a) Have students determine the angle measure of each interior angle of several cardboard triangles or figures of triangles	(a) Draw a triangle to represent any triangle, as $\triangle ABC$

(b) Ask students to make a summary of results

(b) Deduce from previous elements in the postulate system that the sum of the measures of the interior angles is that of two right angles

(c) Ask for a hypothesis or a generalization

(d) Help the student to see if the conjecture made in (c) can be seen to fit into the logical organization of the geometry thus far studied

To lead to the knowledge and justification of the binomial theorem

$$(a + b)^n = a^n + na^{n-1}b + \frac{n(n-1)}{2}a^{n-2}b^2 + \frac{n(n-1)(n-2)}{2 \cdot 3}a^{n-3}b^3 + \cdots + b^n$$

there would be a different methodology under the two approaches. Again, the teacher would use either the inductive or deductive approach.

Inductive Approach

(a) Ask students to multiply

$(a + b)^2$
$(a + b)^3$
$(a + b)^4$
$(a + b)^5$
$\vdots$

(b) Ask the students if they can predict the product for $(a + b)^6$ or for some like power of $(a + b)$

(c) Ask the student to see if he can perceive some pattern of formation of the product

(d) Help the class to come to some agreement upon method of ex-

Deductive Approach

(a) State the expansion for $(a + b)^n$

(b) Ask the student to confirm the validity by the use of several examples, or

(c) Have the student prove the validity of the expression by the principle of mathematical induction if the student is at this level of study of mathematics

pansion which produces all the results thus far

(e) Ask the student to attempt to prove his conjecture by the use of the principle of mathematical induction if he is at this level of mathematical study

In deductive approaches the conclusion or the generalization is given immediately and the proof can be constructed by synthetic or analytic means; in inductive approaches one always starts with examples.

It is important for the teacher to know and for the students to come to realize that the "conclusions" reached by these two methods have widely differing status. These differences may be expressed as follows:

Inductive Approaches	**Deductive Approaches**
(a) A "conclusion" is arrived at by the study of examples	(a) One general case is used; results are arrived at by chains of syllogisms or by reasoning from previous theorems and postulates: the conclusion comes quickly and "at one stroke"
(b) The more cases for which the tentative "conclusion" holds the stronger it is to be regarded	(b) The conclusion is as strong as the premises
(c) The "conclusion" can be rejected as a result of one example which does not support it	(c) The conclusion can be rejected only if the argument is fallacious or if one or more of the premises are undermined
(d) The "conclusion" is at best a hypothesis	(d) The conclusion is a theorem

Classroom discussion should often be directed so the student will feel deeply the difference in character between a *proposed hypothesis* and a *deduced conclusion* or *theorem.* In a way, these two approaches have much in common with the outward and the inward reaches of mathematics for it is often in the service of man that mathematics helps to seek relations from a study of examples, while the inward reach tests this conjectured relation to see if it fits into the logical structure. No one can ever be sure he is "right" from a study of examples, but if one is right as established by deduction, it takes a nearly cataclysmic undermining of the deductive structure to prove that he is wrong. Einstein has said, "Any number of experiments will never prove me right, it takes only one to prove me wrong."

It is important for us that we look at the merits of these approaches relative to their use in the classroom. Inductive approaches give the student concrete experiences and by these one understands better what is being done or what one is trying to find out. In many cases the student can suggest by himself what concrete situations to use to study a given problem. In the inductive approach most students can make a contribution. This helps much to keep the morale of the class at a high level and most students can make the generalization which is sought. Sawyer reminds us that, "In the early teaching of mathematics, there is no danger in making the subject too concrete. The danger is that the subject gets so far from the concrete that it comes to mean nothing at all."[9] Deductive approaches, on the other hand, often depend for their success on a specialized skill in the art of making and writing proofs. Deduction is sometimes difficult for students and it has often happened that the student is learning the proof for a statement which he does not understand. We must "concede that... pupils need to see and to feel mathematics *at work* (italics ours) as well as to think it."[10]

Some teachers may like to think of *inductive methods* as those which employ gradual approaches to harder problems through easier ones, or to more general problems through problems which are somewhat specialized. For the most part, this idea of inductive methods still fits into the picture above, for the examples of simple but increasingly difficult problems may well help a general method to emerge. Simple problems worked by physical means may give answers which will lead to conjectured algorithms on what to do with the numerals themselves to yield these answers. In calculus it is not difficult for the student to discover the usual algorithm for the differentiation of polynomial expressions through the computation of derivatives of many examples by the use of the definition. This is induction, too, and the student can easily deduce that his conjecture is a logical consequence of the definition of the derivative and other theorems which he has had.

The approach by induction in this section is not to be confused with *mathematical* induction. The approach used here could be called scientific induction for it resembles the method of science in studying examples and attempting to make a conjecture or to assemble a theory. Mathematical induction, the reader recalls, is really deduction, for conclusions established

---

[9]W. W. Sawyer, "Abstract and Concrete," *New Approaches to Mathematics Teaching*, edited by F. W. Land. (London: MacMillan and Company, Ltd., 1963), p. 117.

[10]National Council of Teachers of Mathematics, *Emerging Practices in Mathematics Education*, Twenty-Second Yearbook (Washington: The National Council of Teachers of Mathematics, 1954), p. 101.

by the method of mathematical induction are logical consequences of the postulate of mathematical induction and other logical elements of the system.

Exercises

1. You wish to examine the problem of the sum of the measures of the exterior angles of a quadrilateral. Outline carefully some of the different experiences you could plan for your students to make use of *inductive* methods. Point out how each student can make a contribution to the study.
2. A *deductive* approach to the sum of the measures of the exterior angles of a quadrilateral would ask the student to start with some known theorem or theorems and move by a series of logical consequences to a conclusion about the sum. See if you can do this. It will be interesting to learn if this deductive approach was carried out step-by-step in synthetic style, or if an analytic approach was used.
3. You wish your students to suggest the hypothesis that the joins of the midpoints of the sides of a quadrilateral form a parallelogram. Plan experiences to arrive at this conjecture by the method of induction.
4. You are teaching the addition of fractions to freshman algebra students and you wish to construct a series of exercises which will lead ultimately to their adding fractions as

$$\frac{2}{x+4} + \frac{3}{x+7}$$

   Write a list of several simple problems which you hope would lead to such skill.
5. In inductive approaches one might well introduce the idea of sampling. In working with triangles, for example, it might be suggested that one or more should be isosceles, one or more might be right triangles, and so on. How many different kinds of triangles are there?
6. Two points have one line as their join. Three points give rise to three lines and four points can create six lines. How many lines are determined by n points? Describe what can be done to establish a conjecture under inductive approaches. What approach can be made from the deductive point of view?

2.4. Dogmatic and Psychological Approaches

By a "dogmatic approach" we shall mean that a teacher or a text demonstrates or tells how a problem should be worked; the student then imitates this method. Students who study examples to find out how problems of a certain type are worked without inquiring why they are done in this manner are subjecting themselves to dogmatic methods.

Dogmatic approaches in teaching are perhaps the oldest and the most firmly entrenched in our classrooms. As far back as 1650 B. C. the Egyptian

(*Rhind*) papyrus used this method with its eighty-five sample problems to serve as models. Later the Babylonian problem texts were assembled to teach in the same manner. There was little reason or logic in any of these repsentations; they were merely recipes or algorithms. This "tell" method, often followed by drill (many times meaningless drill), seemed to be the accepted method of teaching for a long time. In fact, this approach to teaching is supported by the connectionist theory of learning. Even now in our schools and universities we fear that there is too much blind following of examples. In dogmatic approaches the teacher tells, the student memorizes, and the student finally "gives it back" when requested to do so. The approach is much teacher and text centered.

The approach which we shall call psychological is the antithesis of the dogmatic approach; it is student centered and tends to make use of the best knowledge from the psychology of learning.

In this brief survey of methodology we can suggest contrasting characteristics of these two approaches somewhat better if we summarize them.

Dogmatic Approach	**Psychological Approach**
Rule and rigor are the chief concern; students study models of proofs and examples	Teaching is patterned after findings in psychology of learning
Often abstract generalizations are given first—the student *may* understand later	
Refinements are used immediately	

On rigor, one might say that a development which is rigorous to the teacher may not be so for the student unless the student has been so oriented or unless the student sees a need for such rigor. If an exactness or a preciseness is not understood by a student, then it is not an exactness for him. It is not meant to suggest here that students should not be invited or encouraged to contribute conditions and restraints and other points of good mathematical thinking, but points of rigor which arise in student-teacher discussions are psychologically better for the student than those dogmatically imposed by the teacher.

The psychological approach is very well summarized in "Implications of the Psychology of Learning for the Teaching of Mathematics" by Lankford.[11] In this discussion there appear very practical and fruitful illustrations for teachers who wish to utilize suggestions of psychologists for

[11]Francis G. Lankford, "Implications of the Psychology of Learning for the Teaching of Mathematics," *The Growth of Mathematical Ideas: Grades K-12*, Twenty-Fourth Yearbook (Washington: National Council of Teachers of Mathematics, 1959), Chapter 10.

more effective work in the classroom. There is some learning under all methods, but the advice of psychologists may help us to do a better job. Lankford summarizes his conclusions by saying that effective learning in mathematics results when:

Meaning is emphasized in contrast to manipulation

Teaching is adapted to the variations among individuals

The learner is appropriately motivated

Practice is provided as needed by individuals, but is not relied upon to develop meaning

The learner is helped to discover ideas of mathematics through developmental, reflective activities

A reservoir of success is available as an antidote to the deterrent effects of failure

Reasonable goals are formulated and accepted by the learner

The learner is an active participant rather than merely a passive listener

The learner knows what progress he is making toward his goal.[12]

We would urge that the reader note well Lankford's entire discussion. He skillfully relates current and suggested teaching practice in our classroom with various tenets in the psychology of learning.

The Learning of Mathematics: Its Theory and Practice discusses important aspects which may help guide the teacher in devising effective classroom procedures.[13]

2.5. Lecture and Heuristic or Genetic Approaches

This classification suggests also a teacher-centered versus a student-centered approach. Lectures in mathematics need not have a predominantly dogmatic style, but frequently this is the case. The usual presentation of mathematics by the lecture method consists of the lecturer writing notes on the chalkboard to be copied by the student or of a discussion or reading of reproduced lecture notes. Some lecturers prepare transparencies and reproductions of their lecture notes for the student and then read the lecture from the overhead projector screen; this can all be dogmatic and nearly lifeless. Genetic and heuristic methods, however, use the student as much as possible to promote the unfolding of subject matter.

Briefly written features of each of these methods show interesting differences.

---

[12]Lankford, *ibid.*, p. 429.
[13]Clark and Fehr, *op. cit.*

Lecture Method	**Heuristic or Genetic Method**
Students listen to logically organized materials	Students develop much or at least some mathematics of their own
Much material can be presented	Student become creators
Teachers find their work easier	Leading questions are used by the teachers
Students receiving information about mathematics	Direct information is at a minimum
	Student understands better
The students are *passive* most of the time rather than *active*	Approach is slow, especially at the beginning
The students are followers most of the time—not creators	Method more difficult for the teacher
Students can be motivated deeply	Student interest often high

A method is said to be *genetic* if the class as a whole is considered as a teaching unit; questions and discussion are directed to the class topics are developed as class projects. *Heuristic* methods, on the other hand (although the word means to discover), are regarded usually as those which point to students as individuals. Guides which help a student to develop ideas and generalizations by himself are features of the heuristic method. Of course, such a method is filled with discovery, too. A class in geometry under heuristic approaches might develop thirty different sets or systems or chains of theorems if there are 30 students in the class as opposed to one set of theorems through cooperative (genetic) effort.

In the preceding paragraphs we note that analytic, inductive, psychological, and heuristic and genetic approaches are all somewhat overlapping and that they all have the quality of being student centered. The contrasting approaches, on the other hand, tend to be teacher centered, some of them extremely so. We would not wish to imply some lecturers do not keep in mind the nature and interests of their student audience. It is true that some lecturers are so effective that they gain the interest and attention of their audiences and the students relive, through skillfully presented steps, the discoveries and evolution of ideas concerning a given topic in mathematics. However, this is indeed rare.

It is not difficult to make the claim that, for the most part, the student-centered approaches require more masterful teachers to be effective. This does not mean that teachers who would strive for understanding among their students (which such approaches encourage) should be hesitant to attempt such a methodology. Little by little, well-thought-out approaches which encourage student contribution and participation will develop in the teacher a technique which will produce far more proficiency and understanding on the part of the student as well as a deeper understanding of the subject by the teacher.

Exercises

1. What good features do you see in the dogmatic approach? Are there times or situations in which the dogmatic approach is completely justified?

2. You wish to introduce the idea of the derivative in calculus. How would this be done with a dogmatic approach? How could this be done under a genetic approach? Outline what you could do and what you might expect from the student.

3. You have been studying mathematics. In looking back over your experience try to elect one topic which was presented in dogmatic fashion but which you now see might well have been introduced through a student-centered approach. Outline how you might have introduced the topic with student-centered approaches.

4. The gestaltists and the connectionists have different points of view on how to teach mathematics. Elaborate on each of these points of view and their implications for classroom methodology. You may wish to consult Lankford.[14]

5. Here is a list of some topics or items which appear in courses in mathematics; some of these occur in elementary courses and some in more advanced ones. Select three of these and in each case outline what you would do under a dogmatic approach to introduce the topic. Then for each case outline experiences which you plan for your students with the hope that ideas will emerge from the planned activities. Of course, you may use topics which are not listed here.
   (a) The sum of the interior angles of a triangle
   (b) The sum of the interior angles of the same side of the transversal cutting two parallel lines
   (c) Solving simple algebra problems as "there are two numbers of which one is six more than the other and their sum is twenty"
   (d) Computing the slope of a line
   (e) Extracting the square root of a number
   (f) The algorithm for differentiation of polynomials
   (g) The derivative of the function $\{x, \sin x\}$, or $D_x \sin x$
   (h) Determining "partial fractions" whose sum is a given fraction
   (i) Constructing graphs or plotting points

---

[14]Lankford, *op. cit.*

(j) Multiplication of fractions
(k) $a = b$ modulo m
(l) The area of a triangle (arithmetic)
(m) The area of a triangle in analytic or coordinate geometry—computed in terms of coordinates of vertices
(n) Multiplication of matrices
(o) The idea of a vector space; a vector subspace
(p) Moment of area
(q) The area of an ellipse
(r) Division of fractions
(s) Addition of "signed" numbers—or addition of integers
(t) Solving two simultaneous linear equations
(u) Defining the tangent of an angle
(v) Introducing parametric equations
(w) Infinite series expansions; Maclaurin and Taylor series
(x) The least upper bound of a sequence
(y) An open set
(z) The use of E in computer science language
(aa) Instructing the computer to perform multiplication
(ab) Conditional probability
(ac) Measures or "weights" of sets.

2.6. The Laboratory Approach

In this approach students experiment with concrete objects to develop mathematical ideas. One might regard this definition as including also the use of exploratory numerical activities which help to develop understanding of later desired abstract ideas. For the most part, however, we shall mean by the laboratory approach the use and handling of *things*.

For many students the laboratory approach is nearly indispensable. All our ideas of things are gained from their "sensible effects"[15] (sight, mind, taste) and for many students this means, for effective learning, contact with concrete things rather than reliance on abstractions.

Laboratory approaches have been used by some of our greatest mathematicians. Archimedes explored the possibility of the truth of certain mathematical ideas by using concrete objects. Galileo (1599) demonstrated by weighing thin slabs that the area under one arch of a cycloid equaled three times the area of its generating circle. As the Greeks developed and,

[15]C. S. Peirce, "How to Make Ideas Clear," *Popular Science Monthly,* XII (January, 1878), 293.

later, respected deductive methods so highly, a feeling grew up, even among teachers, that in some way mathematics was cheapened by the use of devices which can be moved, handled, measured, and studied as concrete things. It must be remembered, however, that most mathematics had humble origins in physical situations. If we, then, use physical materials and laboratory approaches to help the student formulate conjectures later to be tested by deductions, we have proceeded much in the same manner as mathematics itself has grown. We would not propose that all great mathematics developed from concrete objects, but it is known that much mathematics developed in investigating physical problems.

Laboratory activity or working with things is receiving increasing attention in the mathematics classrooms to help build ideas and to see mathematics at work in the world. In fact, one entire section of the Twenty-Second Yearbook[16] is devoted to laboratory teaching in mathematics and in journals on the teaching of mathematics there is a growing number of articles on the use of laboratory approaches in teaching. Fruitful sources as the Eighteenth Yearbook[17] and as Rao Sundara's *Geometric Exercises in Paper Folding*[18] are good beginnings for consultation on this subject.

Laboratory work in mathematics can be used in several ways. Experimental activities can be used to

(a) *Help formulate conjectures on properties of figures.* From the use of triangles cut from cardboard or simply from triangles drawn on paper, students may collect sufficient data to suggest that the length of the join of midpoints of two sides of a triangle is half that of the third side. This suggestion is at best a conjecture, but the very seeing and helping in the wording of the conjecture yields deeper understanding of what the conjecture means. Many times a student is asked to prove a theorem of which he does not understand the full meaning. Using laboratory work for this purpose constitutes a facet of effective methodology which helps to make the student a contributor to the growth of the subject.

(b) *Test mathematical ideas.* If a class has just studied the tangent function, the concept of $y = x \tan \theta$ can be put to test by seeing if the height of a pole is given reasonably accurately by this method of indirect measurement. A junior-high-school class which

---

[16]National Council of Teachers of Mathematics, *Emerging Practices in Mathematics Education*, Twenty-Second Yearbook, *op. cit.*

[17]National Council of Teachers of Mathematics, *Multi-Sensory Aids in the Teaching of Mathematics*, Eighteenth Yearbook (Washington: The National Council of Teachers of Mathematics, 1943).

[18]Rao (Row) Sundara, *Geometric Exercises in Paper Folding* (Chicago: The Open Court Publishing Company, 1901).

has been told about the Theorem of Pythagoras may use laboratory methods to check the validity of the theorem.

(c) *Show mathematics "at work."* Mathematics is at work in various instruments: parallel rulers, the Egyptian level, the circle square, proportional dividers, the sextant, the slide rule, and others. Laboratory exercises may be designed to show these tools at work in practical situations. Also student activity can be directed toward the manufacture of some of these instruments with an accompanying mathematical study to substantiate their use. Of course, the depth of all these considerations is dependent on the mathematical level of the students.

There are very distinct advantages to the use of laboratory work in mathematics other than those implied above. From the kindergarten to the graduate school, laboratory work lends concreteness to ideas and helps to make ideas clear. Concrete things help to center interest, help to communicate ideas, and help to make the instruction more rich. Often laboratory work calls for skills and knowledge related to the problem at hand which abstract approaches would never introduce at all. Measurement requires skill in reading rulers or in handling protractors and the use of these and other measuring devices may even give rise to discussions of error. Other laboratory activities may call for the use of levels, plumb bobs, T-squares, and other tools. The use of all these, we insist, is most enriching to the students.

Laboratory work may have shortcomings, however. It is time-consuming and, furthermore, there is a danger that it may become manipulatory to the extent that inward-looking mathematics is neglected—that is, examining how the mathematics learned or used fits into the deductive pattern of the system may not be pursued sufficiently.

If laboratory methods are used, they must be considered as an integral part of the student's activities. To be most effective there must be a smooth and natural flow from laboratory experiences to considerations of rationale or to methods of proof itself. Laboratory work is not an adjunct or an appendage or an "aid"; it is an important facet of the entire learning situation. In arithmetic, for example, cards as those of Fig. 2.7, can be used to reach the conclusion in a concrete way so that

$$\frac{1}{3} + \frac{1}{2} = \frac{5}{6}$$

Algorithms can come later. In calculus, the estimated slopes taken from the graph of the curve whose equation is $\{x, \sin x\}$ can be plotted against x and a new suggested curve obtained, that which appears to be $\{x, \cos x\}$. Hence

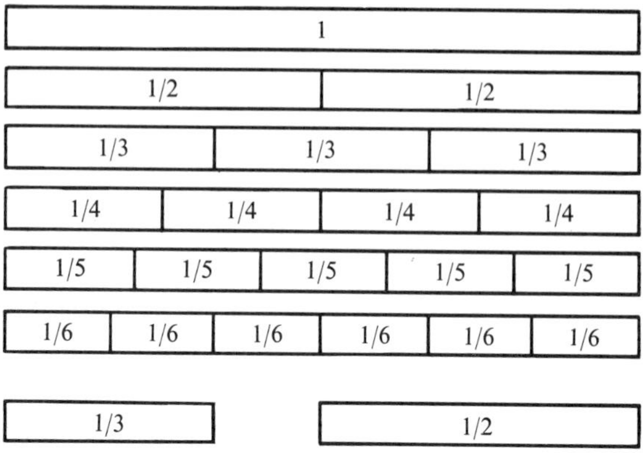

1

1/2 1/2

1/3 1/3 1/3

1/4 1/4 1/4 1/4

1/5 1/5 1/5 1/5 1/5

1/6 1/6 1/6 1/6 1/6 1/6

1/3 1/2

Figure 2.7

students can formulate the conjecture that

$$D_x \sin x = \cos x$$

and use it in problems while more slowly studying the deduction of the conjecture to raise it to the status of a theorem. In both these cases, the student hardly knows that he is doing an experiment; rather in each case, he is using concrete things to help him ponder the answer to a question which is vital at the moment. In algebra, thc students may have added integers with "sliding sticks." From the law that $\log a + \log b = \log ab$ the challenge can well arise to construct sliding sticks which will *multiply* two numbers. In geometry it is highly rewarding to deduce that results from the use of certain instruments as the Egyptian level can be logically substantiated. One does not consider such things as these merely on certain laboratory days or as experiments to be performed sometime during the course; rather, these are done as means of exploring or checking to bring light upon the subject at hand at the moment which is most opportune. Laboratory experiences should be integral parts of courses, not extras.

Exercises

1. Suggest how you might plan laboratory work on triangles to help the student formulate the conjecture that the line joining the midpoints of two sides of a triangle is parallel to the third side. You might also plan laboratory work to study the length of the join of the two midpoints relative to the length of the third side (see Fig. 2.8).

2. Construct an "adding rule" with scales from 0 to 10 placed on cardboard "sticks" as shown in Fig. 2.9. Practice in adding and subtracting on these

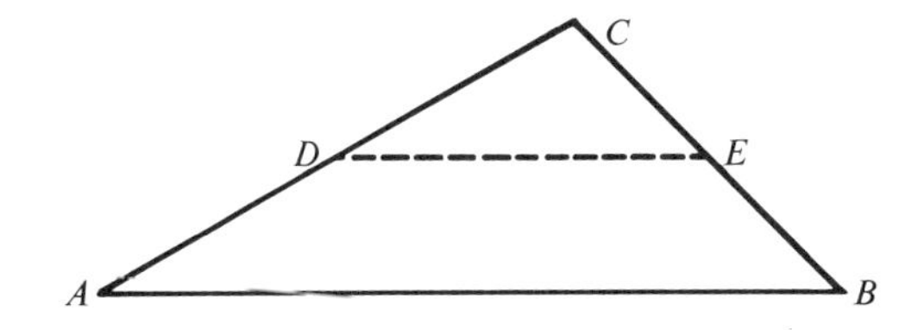

Figure 2.8

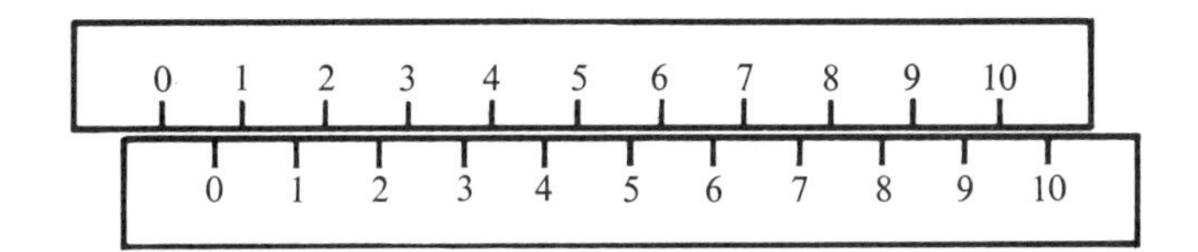

Figure 2.9

sticks gives the student the feel of adding and subtracting by motion. Indeed, this is making use of the number line in adding and subtracting.

3. Now on the adding stick in No. 2, let us make some changes. Let us replace the name 0 by the number whose base ten logarithm is 0; namely 1. Since 3 is at the point which indicates approximately 0.301 of the length of the stick, let us replace the name 3.01 by the number whose logarithm is 0.301; namely 2. Similarly replace 4.77 by the numeral 3. In like manner complete the scale so it contains numerals 1, 2, 3, 4, 5, 6, 7, 8, 9, 10 (or 1). Note the incomplete diagram in Fig. 2.10. Now on this rule

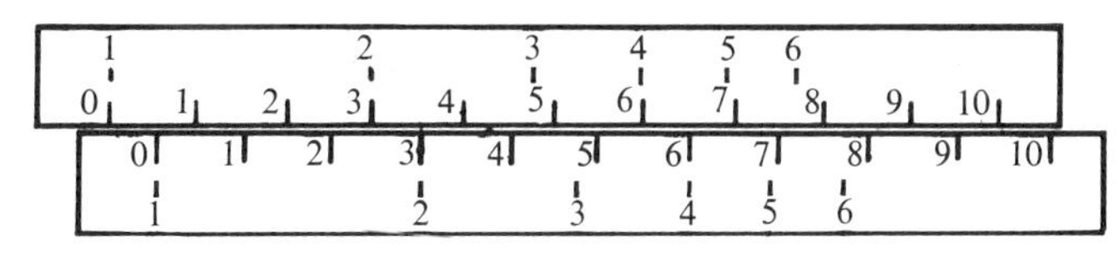

Figure 2.10

go through the "adding motion" with 2 and 3, 2 and $2\frac{1}{2}$ and 2, and others. What "answers" do the "adding motions" produce with the new scales?

4. Construct from cardboard a circular slide rule. Try to determine several numbers in the scale between 1 and 2.

5. When the Egyptian level is used the plumb bob hangs directly over D which is the midpoint of the base of $\triangle ABC$ (Fig. 2.11). This indicates that AB is horizontal. If for a certain surface the plumb bob does not

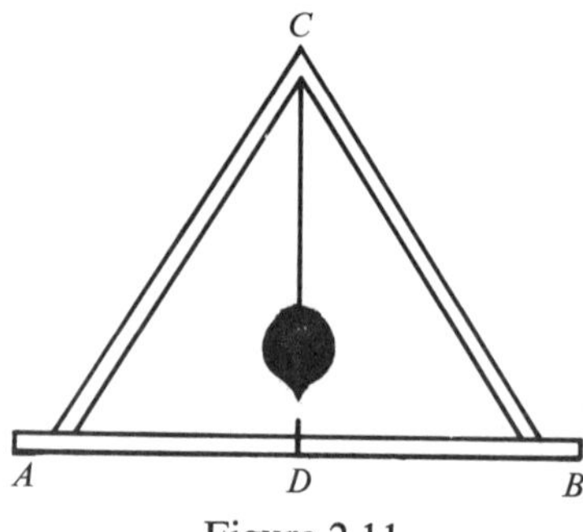

Figure 2.11

hang over D then it is assumed that AB is not horizontal. Translate this physical situation into mathematical hypotheses and then substantiate by deduction that the conclusions made by the use of the Egyptian level are valid.

6. Figure 2.12 shows a "proportional divider." It is used to construct a new figure similar to a given one. It can be set to preserve a certain ratio. Substantiate the use of this instrument by deduction.

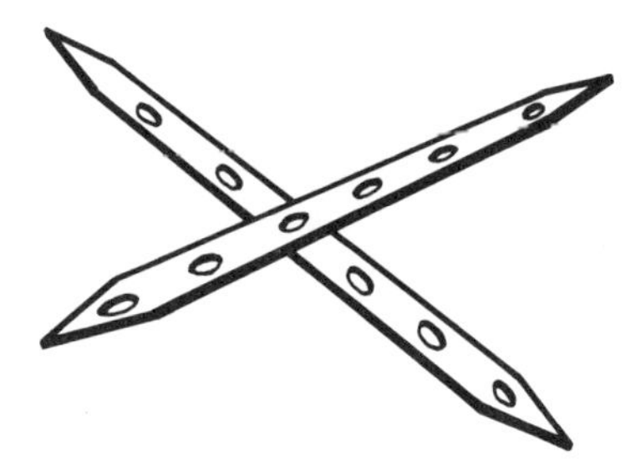

Figure 2.12

7. Suggest an experiment to convince eighth graders that the volume of a cone is one-third that of a cylinder of the same height and radius.

8. Your students doubt that the area of a sphere is $4\pi r^2$. How can you convince them by experiment? You may wish to consult an arithmetic text.

9. How are the laboratory and the inductive approaches related?

10. Suggest three experiments or laboratory studies designed to put mathematics to test; i.e., to see if certain theorems hold in physical situations.

11. Suggest three experiments or laboratory studies or demonstrations to help make ideas clear.

12. Mention three laboratory studies to help the student to arrive at hypotheses which he will investigate further by deduction.

2.7. Discovery Approaches

In the onward-flowing development of classroom methodology, there has emerged a significant and genuine interest in student-centered approaches and especially in those which help the student *discover* as much mathematics as he can for himself. In practice it appears that effective use of analytic, heuristic and genetic, inductive, psychological, and laboratory approaches have much to offer to encourage student discovery.

It would be erroneous to suggest that student-discovery approaches are new. Many great teachers have used methods which have encouraged the student to make use of his tendencies to be creative. Some statements made by teacher-scholars of the past suggest the earnestness and compelling sincerity with which they urged effective use of student-centered approaches. Comenius, known by some as the father of modern educational theory and practice, urges that learning can be helped much when we use *non verba, sed res.*[19] Emmanuel Kant suggests that we must help build ideas by use of specific examples either in mental imagery or by means of physicalization of ideas and processes, for he writes, "Begriffe ohne Anschauung sind leer."[20] John Perry of England, one of the first reformers in the teaching of mathematics, in his famous address before the Engineering Section of the British Association in 1902, challenges us even today when he urges that we should ". . . compel the teacher to take the pupil's point of view rather than the pupil the teacher's; give the student a choice of many directions in which he may study, . . . leave him (the student) greatly to himself."[21] E. H. Moore, speaking before the Mathematical Association of America, on December 29, 1902, raises a goal for which we should strive in his statement:

> . . . the teacher should lead up to an important theorem gradually in such a way that the precise meaning of the statement in question, and further, the practical—i.e., the computational or graphical or experimental—truth of the theorem is fully appreciated; and furthermore, the importance of the theorem is understood, and indeed the desire for the formal proof is awakened, before the formal proof itself is developed. Indeed, much of the proof should be secured by the research of the students themselves.[22]

Felix Klein, one of the great mathematicians of modern time, suggests that,

---

[19]National Council of Teachers of Mathematics, *Teaching of Mathematics in Secondary Schools*, Eighth Yearbook (Bureau of Publications: Teachers College, Columbia, 1933), p. 75.

[20]*Ibid.*, p. 226.

[21]John Perry, "The Teaching of Mathematics," *Science* (November, 1902), 771.

[22]E. H. Moore, "On the Foundations of Mathematics," *Bulletin of the American Mathematical Society*, IX (May, 1903), 419.

"The investigator himself... does not work in a rigorous deductive fashion. On the contrary he makes essential use of his phantasy and proceeds inductively, aided by heuristic expedients."[23] Polya writes, "... mathematics 'in statu nascendi,' in the process of being invented, has never before been presented in quite this manner to the student, or to the teacher himself, or the general public."[24]

It would be equally wrong to say that no student has even experienced "discovery" without being involved in student-centered approaches. It is true that some students have felt the urge to study more in their field by inspiring lecturers or by motives of their own. The point to be emphasized, however, is that, even under this approach the student has taken notes or has read texts and has then digested and reorganized the work in his own way before he really understands. Perhaps we can say that teacher-centered approaches often result in the student being a follower and then, to understand, becoming a creator, if he can, merely to create in some way, or to justify in some way, conclusions by the lecturer. It is the common experience of teachers that justifications made in this way are not always valid ones. This kind of creativity is a far cry from what is meant by "student creativity" in discovery approaches.

Student-discovery methods produce pleasing results. Studies[25~28] indicate that discovery approaches are exceedingly effective in increasing understanding and in improving attitudes toward mathematics. Skill and grasp of content are not impaired; if anything, these are improved, too. Student-discovery methods have an effect on the teacher also; they help his insight into mathematics to deepen as students suggest mathematics "of their own making." Once the teacher has experienced the effectiveness of such methods on the mathematical growth of both the student and himself, it is difficult for him ever again to employ teaching techniques which are highly colored with dogmatic and teacher-centered approaches.

It will be our purpose to examine discovery approaches in more detail in the next chapter.

---

[23]Felix Klein, *Elementary Mathematics from an Advanced Standpoint* (New York: The Macmillan Company, 1932), p. 208.

[24]G. Polya, *How To Solve It* (Princeton: Princeton University Press, 1945), p. vii.

[25]Harold P. Fawcett, *The Nature of Proof*, Thirteenth Yearbook (Washington: National Council, 1938).

[26]Max A. Sobel, "Concept Learning in Mathematics," *The Mathematics Teacher*, XLIX (October, 1956), 425–30.

[27]Oscar Schaaf, "Student Discovery of Algebraic Principles as a Means of Developing Ability to Generalize," (Doctoral dissertation, the Ohio State University, 1954).

[28]Kenneth Cummins, "A Student Experience-Discovery Approach to the Teaching of Calculus," *The Mathematics Teacher*, LIII (March, 1960), 162–70.

2.8. Summary

This chapter has attempted to make the student aware that there are, in general, two widely different approaches to teaching and that in most instances the student-centered approaches are the more effective.

It is important to note that much support and many of the quotations used in this chapter have come from distinguished mathematicians who are convinced that mere telling is not enough. Such mathematicians as E. H. Moore, J. W. Young, George Polya, Felix Klein, and other great teachers urge that future teachers will use classroom experiences in such a way that mathematics will develop through the students' own efforts.

The numbered references and the supplementary reading suggestions should indicate to the reader the immense amount of thinking which goes on concerning the teaching of mathematics. They should help the student to begin now to collect ideas for interesting and fruitful classroom experiences in addition to the ones which he creates by his own inventiveness.

Bibliography

Auer, A., "I Run a Gamut of Gimmicks," *School Science and Mathematics*, LXIII (October, 1963), pp. 576-8.

Bengtson, Ray, "The 'Make 'Em Run' Theory," *The Mathematics Teacher*, LI (January, 1958), pp. 10-12.

Bowen, J. J., "Addition and Subtraction with Linear Scales as an Introduction to Study of the Slide Rule," *School Science and Mathematics*, LVII (November, 1957), pp. 626-8.

Bowie, H. E., "Recent Developments in Mathematics Education," *School and Society*, 93 (April 17, 1965), pp. 252-4.

Butler, Charles H., and F. Lynnwood Wren, *The Teaching of Secondary Mathematics* (New York: McGraw-Hill Book Company, 1960), Chapter 6.

Christofferson, H. C., "Creative Teaching in Mathematics," *The Mathematics Teacher*, LI (November, 1958), pp. 535-40.

Christofferson, Halbert Carl, *Geometry Professionalized for Teachers* (Menasha: George Banta Publishing Company, 1933), Chapter 3.

Coyle, F., "Teaching for Meaning," *School Shop*, 24 (September, 1964), pp. 21-2.

Dadourian, H. M., "How to Make Mathematics More Attractive," *The Mathematics Teacher*, LIII (November, 1960), pp. 548-51.

Davis, David R., *The Teaching of Mathematics* (Cambridge: Addison-Wesley Press, Inc., 1951), Chapters 2, 4.

Dunn-Rankin, Peter, and Raymond Sweet, "Enrichment: A Geometry Laboratory," *The Mathematics Teacher*, LVI (March, 1963), pp. 134-40.

Gager, W. A., "Functional Approach to Elementary and Secondary Mathematics," *The Mathematics Teacher*, L (January, 1957), pp. 30-4.

Gagne, Robert N., "Learning and Proficiency in Mathematics," *The Mathematics Teacher*, LVI (December, 1963), pp. 620-6.

Glennon, V. J., "And Now Synthesis: A Theoretical Model for Mathematics Education," *The Arithmetic Teacher*, XII (February, 1965), pp. 134-41.

Glennon, V. J., "Method: A Function of a Modern Program as Complement to the Content," *The Arithmetic Teacher*, XII (March, 1965), pp. 179-80.

Groenendyk, E., "Mathematics Laboratory," *The Education Digest*, XXVIII (April, 1963), pp. 40-1.

Harmeling, Henry, "Using Historical Stories to Stimulate Interest in Mathematics," *The Mathematics Teacher*, LVII (April, 1964), pp. 290-7.

Harwood, E. H., "Enrichment for All!" *School Science and Mathematics*, LXIII (May, 1963), pp. 415-22.

Henderson, K. B., "Anent the Discovery Method," *The Mathematics Teacher*, L (April, 1957), pp. 287-91.

Holcomb, John D., "A Christman Graph," *The Mathematics Teacher*, LVII (December, 1964), pp. 560-1.

"Improving Mathematics Instruction," *School and Society*, 86 (October 11, 1958), p. 365.

Jackson, R., "Development of a Concept: A Demonstration Lesson," *The Mathematics Teacher*, LIV (February, 1961), pp. 82-4.

Johnson, D. A., "Study of the Relative Effectiveness of Group Instruction," *School Science and Mathematics*, LVI (November, 1956), pp. 609-16.

Keiber, Mae Howell, "A Heart for Valentine's Day," *The Mathematics Teacher*, LII (February, 1959), pp. 132.

Kemeny, J. G., "Rigor vs. Intuition in Mathematics," *The Mathematics Teacher*, IIV (February, 1961), pp. 66-74.

Kluttz, M., "Mathematics Laboratory: A Meaningful Approach to Mathematics Instruction," *The Mathematics Teacher*, LVI (March, 1963), pp. 141-5.

Kovach, L. D., "Unilateral Mathematics Teaching," *The Mathematics Teacher*, LVI (November, 1963), pp. 550-2.

Lankford, F. G., Jr., "Helping Pupils Make Discoveries in Mathematics," *The Mathematics Teacher*, XLVIII (January, 1955), pp. 45-7.

Laycock, Mary, "A Bulletin Board," *The Mathematics Teacher*, LV (April, 1962), p. 301.

Lewis, E., "Role of Sensory Materials in Meaningful Learning," *The Mathematics Teacher*, XLIX (April, 1956), pp. 274-7.

Lowry, W. C., "Pupil Discovery in Junior High School Mathematics," *The Mathematics Teacher*, XLIX (April, 1956), pp. 301-3.

Macarow, L., "Teaching High School Essential Mathematics; Approach Stresses Student Participation," *Chicago Schools Journal*, XLVII (February, 1956), pp. 259-63.

Mayor, J. R., and J. A. Brown, "Teaching the New Mathematics," *School and Society*, 88 (October, 22, 1960), pp. 376-7.

Mock, G. D., "Method Course," *The Mathematics Teacher*, LTV (January, 1961), pp. 17-9.

Moise, E. E., "New Mathematics Programs: What Do They Mean?" *The Education Digest*, XXX (March, 1965), pp. 27-9.

Nemeck, P. M., "Stimulating Pupil Interest," *School Science and Mathematics*, LXV (January, 1965), pp. 47-8.

Payne, Joseph N., "Self-Instructive Enrichment Topics for Bright Pupils in High School Algebra," *The Mathematics Teacher*, LI (February, 1958), pp. 113-7.

Perham, A., "Exercise for the Mathematics Laboratory," *The Mathematics Teacher*, LVIII (February, 1965) pp. 114-7.

Phillips, H. L., "Mathematics Laboratory," *American Education*, I (March, 1965), pp. 1-3.

Ranucci, E. R., "Discovery, in Mathematics," *The Arithmetic Teacher*, XII (January, 1965), pp. 14-8.

Read, C. B., "Use of History of Mathematics as a Teaching Tool," *School Science and Mathematics*, LXV (March, 1965), pp. 211-8.

Reeve, W. D., *Mathematics for Secondary Schools* (New York: Henry Holt and Company, 1954), Chapter 5.

Rummell, F. V., "She Made Mathematics Almost Flunk-Proof," *The Education Digest*, XXIV (October, 1958), pp. 32-4.

Stein, H. L., "How to Make Arithmetic Meaningful in the Junior High School," *School Science and Mathematics*, LIII (December, 1953), pp. 680-4.

Todd, J., "Is Review Really Necessary?" *School Science and Mathematics*, LXIII (January, 1963), pp. 68-9.

Tulock, M. K., "Emotional Blocks in Mathematics," *The Mathematics Teacher*, L (December, 1957), pp. 572-6.

Van Engen, "Breaking Through the Know-How Barrier," *The Mathematics Teacher*, LI (February, 1958), pp. 131-2.

3

Discovery Approaches

> ... there is no more significant privilege than to release the creative power of a child's mind.
>
> Franz F. Hohn[1]

3.1. Introduction

Increasing attention is being given to discovery approaches in modern educational literature and many texts appear which claim discovery-oriented features. Some writers mention employment of the discovery approach and thereby suggest that this is a technique or a classroom practice which is entirely new and completely foreign to other types of teaching. This chapter will show that there are several facets to discovery approaches and that for many years good teachers have been using methods (combinations of heuristic, genetic, psychological, and inductive, all of which are student centered) which encourage student creativity.

In this chapter we will discuss discovery approaches in some detail and describe several kinds of methodology which will be helpful in translating these procedures into action.

---

[1]Franz F. Hohn, "Teaching Creativity in Mathematics," *The Arithmetic Teacher*, VIII (March, 1961), 102-6.

3.2. What Discovery Means

In brief, discovery means that the student recognizes some mathematical properties or relationships through his own effort. To qualify this statement, discovery, as the experienced teacher sees it, involves encouragement of the student to learn, uncover, and devise as much mathematics as he reasonably can by himeself and to organize his findings in a logical manner. Of course, one does not expect a student to discover all the mathematics with which he is immediately confronted, but a spirit of investigation can be encouraged and an atmosphere of making conjectures with subsequent checking can contribute much to the student's growth. Indeed, such a spirit and atmosphere pay large dividends. Discovery, also, is like the call of mathematics: there are several facets and all of them help to synthesize approaches which we might call discovery-oriented.

Discovery means the growing awareness of mathematical principles or relations on the part of the student without being told of them by the teacher or by some other authoritative source. After many trials with concrete objects the student may become aware that another name for

$$\frac{2}{3} \times \frac{4}{5} \quad \text{is} \quad \frac{2 \times 4}{3 \times 5}$$

and that another name for

$$\frac{a}{b} \times \frac{c}{d} \quad \text{is} \quad \frac{a \times c}{b \times d}.$$

He has, through this exploration, discovered a way to compute the product of two fractions in terms of products of other numerals. A discussion of rationale would come later. Or, by suitable exercises, a student can become aware that if one had the differentiable relations

$$x = x(t),\, y = y(t) \text{ then } D_x y = \frac{D_t y}{D_t x}\ (D_t x \neq 0)$$

and he has made a useful discovery. This moment of awarness is an exciting one for the student and for the teacher. This phase of discovery captures some of the thrill experienced usually only by researchers in mathematics and which it should be the right of every student to enjoy. "The joy of discovery . . . is the antithesis of the anxiety of rote: . . . this joy can be experienced at any conceptual level."[2] The student may not be able to tell in good mathematical language what he has discovered since he may lack the vocabulary and the niceties of mathematical expression. The point is that

---

[2]John Biggs, "The Psychopathology of Arithmetic," *New Approaches to Mathematics Teaching*, edited by F. W. Land (London: MacMillan and Company, Ltd., 1963), p. 65.

he is aware of the *principle*, and *he* has discovered it. Such a state of learning is called "non-verbal awareness" by Hendrix.[3] Later the student may experience a second deep thrill when he perceives that there is a *logical connection* between his discovery and the logical organization already built up. Then comes the time for the refinement and precision of language and the attempt at the construction of a formal proof. True, to help the student become aware of certain principles, the teacher as a guide must structure experiences to set the stage for awareness. We are not emphasizing that the student be led around and around until he accidentally says the right thing, but we are suggesting investigation and research on the student's level. Awareness of relations need not come only from examples, laboratory work, and other experiential media, but occasionally discoveries come from deductions themselves. More general relations, having found a place in a deductive system, can lead to discoveries about particular cases.

Discovery means *invention*. Leibnitz has said, "Nothing is more important than to see the sources of invention which are, in my opinion, more interesting than the inventions themselves."[4] Invention may refer to algorithms or it may refer to new mathematical forms or to new mathematical ideas. For the student, the invention of a new idea is most deeply significant and moving. For the teacher, inventions by students effect deep satisfaction, pleasant surprises, and extend horizons. The teacher learns much, too; a common spirit of investigation, growth, and respect, is communicated in the classroom. Left to his own devices, the student originates much. Asked to study the area of trapezoids, one student invented the algorithm of "finding the average length of the bases, and then multiplying this number by the height;" another built a rectangle on one base, built a rectangle on the other base, and then found the mean area—what a revelation it was when it was found that both these inventions were able to be summarized by the same mathematical expression!

Discovery means *creativity*. Proper methodology can do much to release and to cultivate the creative powers of students. The expression of creativity need not await the mature experience of writing research papers or of conducting investigations. Creativity in students can arise in the most elementary situations. Students may create operations of their own devising, they may create explanations for answers received in laboratory experiences, they may create conjectures of all kinds, or they may create new kinds of mathematics. All course in mathematics are rich with possibilities of exploration and creative study: the problem of how many lines are determined by points, no

[3]Gertrude Hendrix, "Learning by Discovery," *The Mathematics Teacher*, LIV (May, 1961), 290-9.

[4]George Polya, *How To Solve It* (Princeton: Princeton University Press, 1945), p. 112.

three of which are collinear, arises naturally in the geometry classroom. While writing abc as the product a and b and c, one can ask, in freshman algebra, how many different ways this product can be written. If he so chooses, one is off to the entirely new field of permutations of letters and to certain considerations in probability. At almost any level one can raise the problem of the slope of the tangent to a curve and students may create a method for answering this question. As much as possible, topics which are to be considered later should be started in very humble, explorable ways to capitalize on the creative powers of the student.

Discovery means *testing and proving*. Here we have in mind both the original meaning of proving (from the Latin *probare* meaning to test) with more examples and testing to see if a conclusion can be found by deduction to be a logical consequence of foregoing material. Along with every discovery of what seems to be a mathematical principle there is the intellectual obligation to see whether or not the suggested principle can withstand the refining fires of counterexamples and the test of placement in a deductive system. There is abiding value in a student's defense of his own conjecture before others and a sense of intellectual humility arises when he realizes that perhaps time or ingenuity does not permit all conjectures to be tested successfully by deduction, This realization emphasizes the difference between a conjecture and a theorem and helps the student to recognize that mathematics is a growing field. Most important, he has been responsible for at least one step in its growth.

Discovery means *growth*; a kind of growth which carries one far beyond the confines of a text or of a prescribed course of study. The student who wishes to devise a means of computing the area of a triangle in terms of coordinates of the vertices may not devise a determinant, but he may invent other forms whose study takes him into unpredicted directions. Discovery approaches invariably open new fields and encourage the construction of new special disciplines. What is mathematics, but a way of thinking and a manner of organizing material?

Discovery means *using the student*. Perhaps the discussion above has caused the reader to become aware that discovery approaches are largely student centered and that countributions by students are of paramount importance. In fact, we agree with Ivey when he says ". . . education has a great teaching facility which as yet is unused—the student."[5] Hutchinson says, "May it be that we have overlooked our best allies in our fight for a sound education based on fundamentals—the students themselves?"[6] Making use

[5]John F. Ivey, Jr., "Matching Quantity with Quality in Education," *The Ohio State University Educational Research Bulletin*, XXXIX (September 14, 1960) No. 6, p. 152.

[6]C. A. Hutchinson, editor, Section on Mathematics Education, *The American Mathematical Monthly*, XLVI (May, 1939), 280.

of student contributions simply means that the teacher's method takes on the predominant characteristics of those found in the previous chapter which were student centered. Using the student does not mean that there is no direction, no practice, no demands, no obligations, no discipline of study. It means that, as much as possible, the mathematics is the result of student investigations. The teacher serves as an engineer to structure experiences, as an interested learner, and a referee in the right of his more mature mathematical background. At one time he is a drill master; at another time he helps with investigations; and at still another time he is a creative mathematician along with his students. Instead of being responsible for covering the ground he is responsible for guiding the student in uncovering ideas.

Discovery in the classroom, then, is a concept with several facets; the reader could perhaps add more than those discussed here. Above all, discovery is a spirit which pervades the classroom. To have a dogmatic teacher-centered presentation of material and then a few discovery exercises at the close of a chapter is not discovery in the sense used here; nor is it the solution of a few puzzle problems, or the discovery of a few relations among numbers, or finally guessing at the right word to say, as a teacher gives hint after hint to the wondering student. Rather it is a kind of methodology which pervades the whole process of teaching and which encourages student contributions from most experiential and simple beginnings to considerations of a more profound nature, and then seeks the cooperative help of the students to organize the knowledge into a logical system. With such a methodology passiveness is at a minimum while activity and creativity are at a maximum. The mind of the student is not a cistern to be filled, but an instrument to be used.

We urge the student to read more on discovery procedures. Some references are found in the exercises below and at the close of this chapter.

Exercises

1. Read one of the articles on discovery as Reference 28 in Chapter 2, References 1 and 4 in this chapter, or one suggested below or one found in more recent journals. Summarize in several succinct statements. Be sure to include the reference data for the article.

   Humphrey Jackson, "Creative Thinking and Discovery," *The Arthmetic Teacher*, VIII (March, 1961), 107-11.
   Brenda Lansdown, "Creating Mathematicians," *The Arthemetic Teacher*, VIII (March, 1961), 98-101.
   Harold Lerch, "Arithmetic Instruction Changes Pupil's Attitudes Toward Arithmetic, *The Arithmetic Teacher*, VIII (March, 1961), 117-9.
   Luch Nulton, "Arithmetic: Arthritics or Adventure?" *The Arithmetic Teacher*, VIII (November, 1961), 345-9.

2. Consult *The Education Index* to learn of articles on discovery not found in

periodicals of the National Council of Teachers of Mathematics, but preferably in journals for teachers, administrators, or school-board members. Summarize in several sentences.

3.3 Discovery Through Situation Approaches

This classification is used to indicate approaches which involve placing students in situations which lead them to develop or to create mathematics new to them. Sometimes this means that the student will merely see the desirability of having some new mathematics. The arithmetic student who finds himself in a situation which requires the adding of fractions soon feels the need for inventing some algorithms which will help him to add fractions quickly; the student who has designated positions on a map by ordered pairs of numbers soon raises the possibility of computing distances between points in terms of these ordered pairs. All of this is mathematics new to him. The student who becomes aware of some of the problems of transportation (e. g., costs and control of traffic) soon feels that there is a need of creating some kind of mathematics to study these problems.

Another meaning which we assign to the situation approach is that the student experiences a story out of which mathematics emerges, often nearly unknowingly. Such a story may concern John's method of keeping score where (8, 6), let us say, means that he won eight points but lost six and that he knows nothing about positive and negative integers; he does know that if he won five and lost nine then he writes (5, 9). Out of this story or situation can emerge an ordered-pair approach to the integers where the elements of the ordered pair are natural numbers; indeed the student moves into the more sophisticated postulational development very easily, and, with guidance, much on his own power. We will examine and develop such approaches as these in the following discussion.

It is often not difficult and it is challenging to the teacher to try to contrive stories or situations out of which fundamental mathematical ideas can emerge. Through such experiences student creativity rises to a high level and student understanding becomes much deeper than it does when the student is expected merely to be an acceptor of what the teacher gives. In such situations, the student is interested, he enters into the situation with an open and unprejudiced mind, and he often identifies himself with one of the characters in the story. Above all, there is no anxiety and the student has no fear. Often the student contributes potential solutions to the problem at hand with a zeal and intensity not found elsewhere. Many times the student begins to use concepts and techniques of which, at the time, he may not know the name; indeed, he may even help in their classification and naming. Rationale and abstractions come later.

Of course, the situation approach is student centered and already our discussion in this chapter has included features of this approach. We shall elaborate on it by the use of further examples.

(a) Here is a situation which helps the student eventually to develop the concept of adding signed numbers. Yet the idea grows from a setting for which the teacher does not mention that, "Today we are going to study signed numbers." Concepts come first, names afterward. Note that there would be no fear on the part of the student; he is drawn into the story or the situation.

Johnny is playing a game in which he records his score as 3 G if he has *won* three points or if he has "three to the good." When he loses or *owes* points, say 4, then he writes 4 σ. In one game he had 4 points to the good and then he lost 5. He wrote

4 G (beginning score)

5 σ (change in score)

so, in the end what was his standing and what symbol would represent it?

In other parts of the game Johnnie had beginning scores and changes as follows. What was the score after each of these parts of the game?

Beginning	Change	End Result
3 G	2 G	
4 G	5 σ	
10 G	12 σ	
12 G	10 σ	
14 G	10 σ	
6 σ	3 σ	
10 G	9 G	
3 σ	4 σ	
6 σ	3 G	

Teachers may prefer to write these situations in vertical form too, as

Beginning	3 G	4 G	10 G	12 G	14 G	6 σ	10 G	3 σ	6 σ
Change	2 G	5 σ	12 σ	10 σ	10 σ	3 σ	9 G	4 σ	3 G
End Result									

It is not long, however, until the student can use *positive* and *negative* language instead of G and σ, and the concept becomes deeply rooted. Other situations from which the addition of *signed numbers* can be clarified will be discussed in a later chapter.

We wish to emphasize that any algorithms for adding signed numbers can and should emerge from actual or contrived experiences of the students and that student-invented algorithms should come before rules.

(b) To introduce the idea of parametric equations, a story like this has been found effective:

John and Tom decided to play a game. John would tell the value of $3t^2 + 10$ (which he labeled x) and Tom would tell the value of $t^3 - 2$ (which he labeled y). Now Tom's little brother Harry wanted to play too, so they gave Harry the task of selecting cards containing values of t.

When Harry called out $t = 1$
What should John (x) say?
What should Tom (y) say?
When Harry called out $t = 2$
What should John (x) say?
What should Tom (y) say?

Here it is seen clearly that as the value of t is named, x and y both become activated and produce values. Indeed t is a variable also, and it seems to act as the control for the values of x- and y-variables in the discussion. t is called a parameter.

(c) In introducing experiences in various numeral systems, one might suggest a story like this:

Johnny went with his parents to a country called Fivonia. The people there were about the same as Johnny's friends, but they had a strange way of writing numerals for the numbers. In fact, when a Fivonian wished to write the numeral symbolizing how many objects there were in the pile in Fig. 3.1 he would

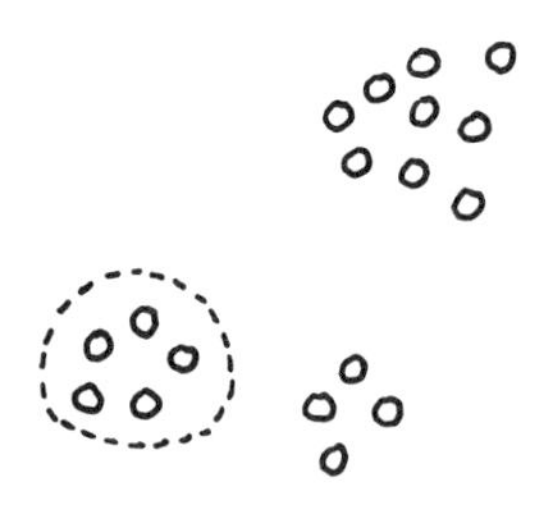

Figure 3.1

gather them up by *fives* and write 1 4 to mean *one five* and *four* more. The collection shown in Fig. 3.2 whould have its number

Figure 3.2

of elements indicated by the numeral 3 2. Tommy was told that once there were five packages of five each, then a numeral was placed in the third position from the right. The numeral 2 4 1 meant one and four groups of five each and two groups of twenty-five each. Johnny became quite interested in this strange system and he decided to try his hand at writing numerals in the Fivonian style. What numerals should he write to communicate the "how-many-ness" of these sets shown in Fig. 3.3?

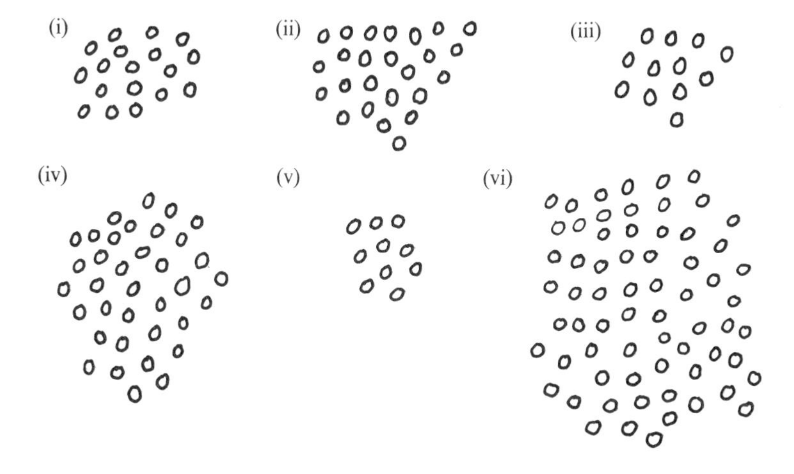

Figure 3.3

The teacher may prefer to use numerals which are not Hindu-Arabic, but rather different symbols and word names as different systems are used. Also, the teacher could well have the student act out this situation in the classroom with lima beans or other concrete objects.

Longer situation approaches may well continue for several days and the story accordingly becomes less prominent as the development proceeds. We give here an example on developing the idea of a matrix and the addition of matrices.

Tommy was a basketball player and he was interested in keeping a record of the baskets made in various quarters and games. His successes were tabulated as follows:

Quarter	1958 Season Game Number 1	2	3	4	5		1959 Season Game Number 1	2	3	4	5	
1	3	2	1	0	1		0	1	3	10	4	
2	4	1	3	2	1	$= A;$	10	1	3	1	2	$= B$
3	1	3	2	4	5		10	10	3	10	8	
4	3	0	0	2	1		9	8	7	1	3	

$$A = \begin{pmatrix} 3 & 2 & 1 & 0 & 1 \\ 4 & 1 & 3 & 2 & 1 \\ 1 & 3 & 2 & 4 & 5 \\ 3 & 0 & 0 & 2 & 1 \end{pmatrix}; \quad \begin{pmatrix} 0 & 1 & 3 & 10 & 4 \\ 10 & 1 & 3 & 1 & 2 \\ 10 & 10 & 3 & 10 & 8 \\ 9 & 8 & 7 & 1 & 3 \end{pmatrix} = B$$

Such an arrangement of data is called a matrix. Once we have the form in mind (what the rows and columns mean), then the labels need not be written. Hence, for the 1960 season Tom might have

$$\begin{pmatrix} 1 & 3 & 2 & 1 & 4 \\ 1 & 3 & 2 & 9 & 8 \\ 10 & 4 & 2 & 1 & 6 \\ 0 & 0 & 3 & 5 & 5 \end{pmatrix} = C$$

The matrices in the examples are called 4 by 5 matrices, for there are four rows and five columns and the numbers 4 and 5 are called the dimensions of the matrix. The entry in the ith row and the jth column is indicated by a_{ij}.

Exercises

1. In Matrix A what is a_{12}? a_{13}? a_{24}? a_{35}? a_{45}?

2. Suppose in Matrix A the term a_{15} is changed from 1 to 0. Is this now the same matrix? Hence what definition do you suggest for two matrices to be equal?

3. Let us suppose now that Tom wishes to construct a new table or a new matrix to show the total scores made by him in each game and in each quarter for a period of two years, those of the 1953 and the 1959 seasons. What is the sum of the scores for the two games Number 5 for the first quarter and hence the entry in the first row and fifth column of the new matrix? Construct the new matrix which summarizes the results of the two seasons.

4. Hence how might one define the addition of two matrices with like dimensions?

5. Hence given the matrices

$$A = \begin{pmatrix} a_{11} & a_{12} & a_{13} \\ a_{21} & a_{22} & a_{23} \end{pmatrix} \quad \text{and} \quad B = \begin{pmatrix} b_{11} & b_{12} & b_{13} \\ b_{21} & b_{22} & b_{23} \end{pmatrix}$$

what matrix defines $A + B$?

6. Is the addition of matrices which have the same dimension a commutative operation with addition defined as you suggested in Exercises 4 and 5 above? Let us assume that the entries in the matrices are real numbers and then see if you can deduce your conjecture about commutativity of the addition of matrices from the properties of the addition of real numbers.

7. What other questions might we raise about the addition of matrices? See if you can deduce the properties which you suggest from the properties of real numbers and the definition of addition of matrices.

8. What must be the structure of a matrix which when added to Matrix A would yield as the sum Matrix A? What might such a matrix be called?

9. If

$$A = \begin{pmatrix} 2 & 3 & 4 \\ 1 & 2 & 1 \end{pmatrix} \quad B = \begin{pmatrix} -3 & +5 & +7 \\ -1 & -1 & -1 \end{pmatrix} \quad \text{and} \quad C = \begin{pmatrix} +4 & 0 & +3 \\ -4 & +1 & +6 \end{pmatrix}$$

what matrix expresses

(a) $A + B$? (b) $A + C$?

10. From the matrices given in Exercise 9, what Matrix X added to A would yield B? That is, if $X + A = B$ then what is Matrix X?

11. Given this statement on the equality of two matrices, we can write four equations in x and y. See if you can solve these four equations simultaneously for x and y values.

$$\begin{pmatrix} x - y & 2x + y \\ x - 3y & x + 4y \end{pmatrix} = \begin{pmatrix} 4 & 14 \\ 0 & 14 \end{pmatrix}$$

12. Find values of x, y, and z which satisfy the relation

$$\begin{pmatrix} 2x - 5 & z - 4 & 3 - 5y \\ x + 6 & y - 4 & 6 - 4y \end{pmatrix} = \begin{pmatrix} -1 & 0 & -12 \\ 8 & -1 & -6 \end{pmatrix}$$

In the chapter on the teaching of algebra, we shall continue this situation approach to encourage the emergence of what mathematicians regard as the definition of the multiplication of matrices.

Note that in the situation approach the student is not always informed of the topic which he is going to study. Often an announcement of the topic may cause anxiety and possible fright, especially if parents or older brothers and sisters give unfavorable impressions about their mathematical experiences. A boy who relates that, "Tomorrow our teacher said that we would start logarithms" and is then advised that, "you will find these difficult and mysterious—I never did understand them," is beginning to develop a block against logarithms at that very moment. Yet, through situation and other approaches the student may find himself helping to invent logarithms and he is likely to understand their nature and the usefulness of these interesting numbers, perhaps before he even knows that they are called logarithms!

Exercises

1. Work through the examples and exercises in the situation approach above on matrices.
2. Note again the topics listed under Section 2.5 in Exercise 5, and use one of these or another of your interest to develop a short situation approach to encourage formulation of ideas by the student himself. Include questions and exercises.
3. A good situation approach to the meaning of numeral as opposed to that for number is found in Beberman and Vaughn's *High School Mathematics Course 1*.[7]
4. Denbow and Goedicke, in their *Foundations of Mathematics* for college-level classes, employ a situation approach with "the furniture mover's story" to develop the idea of a group. It is largely descriptive but it is a method of introduction which differs from the formal definitions of groups.[8]

3.4. Discovery Through Closely Guided Developmental Approaches

Perhaps the critic of discovery approaches will contend that no student or group of students has the time or the capability of discovering all the mathematics even in a given course of study. It is not claimed that the student can discover all, but there is convincing evidence that the student learns much more when he is encouraged to discover and to formulate what he can in his own way than if the material were presented in a dogmatic fashion.

It is true that there are topics in mathematics to which a critic of discovery approaches may point because the student might not be expected to devise ways of handling the problem; that is, there are some topics in which the student needs much guidance. In such cases one can use what might be called *closely guided developmental approaches*. In actual practice, we may use the technique of inductive quizzes or of written class exercises written by all students, not in the sense that the teacher is an examiner and

---

[7]Max Beberman and Herbert F. Vaughan, *High School Mathematics, Course 1* (Boston: D. C. Heath and Company, 1964), p. 1.

[8]Carl H. Denbow and Victor Goedicke, *Foundations of Mathematics* (New York: Harper, 1959), pp. 118 *et seq.*

that these exercises are graded, but in the spirit of developing an idea. Such an approach guides the student in his thinking by means of a series of questions with short answers, with the advancing steps so small that the student passes easily from one to the next. Answers are checked immediately. This procedure provides a very effective reinforcement in the learning process and one notes that it is quite close to programmed learning.

We give hcre some examples of inductive quizzes, or closely guided developmental approaches, in algebra, geometry, and calculus.

(a) *Problems on consecutive numbers* (answers are given in parentheses)

1. I am thinking of an integer n. Write n on your paper as the numeral for this integer.
2. What numeral would represent the next consecutive integer? $(n + 1)$
3. What numeral would represent the second consecutive integer? $(n + 2)$
4. We now have $n, n + 1, n + 2$ as the numerals for three consecutive integers. Suppose now I tell you that the first integer plus three times the second integer plus six times the third integer is 65. See if you can write a relation which says this:

$$n + 3(n + 1) + 6(n + 2) = 65$$

5. Now see if you can determine the integers from this relation. (5, 6, 7)

(b) *A problem in geometry*

In testing the hypothesis, or discovery made by other means, that the "Join of the midpoints of two sides of a triangle is parallel to the third side, and the measure of its length is one-half the measure of the length of the third side" the student would hardly be expected to devise by himself the usual procedure of producing

$$\overline{DE} \text{ to } F \text{ so that } \overline{DE} \cong \overline{EF}$$

An inductive quiz, after the assumed data are clearly recognized, might appear as follows (refer to Fig. 3.4):

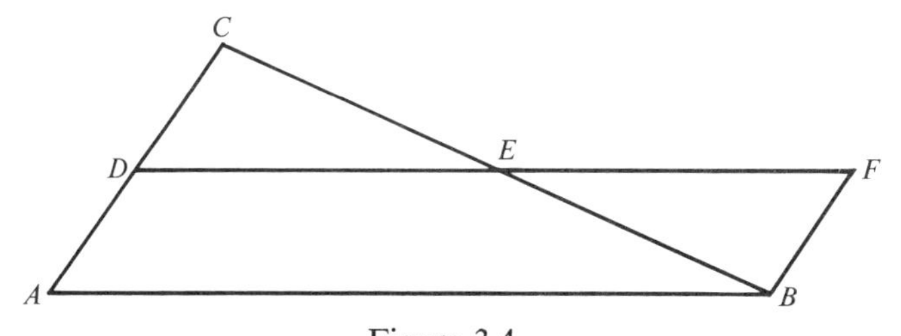

Figure 3.4

1. If DE is produced to F so that $\overline{DE} \cong \overline{EF}$, do you note any triangles which might be investigated for a useful relation? Deduce this relation, if possible.
2. Is a relation between $\overrightarrow{\mathbf{CD}}$ and $\overrightarrow{\mathbf{FB}}$ readily apparent? Test this hypothesis by deduction.
3. Can you now deduce that $ABFD$ is a special kind of quadrilateral? Give reasons.
4. Hence, what can be said about $\overrightarrow{\mathbf{DE}}$ and $\overrightarrow{\mathbf{AB}}$? Why?
5. See if you can deduce a relation between $m\overline{DE}$ and $m\overline{AB}$.
6. State again the hypotheses which you made in your study of this figure and which can now be deduced and hence can fit into our deductive system of geometry. This leads to an assignment for the student to organize a proof in good synthetic style. The above merely suggests questions for the student. The teacher may well use invitations as "I wonder if. . ." or might well stop at any time and ask, "Does anyone see a way to complete this argument?" Students may also suggest the next item which can be proved. The writers have found such classroom approaches very effective in introducing proofs.

(c) *A problem in calculus*

Let us suppose that you wish your students to suggest the usual method of defining the area of a region described in polar coordinates

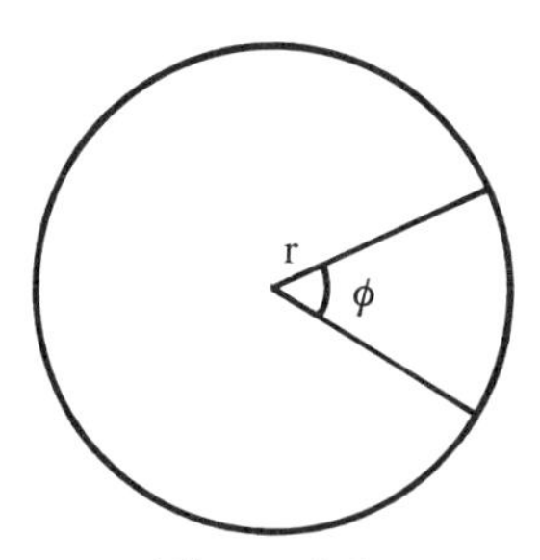

Figure 3.5

(see Fig. 3.5). Your students

(i) have already developed the formula

$$A = \frac{r^2\phi}{2}$$

for the area of a circular sector where r is the radius and ϕ is the angle in radians between two radii which form the sector;

(ii) have had previous experience with the definition

$$\lim_{n\to\infty}\left\{\sum_{i=2}^{n} y_1\Delta_1 x\right\} = \int_a^b ydx$$

where

$$\sum_{i=2}^{n} \Delta_i x = b - a$$

(iii) have had previous experience with setting up and examining sequences of approximations leading to the definition of area under curves in rectangular coordinates.
Suppose the problem is to define the measure of the area of the region bounded by the curve whose equation is $r = 1 + \sin\theta$ and the lines $\theta = \pi/6$ and $\theta = \pi/2$. One might use questions as follows:

1. What is the length of r when $\theta = \pi/6$? (Answer: 1 1/2)
2. If we use this value of r as the radius of a circular arc (Fig. 3.6) to

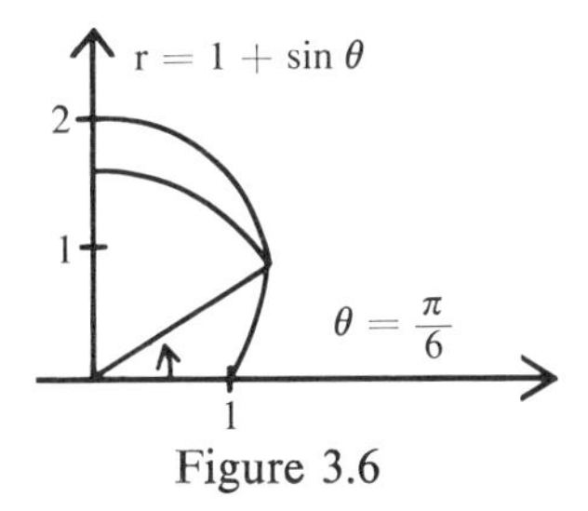

Figure 3.6

approximate the measure of the area under the given curve, what will this approximate measure be? (Answer: $3\pi/8$)
3. Is this approximation to the measure of the required area too small or too large?
4. How could we get a better approximation to the measure of this area?
5. Let us suppose that we use two circular sectors and consider their sum as a better approximation (Fig. 3.7). Let us use the radius at $\theta_1 = \pi/6$ as that of one of the sectors and the radius at $\theta_2 = \pi/3$ as the radius of the other sector. What would be the value of the approximate measure now? [Answer: $\pi/12(4 + \sqrt{3})$]
6. How could we get a better approximation? Using three circular arcs with equal central angles taken from the partition $[\theta_1\theta_2\theta_3\theta_4]$ see if you can write a third approximation in $\sum$ notation.

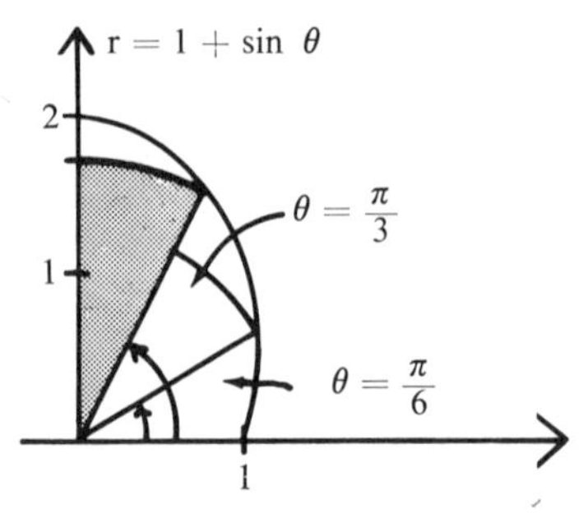

Figure 3.7

7. See if you can write in $\sum$ notation an approximation using n sub-angles each of size $\Delta_i\theta$.
8. Now let us see if we can define the area measure-number as the limit of a certain sequence.
9. Define the limit in No. 8 as an integral.
10. See if you can evaluate this integral by the use of the Fundamental Theorem of the Calculus.

Many students achieve the desired results with sets of class exercises as those shown above. Some students may not need some of the intervening steps, while a few are able to make progress from one step to another from the merest suggestions. Experience has shown, however, that many students need the kind of guidance outlined and that this guidance helps to emphasize the place of definition, primitive statements, previously proved theorems, and the logical organization of mathematics. The reader needs only to reflect on the difference between a student who studies illustrative examples of the use of algorithms and methods and the student who sees a method grow as a result of his own writing and thinking. The latter becomes a partial creator of the method.

Developmental approaches may lack some degree of complete creativity on the part of the student yet he is not just an onlooker or a follower.

Approaches of this kind have many benefits for the teacher in the classroom who may use such approaches as class exercises or as material for study guide sheets to be discussed in the ensuing class period. The writer has found such short developmental exercises useful in the class period itself. Further discussion develops from which students may be strengthened mathematically to begin on work for tomorrow. Such procedures make the student more confident in that he understands because he has helped to develop parts of the subject. Such procedures help to keep the class interested and industrious on worthwhile creative and constructive activity. A student-as-a-participant approach of this kind might well be used to develop ideas or methods of proof or algorithms which the student could hardly originate heuristically in the time available. Indeed, developmental approaches might well take care of some of the "imperfections" mentioned by J. W. A. Young when he says:

> But it must always be remembered that the heuristic method, with all the imperfections which it may have, comes much nearer to realizing the aim and ideals of mathematical instruction than any mere passive ingesting of a body of mathematical facts, however voluminous.[9]

In the constructing of developmental approaches we might well take more advice from Young:

> The work must be broken up into simple steps, well within the child's power. And it is easy to be mistaken in thinking the steps are sufficiently simple. The only proper test of this is success of the pupil in doing what he is asked. . .[10]

Exercises

Select three topics in mathematics which you feel might profitably be developed by methods suggested above. Write five or six appropriate questions for each as you would on a developmental quiz.

3.5. Discovery Through "Mathematics to the Rescue"

"Mathematics to the rescue" means that the student is placed in a situation or in a predicament which requires much tedious work to resolve and, in order to emerge successfully or with greater efficiency, he sees that he can use mathematics which he knows or mathematics which has been previously developed in another connection. True, this may appear to be close to situation approaches, and, indeed it is, but we believe that this approach has a slightly different tinge. Some examples will be taken from algebra and from calculus.

Mixture Problems in Algebra

Consider the following problem: "How many grams of pure salt must be added to 50 grams of 4% solution of salt to yield a 6% solution?" The mathematics-to-the-rescue approach would encourage the student to make guesses at the answer and to check each until he arrives at a desired result.

[9]J. W. A. Young, *The Teaching of Mathematics in the Elementary and the Secondary School* (New York: Longmans, Green and Company, 1924).

[10]*Ibid*, p. 73.

In guessing and subsequent checking by arithmetic, the student is assured of understanding what the problem is about. For example, if the student guesses at 3 grams, then he checks by

Computing how much salt is in the solution (2 grams)
Adding the guessed 3 grams of salt to the original 2 grams (5 grams)
Dividing 5 by the total number of grams, 53, to see if he obtains 6% (9.4%).

Mathematics comes to the rescue when, after several such guesses, the student becomes aware of or originates the idea that he could use a letter in place of the guessed answer and try to solve an algebraic equation! The latter procedure gives the answer on the first try and illustrates the power of algebra and of deduction. Often this means some exciting and dramatic moments in the classroom. More examples of this approach in algebra will appear in our chapter on the teaching of algebra.

Computing the Slope of a Curve at a Given Point on the Curve

Using Fig. 3.8, let us suppose that we wish to arrive at some conclusion about the slope at $a = 3$ of the curve whose equation is $y = x^2$. In begining discussion the slop of a curve might not be precisely defined and so we shall depend on an intuitive notion of slope. Indeed, a precise definition of slope of a curve at a given point might well emerge from the following discussion. We seek a method to approximate what we might intuitively regard as

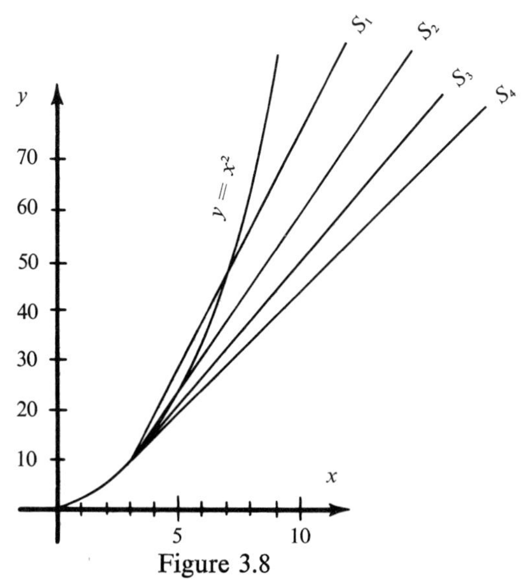

Figure 3.8

the slope of the curve by using the slope of the secant S_1 from $x = 3$ to (let us say) $x = 7$. This approximation yields $m_1 = (49 - 9)/4 = 10$, where m_1 is the slope of the secant S_1. A better approximation is gained by using the slope of the secant S_2 from $x = 3$ to $x = 5$, or $m_2 = (25 - 9)/2 = 16/2 = 8$. A still better approximation is obtained by using the slope of the secant S_3 from $x = 4$, or $m_3 = (16 - 9)/1 = 7/1 = 7$. Using the slope of the secant S_4 from $x - 3$ to $x = 3\frac{1}{2}$, one secures

$$m_4 = (12.25 - 9)/\tfrac{1}{2} = 3.25/\tfrac{1}{2} = 6\tfrac{1}{2}.$$

Hence, we get the following sequence of approximations:

$$10, 8, 7, 6\ 1/2, \cdots$$

How can we find the limit of this sequence as the lengths of the abscissa change become smaller and smaller or as the number of terms becomes larger and larger? Here is a situation in which one can make only conjectures and the student may be at a loss concerning how to proceed. For the moment we are at a stopping point. Some students have rewritten this sequence as follows:

$$10, 10 - 2, 10 - 2 - 1, 10 - 2 - 1 - 1/2, \cdots$$

with the terms finally appearing as

$$10, 10 - 2, 10 - (2 + 1), \cdots, 10 - (2 + 1 + 1/2 + \cdots)$$

and now we can use the expression

$$S_\infty = \frac{a}{1 - r}$$

for the sum of an infinite geometric progression, an old expression useful in a new setting. We would not expect this result to be used at the close of a lecture; rather we would hope that some student would suggest this or, if the use of this expression did not materialize from the class, then the teacher would direct the students to its use. Mathematics comes to the rescue to give an answer of $10 - 4$, or 6, as the limit of the sequence of slopes of secants. The answer may now be taken as the definition of the slope of this curve at $x = 3$. For a more complete discussion on this approach to the teaching of calculus the reader is referred to Reference 32.

Such an approach, as all such approaches do, helps the student to understand what the problem is all about. The concreteness of the numerical approach coupled with the experience of using "old" mathematics in a new setting gives insight into how mathematics grows as well as appreciation of the power and the consistency of mathematics as a discipline. Mathematics emerges in a dramatic way to rescue the student from tediousness and from situations which one might not otherwise be able to resolve. Such an approach helps to keep the student aware of man's continual groping for mathematics to serve his ever-increasing need and the fact that uses of mathematics appear in most unexpected places; *a priori*, what relation might there be between the infinite sum of a geometric progression and defining the slope of a parabola at a given point? What might have appeared to be relatively unimportant in the algebra classromm now looms into fundamental significance. This approach, too, emphasizes the *outward* call of mathematics, and, at the same time, makes the student conscious of how various tools in mathematics advance the progress of mathematics itself and how each step in the process is a logical consequence of a previous step.

The spirit in the classroom which is present when discovery approaches are used (e. g., situation, mathematics to the rescue) is markedly different from that which arises from the text description or from receiving generalities which are often not understood. Generalizations which the students perceive and attempt to state are much more vital to the student's development than generalizations which are dictated for him to accept in a passive manner.

In practice one should not expect these dramatic moments of "rescue" to come from experiential beginnings limited to one class period. Time is required for exploration, experimentation, and reflection. Hence, in this approach, the student may explore a proposed new topic for several days by means of concrete (which may mean numerical) examples while other work in more advanced stages is currently in progress. The time factor and organization of class work will be discussed more fully in Section 3.7 and in the various succeeding chapters on the teaching of special subjects.

Exercises

1. Note again the mixture problem (a) and make several more guesses to the answer. Now in all those positions in which you used a guessed answer to make the check, place a letter, as x, and impose the condition that the resulting expression must equal 6%. What value of x makes the statement valid?

2. Using the method of (b) arrive at a number which will define the slope of the curve $y = 3x^2$ at $x = 4$. If you prefer you might use the point at which $x = a$.

3. John passed a certain point going west at a uniform rate of 40 miles per hour. Three hours later Harry passed this same point going west at a uniform rate of 50 miles per hour. After how many hours will Harry overtake John? Guess at the answer and check by arithmetic.

4. Consider the region bounded by the curve $y = x^2$, the x-axis, and the line $x = 8$ (Fig. 3.9). You compute a sequence of approximations to what we would intuitively call the area of this region. Let us suggest that the initial approximation be that of the area of a right triangle and that succeeding

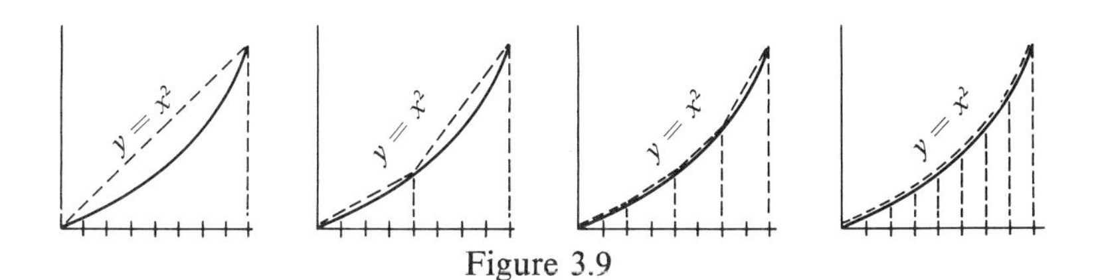

Figure 3.9

approximations come from the use of one right triangle and trapezoids. See if you find the limit of the sequence of sums of areas of rectilinear figures as suggested above. Note the simple arithmetic beginning which uses again

$$S_\infty = a/1 - r.$$

3.6. Discovery and Student Organization of Materials

The organization of materials into a logical system or the proposal of a rationale by the student is a most effective means of giving the student insight into what mathematics is as defined by Cassius Jackson Keyser when he says that

> As an enterprise, mathematics is characterized by its aim, and its aim is to think rigorously whatever is rigorously thinkable or whatever may become rigorously thinkable in course of the upward striving and refining evolution of ideas.[11]

In fact, we would be shortchanging the student by denying him one of the most valuable experiences in mathematics if we were always encouraging him to discover, to originate, and to devise mathematics new to him and not, at the same time, be asking him to organize his findings in some systematic way. Thorndike urges that "Knowledge which is useful . . . is organized knowledge . . . organized by the learner."[12] Meland suggests that "Learning can be deepened and be made more genuinely human as well as beneficial as a human act if it appropriates the procedures of both creative artist and critic."[13] The last facet may well mean logical organization by the student himself.

It is a highly pressing obligation, therefore, on the part of every teacher to ask and to help his students to organize their findings in logical order.

---

[11]Cassius Jackson Keyser, *The Human Worth of Rigorous Thinking* (New York: Scripta Mathematica, 1940), p. 3.

[12]Robert L. Thorndike, "How Children Learn the Principles and Techniques of Problem-Solving," *Learning and Instruction*, Forty-Ninth Yearbook (Chicago: The National Society for the Study of Education, 1950), Chapter VIII, pp. 211-2.

[13]Bernard Eugene Meland, *Higher Education and the Human Spirit* (Chicago: University of Chicago Press, 1953), p. 81.

Herein lies another teaching advantage of methods which make use of student contribution. What terms shall we regard as primitive? How can we state our definitions carefully, subject to refinements by the student and teacher as the work proceeds and as students become more mature in the subject? What are our primitive statements? Do any of the hypotheses made from results of experiential exploration belong in the system as logical consequences of the primitive terms and statements and previously deduced theorems, or must some of these be regarded as primitive statements? Which hypotheses, if any, remain as hypotheses? The student, through such a consideration, becomes an organizer of materials; now he develops, in a way which would be otherwise difficult, an insight into the meaning of a deductive system. Interestingly enough, what is one student's theorem may be another student's postulate—a striking illustration of "what is one man's meat is another man's poison!" Now is the time for students to compare their systems with these of others and with those of standard text writers. Experience has shown that students find it very instructive and revealing of the nature of mathematics when they are called upon to defend their deductive system against the critical examination of others.

Student organization of the mathematical ideas and principles developed is a compelling *must* in discovery approaches.

Exercises

1. In Section 3.3 there is illustrated a situation approach to matrices and their addition. Study this material again and organize the materials developed there into a deductive or axiomatic system. Select the terms and statements which you regard as *primitive* for your study; select the terms which should be defined and then list properties which can be deduced as theorems.
2. Suggest several advantages of student organization of materials, subject to the scrutiny of his classmates and the teacher. Suggest some disadvantages.

3.7 Discovery and the Time Factor

The experiential, developmental, experimental, concrete, student-discovery approaches obviously require more time than the lecture method. Even an analytic approach to problems, under genetic classroom methodology, requires more time than the presentation of an elegant synthetic treatment unless the lecturer stops to ask what step might come next; then the treatment is not strictly synthetic anymore.

Working under pressure of time in classes one can develop what might be called an oblique organization of materials. In this plan, seeds of larger ideas are sown in study guide sheets while other topics are being studied.

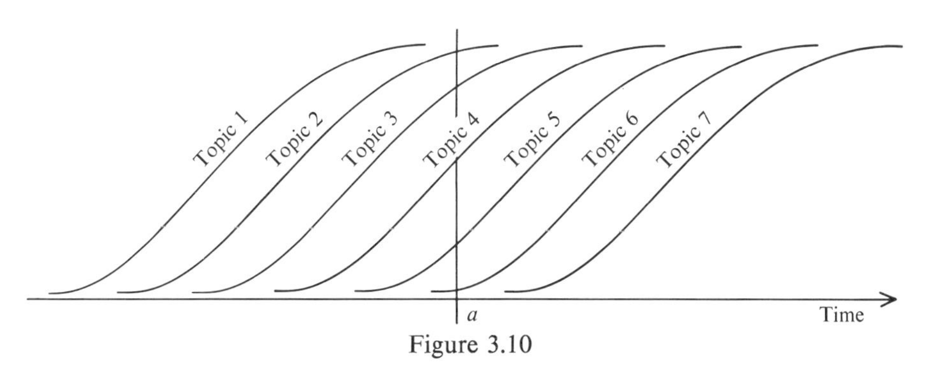

Figure 3.10

The organization appears graphically as shown in Fig. 3.10. Consider Day *a*. A study sheet for this day may contain a very rudimentary exploration of Topic 6, some growth on Topic 5, emphasis on Topic 4, review of Topic 3; and, of course, one might review a number of previous topics.

This methodology in teaching has been found to be effective in covering material and at the same time productive in encouraging a maximum amount of exploration and discovery under existing instructional systems. This plan requires constant arranging of exercises and writing of guide sheets on the part of the teacher or the use of other methods to encourage student contribution, but the rewards for such work (which is not as time-consuming as one might think after the teacher has had experience) are deep understanding, interest, and joy on the part of the student. Moreover, a set of guide sheets can be used again as minor points are improved and made more effective. Texts can still be used under this oblique organization.

The more traditional organization might be given the term "vertical." In graphic form it appears as shown in Fig. 3.11. It is not claimed that in a topic-by-topic organization there is no preparation for the next topic, but it is most often unfortunately the case that there is grave tendency for student exploration and the inductive approach to be seriously neglected or completely ignored.

In this text it is planned to give the student practice in writing study guides as the course progresses, but a few are given here as samples. Even these would need improvement to be effective in given classroom situations,

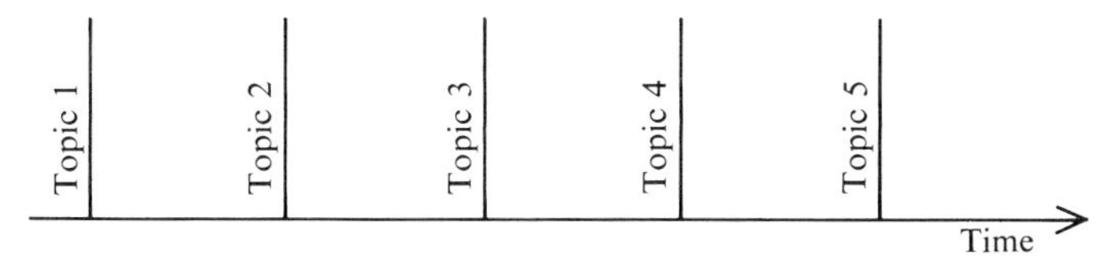

Figure 3.11

but these are guides representative of those which have been developed through practice with classes in the past.

Let us suppose that our algebra class is studying the solving of equations of one unknown which have fractional coefficients. Soon the teacher plans to have the class do work on solution problems. To have a study guide which is patterned after oblique organization suggestions, let us propose something such as the following:

Algebra I

1. *Review*. Compute as requested:
   (a) $(3x - 5)(2x + 4) =$ (d) $(a - 3b)(a + 3b) =$
   (b) $(6x - 7)(2x + 3) =$ (e) $(a - 3b)(a + 2b) =$
   (c) $(2x - 3)(2x + 3) =$ (f) $(y - 3)(y + 7) =$
2. *Review*. What value of the letter used makes each of the statements true?
   (a) $4 - (-2a + 7) = -31$
   (b) $-2(a + 6) + 3(-a + 7) = 4(-2a + 3)$
3. *Text* (appropriate exercises from the textbook being used)
4. John had a chemistry set. He took 40 grams of water and mixed 10 grams of alcohol with it. How many grams of solution did he have? What fractional part of it was alcohol? The solution is what percent alcohol?
5. John needed this alcohol solution for an experiment so he set it aside while he went to town for his mother. While he was gone. Charles, his little brother, put 10 more grams of H_2O into the solution. How many grams of solution were there now? What fractional part was alcohol now? The solution is now what percent alcohol?
6. If Charles had added 10 grams of alcohol rather than water, what would be the answer in No. 5?

Note the exploration through numerical approaches to some of the ideas about percent composition of solutions which, as teachers know, must be reviewed and sometimes relearned by the algebra student. Note, furthermore, the use of a text if it has problems suitable to the students' needs and to the teacher's purposes.

The study guide on the next day would develops further the ideas begun in Nos. 4, 5, and 6.

A study guide in geometry might appear as follows. Let us assume that the class is concentrating for the moment on the study of parallelograms.

Geometry

1. *Review.*
   (a) Lines ℓ and ℓ^1 are given. Name six tests to see if they are parallel.
   (b) By definition two lines ℓ and ℓ^1 are parallel if _______.
   (c) What is the usefulness or "efficiency value" of the theorems in (a)?
2. *Review:* Compute the measure of the angle requested.
   (a) $\overline{AC} = \overline{BC}$
   $\overrightarrow{AF}, \overrightarrow{BF}$ are angle bisectors
   Compute $m\angle 1$.
   (b) $\overline{AC} = \overline{BC}$
   $\overrightarrow{CD} \parallel \overrightarrow{AB}$
   Compute $m\angle 1$.

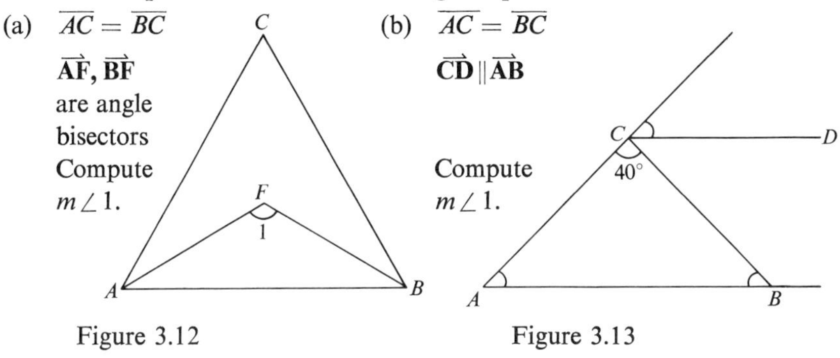

Figure 3.12 Figure 3.13

3. *Text* (problems on parallelograms or other topics being studied at the moment)
4. In each of these triangles in Fig. 3.14 locate the midpoints of $\overline{AC}$ and $\overline{BC}$; call these D and E. Now in each case draw the join $\overline{DE}$.

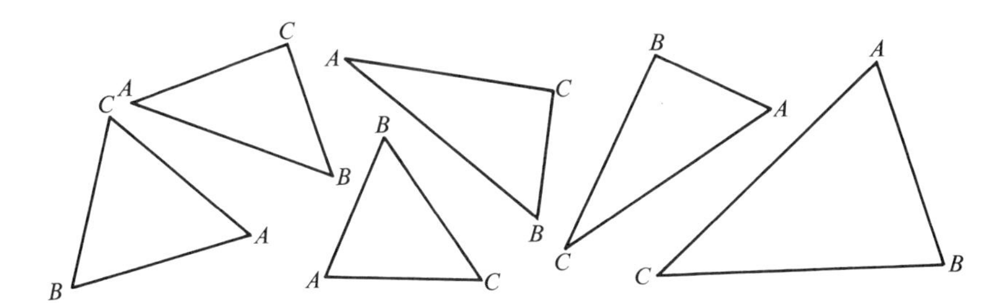

Figure 3.14

What relations, if any, seem to appear between $\overline{DE}$ and $\overline{AB}$ or between other parts of the triangles due to the presence of $\overline{DE}$.

5. How could you be more sure of the truth of your conjectures? Suggest what you could do.

6. See if you can devise a way to deduce one of your conjectures from previous theorems and some possible new definitions.

In the geometry study sheet above, note that first of all there is a rapid review. In No. 3, there is utilization of the text to help focus the energy of the student on the mainstream topic of parallelograms. Numbers 4, 5, and 6 consist of an experiential approach to new considerations. This approach, along with subsequent class discussion, helps greatly in causing mathematics to emerge from student thought.

Next appears a sample study guide for a class in calculus.

Calculus

1. *Review.* Compute

$$\text{(a)}\quad D_x(x^2 + 3x) \qquad \text{(b)}\quad D_x\left(\frac{1}{x^2}\right) \qquad \text{(c)}\quad D_x\left(\frac{x^2 + 4}{x^3}\right)$$

2. *Review.* Compute in each case
   (a) the slope of the curve $\{x, x^3 - 3x^2\}$ at $x = 2$;
   (b) the equation of the line tangent to $\left\{x, \dfrac{x^2 + 3}{x}\right\}$ at $x = -2$;
   (c) the equation of the line parallel to the tangent to the curve $\left\{x, \dfrac{x + 7}{x + 5}\right\}$ at $x = 4$ but which goes through the point $(-6, 8)$;
   (d) the equation of the line perpendicular to $\left\{x, \dfrac{x^2 + 3}{x}\right\}$ at $x = -2$.
3. (Exercises on a topic in calculus which is being studied in detail at the moment.)
4. Suggest a first approximation to the number which we might define as the area of the shaded portion under the curve in Fig. 3.15 whose equation is $\{x, x^2\}$ above the x-axis and between $x = 0$ and $x = 4$. Keep in mind that, at present, we can compute the areas only of regions with rectilinear boundaries. Show computation.

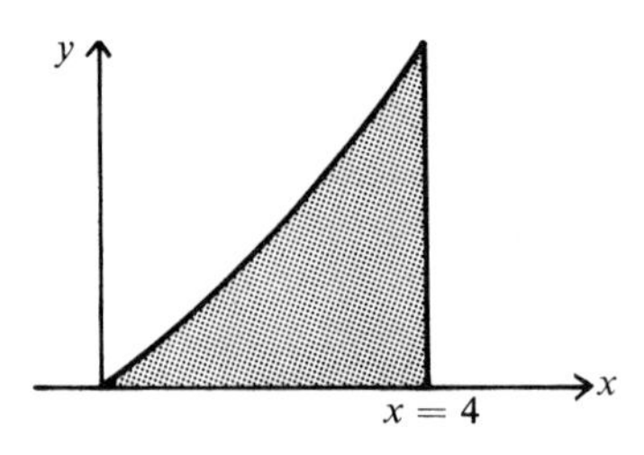

Figure 3.15

5. Suggest now a better approximation to the number which might define the area in No. 4. Show computation.

The above study guide has the same pattern as those in algebra and in geometry. There is opportunity in the guide for the students in the calculus class to strengthen recently acquired skills, to review fundamental concepts, to work in the topic currently being studied, and to explore in a simple way the problem of area under a curve. Just what method the student will suggest in Numbers 4 and 5, one does not know—this is exactly the reason why teaching with such methodology is exciting, challenging, and instructive to the teacher as well as to the student. In some way the teacher will use the suggestions of the students to develop a sequence of approximations leading to a number which will be agreed upon as defining well the area under the curve. In later study guides this idea would be developed as guided by student and teacher contribution.

Some day the finding of areas will become the major topic and then other subjects will be introduced for exploration and for developing rudimentary beginnings on those particular ideas. With such a daily plan for class work, a pace is set which encourages students to explore, to create, to attempt to build a logical system, and to review in a constructive and purposeful way much which they have learned before. Big dividends result when the teacher plans and structures study guides which encourage student contributions to the emergence of mathematics. With some experience and thought on the part of the teacher on "how he would propose to tackle the problem if he did not know" and with a sensitive observation of class discussion, the guide for the next day is most often determined largely by what goes on in the classroom. Actually, many times the study guide outline to be given out the next day is in the mind of the teacher before he leaves the class. Of course, the teacher is guided also by his objectives and the outline of his course; one would not expect the study guides to meander, without an apparent goal.

Exercises

1. Write a study guide which you migh use in seventh grade arthmetic in the following situation: you wish to review the topic of the addition of fractions, you are currently studing the computation of areas of rectangles and triangles, and you wish your students to examine trapezoids to see if they can devise a way to compute areas of such figures.

2. Write a guide sheet in the pattern of those above which you might give to your class as a next day's assignment in one of these subjects: geometry, arithmetic, linear algebar, statistics, calculus, differential equations, set algebra.

3.8. Discovery and the New Age in Mathematics

The advent of the computer is the beginning of a new era in mathematics and makes new demands on the student and the teacher of mathematics. The use of the computer has accelerated the coming of numerical analysis, which has changed the point of view somewhat for mathematics in the schools.

The very approaches which have been discussed in the preceding pages help the student to develop a numerical-methods philosophy and "we firmly believe that this can and should be woven into the mathematics course at all levels, without needing expensive equipment and without it being regarded as yet another special 'topic'."[14]

Many examples can be given in which opportunities for introducing numerical methods can be utilized. Early in one's arithmetic study one can have this problem: If the area of a square has a measure of 19, what is the length of each side? "The length is larger than 4, but smaller than 5", will be the first comment in exploring what the answer might be. Various suggestions may be given but finally the students can suggest, with proper stage setting by the teacher if necessary, that an estimate e_1 divided into 19 yields $19/e_1$ and the mean of these two numbers is a better approximation, and so on. This trial-and-error or guess-and-check method emphasizes (a) the type of answer wanted, that is, the meaning of square root; (b) that a few trials or approximations may give an answer suitable for the purpose; (c) that it might be well to develop a plan or a pattern for using one approximation to make the next. The idea of evolving a plan to make successive approximations is most important. Graphical representations of the progress of making approximations is an enormous aid to the student, too, and this helps in a pictorial way to introduce such ideas as convergence and limits. All mathematics abounds with examples which can be looked at from the standpoint of approximations. Indeed, most problems, in practice, are of the nature that about the best we can hope for are approximate answers.

Another example which might occur in an arithmetic class is this: The perimeter of a rectangle is 30 and its area is 45. What are the length and the width of the rectangle? Long before the student has heard of simultaneous equations leading to a quadratic equation he can set up the sentences

$$2\ell + 2w = 30$$

and

$$\ell w = 45$$

and proceed to try to find a pair of numbers which will make these sentences

---

[14]T. J. Fletcher, editor, *Some Lessons in Mathematics* (Cambridge: University Press, 1965), p. 91.

true. Of course, the traditional mathematician would claim that one has the solution as soon as the equations have been written, but this does not satisfy the arithmetic student who wishes to know the answers and who cannot solve quadratic equations. One can prepare a table and try to find values which satisfy the given conditions; arrangement in the form of a table helps to guide the student toward making better choices.

$\ell + w = 15$

ℓ	10	9	11	10.5	10.9	$\cdots$
w	5	6	4	4.5	4.1	$\cdots$
ℓw	50	54	44	47.25	44.69	$\cdots$

We see that we are getting close and, although this example is extremely simple, the student acquires the idea of accepting approximations for answers. He will soon develop the point of view that such work is not trivial in nature and, in addition, he will gain a kind of understanding of mathematics which is most significant and demanding in this modern day. The approaches which have been described in this and preceding chapters go far in providing this kind of understanding.

Our discussion of the new age in mathematics and its impact on teaching would not be complete if it were not mentioned that giving instructions to a machine requires a point of view which may be somewhat new to the student, but which may well be a part of student-centered approaches in teaching. Again, creativity and understanding on the part of the student are necessary to construct flow charts and sets of directions to accomplish solutions to problems. Flow charts for arithmetical processes can be used to increase understanding and to find errors in reasoning or in decisions of what to do next. If a method is understood, a flow chart follows easily; if it is not understood then to construct a flow chart will require deeper and continued consideration of the processes.

The reader will find materials for further reading on computer-oriented mathematics in References 15 and 16. This is a point of view which we should consider also in our teaching.

Exercises

1. Approximate the square root of 27 by using 5 as a divisor and by finding the quotient and then the arithmetic mean of the divisor and the quotient. Now use this mean as the divisor into 27 and compute another quotient

---

[15]*Ibid.*

[16]National Council of Teachers of Mathematics, *Computer Oriented Mathematics* (Washington: The National Council of Teachers of Mathematics, 1963).

and another mean as the second approximation. Compare this approximation with that obtained by other means.

2. Approximate in the same manner as No. 1 the square root of $a^2 + b$ by using a as the first approximation; that is, use a as the divisor, compute the quotient, and take as a second approximation the arithmetic mean of the divisor and the quotient. Compare this second approximation with the first two terms of the expansion of $(a^2 + b)^{1/2}$.

3. Using Fig. 3.16, consider the quadratic equation

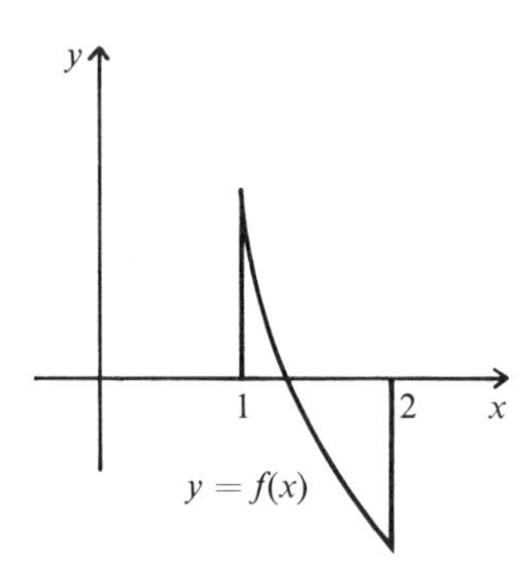

Figure 3.16

$$x^2 - 11x + 13 = y = f(x)$$

which has one root between $x = 1$ and $x = 2$. Try to approximate the root between 1 and 2 by assuming that the curve is a straight line in this interval and by using similar triangles. Now make another approximation. (Ans. The first approximation is 1 3/8.) Compare your results with those obtained by the use of the quadratic formula. Can you suggest any other method of approximation?

4. Consider the simultaneous equations

$$3x + 2y = 17$$
$$x - y = 2$$

Of course, one can solve those by methods which we have learned in algebra, but let us look at the problem in this manner: write the equations as

$$3x + 2y - 17 = R_1$$
$$x - y - 2 = R_2$$

where R_1 and R_2 are called residual. See if you can find values of x and y which make these residuals simultaneously very small. Scholars in the field of numerical analysis have refined such methods greatly, but here we have a glimpse into a new approach to solving equations. It might be helpful to construct a table of results. Compare your approximations with answers obtained by the methods learned in algebra.

3.9. Discovery and Teaching

The preceding development of methods of teaching has helped us to realize that the teacher is and should be several things. He is a many-valued person who can use several kinds of methodology.

First of all, the teacher is an *engineer*. It is a challenge to the teacher to structure experiences for students so they learn with a maximum of creativity, discovery, and independence, and with a minimum of telling (accepting "because it's so") and meaningless manipulations.

Secondly, the teacher is like a *physician*; he is a diagnostician and a therapist. He is a master in various type of methodology, but always with an understanding of how students learn best. He is able to use laboratory methods when it seems advisable; he is able to encourage physicalization of ideas when it seems helpful. The teacher, like a physician, has different helps for different students and for different "illnesses." No teacher uses one method only, but changes his approach depending on the previous experiences, maturity, and difficulties of the learner.

The teacher is at times a *drill master*. This means that the teacher must provide means to make certain responses on the part of students quite automatic, Drill and practice sould come, however, after the student understands and after, in most cases, he has developed the process himself. Practice for proficiency on what the student understands requires less time than if the student is attempting to become proficient when, because of lack of understanding, he is trying to follow a procedure blindly with meaningless processes. This fact should lessen the concern for increased time requirements necessary for discovery learning.

The teacher is like a drama *properties man*, or like a *host*, as he hopes to keep the subject moving among the students. The teacher tries to help the students explore and to create by maintaining an atmosphere of creativity; he must keep enthusiasm alive. Keeping mathematics growing and developing in the classroom, one finds, has much in common with the fortunate circumstance (or fate) of language study: children played in Latin, they sang in Latin, Latin helped to express the joys and sorrows of life—it was exciting until it was placed in the grammar book! Mathematics vibrant and growing is one thing; studying lifeless recordings with little understanding is another. The enthusiastic mathematical host in the classroom keeps the development of mathematics by students in a state of contagious creativity. Must the teacher, therefore, encourage agreeable and pleasant surroundings in learning? Obviously, the answer is "Yes!" Psychologists tell us this; common sense does, too. Should there be any shame in not understanding? No. Methods which help the student to understand and develop or discover ideas make learning more pleasant, and he learns better by its being so. "Nothing breeds success like success itself."

To the students and teachers who are interested in programmed learning or in self-instructional devices this chapter has contained some helpful discussion. It can be recognized that principles of the student-centered approaches have some common elements among principles underlying programmed learning. Some of these common elements are (a) progress in small steps arranged in order, (b) provision for answers (especially in inductive quizzes and study guide sheets); (c) active participation on the part of the learner. One might say that many of the teachers who are best skilled in student-centered approaches or student-participation approaches have already used programmed learning principles.

There remains the question of creativity and discovery in the present programmed learning devices. It has been emphasized by some that the higher learning efficiency under programmed instruction may give more time to the teacher and student for creative work. Discovery and creativity, organization of materials, and differences of opinion deserve attention at *all* times in the classroom. It is doubtful that machine learning as we know it can replace the lasting personal attention and interest of the teacher or can be a substitute for experiencing the growth of ideas by cooperative efforts and fruitful contributions of class investigations.

Indeed, writers on the subject of machine teaching still assert that the machine can be no more than a help to good teaching. Kvaraceus points out that

> . . . the more crucial and far-reaching outcomes of learning will always be found at the level of interpretation, application, appreciation and invention. These levels are still outside the reach of most self-learning devices and thereby place a low ceiling on what is to be mastered.[17]

Exton suggests that

> It may be that the teaching machine most effectively and efficiently teaches the teacher how to teach by providing him information as to the success with each student at every point in the program of instruction.[18]

Cook[19] mentions both advantages and disadvantages. Student participation, reinforced learning, progress at individual rates, along with the unending patience of the machine, are factors in favor of machine learning. The question of readiness and individual differences are raised by some (does

---

[17]W. C. Kvaraceus, "Future Classroom: an Educational Automat?", *Educational Leadership*, 18 (February, 1961), 289.

[18]E. Exton, "Teaching Machines: Fad or Here to Stay?," *American School Board Journal*, 141 (September, 1960), 19.

[19]F. S. Cook, "Some Advantages and Limitations of Self-Instructional Devices," *Balance Sheet*, 42 (December, 1960), 154-6.

everyone learn in the same manner?). Arrowsmith has suggested that we need "men, not programs; galvanizers, not conductors" and that the teacher should be a "radiant exemplification to which the student contributes a correspondingly radiant hunger for becoming.[20] The student has positive learning experience if he has met a teacher who will motivate him to search and to study more. Certain virtues and benefits "'come not from a course, but from a teacher; not from a curriculum, but from a human soul'—could they possibly come from a machine, or from a long dreary series of machines?"[21]

Bibliography

Casmir, F. L., "Human Teacher Is Still Best," *California Teachers Association Journal*, LVI (September, 1960), pp. 41-4.

Crescimbeni, J., "Developing Arithmetical Inquiry with Enrichment Aids," *The Arithmetic Teacher*, XIII (January, 1966), pp. 49-51.

Fleckman, B., "You Have a Role in Discovery Learning," *The Instructor*, LXXV (October, 1965), p. 88.

Frazier, A., "Opinions Differ; Technology in Education," *National Education Association Journal*, XLIX (September, 1960), p. 73.

Gold, S., "Graphing Linear Equations: a Discovery Lesson," *The Arithmetic Teacher*, XIII (May, 1966), pp. 406-7.

Griffin, H., "Discovering Properties of the Natural Numbers," *The Arithmetic Teacher*, XII (December, 1965), pp. 627-32.

Heimer, R. T., "Some Implications of Programmed Instruction for the Teaching of Mathematics," *The Mathematics Teacher*, LIV (May, 1961), pp. 333-5.

Hendrix, Gertrude, "Learning by Discovery," *The Mathematics Teacher*, LIV (May, 1961), pp. 290-9.

Hendrix, Gertrude, "The Psychological Appeal of Deductive Proof," *The Mathematics Teacher*, LIV (November, 1961), pp. 515-9.

Hirschi, L. Edwin, "Encouraging Creativity in the Mathematics Classroom," *The Mathematics Teacher*, LVI (February, 1963), pp. 79-83.

Houston, J., "Programmed Math Passes Three-Year Test," *Nations Schools*, LXXVI (August, 1965), pp. 40-3.

Johnson, Donovan A., "Enriching Mathematics Instruction with Creative Activities," *The Mathematics Teacher*, LV (April, 1962), pp. 238-42.

---

[20]William Arrowsmith, "The Future of Teaching," *Journal of Higher Education*, XXXVIII, 3 (March, 1967), p. 134.

[21]Donald W. Robinson, "But THIS a Machine Can NOT Do!", *California Teachers Association Journal*, 56 (September, 1960), p. 17.

Kaplan, Jerome D., "An Example of Student-Generated Sequences in Mathematics Instruction," *The Mathematics Teacher*, LVII (May, 1964), pp. 298-302.

Kersh, B. Y., "Learning by Discovery: Instructional Strategies," *The Arithmetic Teacher*, XII (October, 1965), pp. 414-7.

Lerch, H. H., and H. Hamilton, "Comparison of a Structured-Equation Approach to Problem Solving with a Traditional Approach," *School Science and Mathematics*, LXVI (March, 1966), pp. 241-6.

Lysaught, J. J., "Programmed Learning and the Classroom Teacher," *New York State Editor*, XLVIII (February, 1961), pp. 9-11.

Marks, John L., and James R. Smart, "Using the Analytic Method to Encourage Discovery," *The Mathematics Teacher*, LX (March, 1967), pp. 241-5.

May, K. O., "Programming and Automation," *The Mathematics Teacher*, LVIII (December, 1965), pp. 705-8.

Miller, E. E., "Don't Be Afraid of Programmed Learning," *Grade Teacher*, LXXXIII (May, 1966), pp. 180-2.

Polya, G., *Mathematics and Plausible Reasoning* (two volumes) (Princeton: Princeton University Press, 1954).

Salzen, R. T., "Discovering What Discovery Means," *The Arithmetic Teacher*, XIII (December, 1966), pp. 656-7.

Skinner, B. F., "Something Good Happens to a Child," *Phi Delta Kappan*, XLII (October, 1960), p. 27.

Smith, M. D., "Points of View on Programmed Instruction," *Science Editor*, XLVI (October, 1962), pp. 302-3.

Smith, W., "Use and Abuse of Programmed Instruction," *The Mathematics Teacher*, LVIII (December, 1965), pp. 705-8.

Snyder, Henry D., "An Impromptu Discovery Lesson in Algebra," *The Mathematics Teacher*, LVII (October, 1964), pp. 415-6.

Szabo, Steven, "Some Remarks on Discovery," *The Mathematics Teacher*, LX (December, 1967), pp. 839-42.

Wehrman, K., "Math Gets Exciting When They Make Their Own Discoveries," *The Grade Teacher*, LXXXIV (January, 1967), pp. 88-91.

Wright, Frank, "Motivating Students with Projects and Teaching Aids," *The Mathematics Teacher* LVIII (January, 1965), pp. 47-8.

4

On Teaching Arithmetic

Begriffe ohne Anschauung sind leer.*
Kant

4.1. Introduction

Although we are preparing for the teaching of secondary school mathematics, there is always a demand for a deeper understanding of principles of arithmetic, since many topics at the secondary and early college level have their origins in arithmetic. Too, in early years of secondary instruction there is an unequalled opportunity to take advantage of the growing up, or maturing, process of the student to help him look at arithmetic in a light and with a more critical eye. The teacher at the secondary level may emphasize the common sense and physical origins of arithmetic and may help the student feel that many arithmetic processes can be invented or discovered through his own efforts as opposed to dogmatic or rote beginnings which the student may have had. The teacher in the secondary school has numerous opportunities to help students reexamine, rediscover, enrich, and continue to develop, systematize, and organize understandings and processes in arithmetic. It is hoped that the material in this chapter may yield an insight into arithmetic the reader may not have at the moment.

*Concepts without visualization are empty.

Emphasis on arithmetical origins of many fundamental ideas and topics, and even of entire branches of mathematics helps to keep alive an ever-growing study of arithmetic. One need only to refer to numerical approaches (see Sections 3.4 and 3.5) to the idea of a derivative or to the concept of an integral, with accompanying discussion in the arithmetic of converging sequences, to perceive how arithmetic is found at the beginning of many great ideas and entire disciplines in mathematics. Such beginnings not only give the student security in basic understanding of new topics, but help him to recognize that arithmetic is a vital subject. During all these experiences his arithmetic skills are being strengthened in new settings.

Moreover, a reexamination of arithmetic in the secondary school may help the student to perceive the wide gap between humble beginnings in arithmetic and ultimate algorithms which we use now. He may see that understanding is often much more important than manipulation, especially meaningless manipulation. Indeed, since many students who have learned by rote have a dislike for and fear of arithmetic, the casual, yet purposeful, restudy of this subject might well help to dispel this lack of confidence and antagonism.

Although the need for attention to arithmetic is often very urgent in the secondary as well as in the elementary school, we cannot hope to give in this one chapter a course in theoretical arithmetic or a full discussion of arithmetic for teachers. We shall attempt to discuss points which seem important for a chapter of this kind; indeed, we might suggest as a subtitle "Some Arithmetic Which the Secondary Teacher Should Know." We shall assume that the reader is well acquainted with sets and cardinals and with the various systems of numbers, including their properties and logical origins.

4.2. What Is Arithmetic?

Basically, arithmetic is the development and study of algorithms for numerals which give the same answer to a problem as we would have obtained had we worked out the problem by physical means. For example, let us suppose that we received an order for 36 shirts and then we received an order for 35 more. What is the total number of shirts on order? We could determine this number at several levels of consideration. The most elementary (physical) way would be to secure a set or a collection of 36 shirts and then another collection of 35 shirts and assemble them—or form the union of the two sets—and then determine how many shirts these are in the new set, either by counting or by comparing the new set with representative sets until an equivalent one is found. If the latter is done, then the cardinality of the equivalent representative set is assigned as the total number of shirts on order. Presently we arrive at the total number of shirts on order by use of an algorithm performed on the numerals 3 6 and 3 5.

3 6
3 5

Our algorithm is this: 6 + 5 makes 11. We place the right-hand 1 of 11 below the right hand column and consider the left hand 1 of the numeral 11 added to the 3 and 3 and write 7 below the left-hand column. This operation on the numerals is conceptually different from determining the cardinality of the union of two sets of shirts! Yet here we have performed an operation on the numerals 3 6 and 3 5 to obtain the same answer as we would have obtained by physical means. This process is called an *algorithm.* Arithmetic is the study of algorithms on numerals or marks to give answers rather than acting out the problem itself. Another well-known algorithm is that performed on $\frac{1}{3}$ and $\frac{1}{2}$ to compute their sum:

$$\frac{1}{3} + \frac{1}{2} = \frac{2+3}{2 \cdot 3} = \frac{5}{6}$$

—a process very different conceptually from finding the sum by physical (original meaning) methods. Levi[1] discusses this further.

It is often a long road from the use of concrete objects in solving problem situations to the use of algorithms on numerals and careful nurturing and patience are required to help the student progress from one level to the other and have him understand what is going on. It is not always easy for the student to translate a situation within his own experience into an arithmetical problem involving the use of algorithms and then to translate the result of the algorithm back to the real-life situation. Indeed, this is a difficulty which older students experience in algebra, in calculus, and in differential equations when they are asked to translate physical situations into mathematical language; it is a skill which requires insight on the part of mathematicians in applied fields. In this journey from the concrete to generalizations and algorithms, the teacher must exercise great care lest the pressure of instruction in manipulation by algorithms crowd out understandings of numerals and what the algorithms really do. There is much opportunity here for students to develop their own algorithms, which should be encouraged.

Exercises

1. An auto dealer estimates that he will place new tires on 17 autos during the next month. How many tires should he order? Show several ways to determine the answer by physical or close-to-physical means. How is the answer determined by use of algorithms?

[1]Howard Levi, "Why Arithmetic Works," *The Mathematics Teacher*, LVI (January, 1963), 2-7.

2. One-hundred-thirty-five dollars are to be partitioned or distributed equally to nine people. What amount does each person get? Suggest several ways to determine the answer by physical means. What algorithms have you learned by which to work this problem?

4.3. Points of View on Curricula in Arithmetic

The aims of good teachers in mathematics are perhaps much alike: pupil understanding of arithmetic principles and processes and their intelligent use in problem situations. Differences of points of view arise in how these aims can be accomplished. Although we would expect the points of view to have features in common, we give names according to the emphasis made.

Intuitive Experiential Point of View

In this approach there is much use of laboratory work and activities out of which arithmetic emerges. Arithmetic arises as a result of experiences with concrete objects or with numerical experiences. Laws in arithmetic are found by the use of objects, rods, abaci, and devices of various kinds. Most of the time students have concrete representations of the problem; the laboratory approach is emphasized and operations are often acted out to help develop new ideas about numbers and operations on them. This approach answers essentially the outward call of mathematics.

With this point of view the student will develop algorithms on the numerals to yield answers which he would have obtained by the use of concrete things. Also as a result of reflection on his findings, certain properties of operations will evolve and his observations can be summarized in the form of a *system* of mathematics in which certain properties may appear as logical consequences of others. Whether or not this system as first conceived is recognized as one of the well-known structures in mathematics (such as a field) depends on the kind of concrete things used. One must nurture in the student the idea that mathematics is man-made. Algorithms for the addition of fractions, for example, are different for different meanings assigned to the fraction. If the numeral $\frac{3}{4}$ means that a player made 3 hits while 4 times at bat and $\frac{5}{6}$ means 5 hits while 6 times at bat, then the correct way to compute the present standing of the player is to say that

$$\frac{3}{4} + \frac{5}{6} = \frac{8}{10}$$

If $\frac{3}{4}$ and $\frac{5}{6}$ mean parts of a whole or have to do with lengths of segment on a number line, then the definition of their addition is different and the algorithm is different; the sum is $1\ \frac{7}{12}$. Our numerals take on the meaning assigned to them; resulting algorithms to perform defined operations are thus expected to be different. The emphasis on the laboratory approach helps us to keep sight of the fact that the child is an operationalist and that the physicalization of ideas plays a most significant part in learning. Indeed, even older learners become as little children again when they begin to study something new: one does not gain full insight into what a group is by studying the postulates but, rather, by working with many examples of groups themselves, even the physical rotation or overturning of geometrical figures.

Formal "System-Emphasis" Point of View

Features of this point of view are less use of laboratory and experimental work as well as more consideration of mathematics for its own sake to discover, study, and perceive properties as a part of a structure (or system) of mathematics. Such an approach seems to start early with an attempt to use mathematics as a unified system. The *number line* is used more and other physical things which the student can handle are used less. In this approach the content is influenced directly by the more sophisticated origins of numbers: cardinals arising as symbols representing equivalence classes of sets, natural numbers as finite cardinals, integers as names of equivalence classes of ordered pairs of natural numbers, rationals as names of equivalence classes or ordered pairs of integers, each of these involving different definitions of equivalence, along with the language used to describe fields. Of course, there is an attempt to use good methodology and, indeed, many of the student-centered approaches discussed in Chapter 11 are being given attention in these curricula, but the emphasis is on structures of systems. Even the Cambridge Report, suggesting a curriculum which leans closest to a purely systems-emphasis point of view, urges that "children can study mathematics more satisfactorily when each child has abundant opportunity to manipulate suitable physical objects"[2]; but at the same time in grades 3 to 6 they will compare properties of the real numbers "with the properties of modular systems, finite fields, and the system of 2×2 matrices."[3] The Cambridge Report invites mathematics teachers to consider possibilities and paths in mathematics teaching so our students can accomplish as much as possible under sound methodology. The writers are sincere in the belief that such programs are possible. Some modern programs extend the number line in both directions almost immediately and one course of study introduces the ordered-pair approach to integers through the use of vectors in early junior high school.

[2]Irving Adler, "The Cambridge Conference Report: Blueprint or Fantasy?", *The Arithmetic Teacher*, 13 (March, 1966), 180.

[3]*Ibid.*

There is some feeling that this emphasis might deprive the student of seeing mathematics at work in the world and of learning from experience that arithmetic and mathematics originated from physical things in the environment of man. A curriculum which emphasizes mathematics as a postulational system is looking to the inward call of mathematics.

There are several gains to be made, however, when we look at various mathematical systems and realize that different kinds of numbers are elements of different systems. One of these gains, as has been pointed out above, is that as new and different kinds of numbers arise, one may expect different algorithms for operations. *Natural* number addition and *rational* number addition, for example, can be expected to be performed in a different manner. An expectancy of this kind will suggest to the student that he examine and formulate the meaning which will be assigned to the sum

$$\frac{2}{3} + \frac{4}{5}$$

and what algorithms can be developed to yield this sum.

Sometimes, of course, it may happen that algorithms for operation with different kinds of numbers are similar, but this is a consequence of the system and need not be generally expected. Too, it is more characteristic of the postulational approach to help the student be ever mindful to delineate carefully what terms are being defined, what ideas are being postulated, and what facts are consequences of others. This viewpoint encourages also a pattern for the study of new operations: is the operation commutative? is it associative? and so on. This approach brings out into the open some troublesome questions in arithmetic: why one inverts the divisor when dividing one fraction by another, how one can handle $0/0$ and $5/0$, why $1/a \cdot 1/b = 1/a \cdot b$, and other questions.

A Series-of-Skills Approach

This point of view for arithmetic programs is outmoded. There was a time at which arithmetic was considered as an accumulation of skills on how to do certain problems. In such programs the essential features of texts were sample problems and solutions along with some explanation of certain algorithms, followed by exercises for the student. There was little attempt at logical organization and rationale for what was being done and, if this was developed, the presentation was dogmatic. Certainly there was a minimum of concern about how students learned except through repetition.

This "series-of-skills" approach is mentioned here for, although we may pay homage to new discoveries on learning, yet there is danger of continuing or reducing our teaching to the developing of manipulative skills without understanding.

One must always remember that the curriculum is not that program which is written on paper and found in the files for reference, but it is the record of both topics and methodology which are going on in the classroom along with the resulting attitudes and abilities of the students in mathematics. The three points of view, therefore, are merely guides to the various schools of thought and are not mutually exclusive. The different texts in mathematics which emerge from these schools have ideas in common, but have different emphases. A point of view on curriculum is perhaps more nearly a *spirit* in which the teacher believes and with which he can work most effectively for the good of the student.

This section should not be concluded without mentioning a few of the forces and movements which are influencing curricula at the present time. Of course, text publishers and their staffs help to shape programs of study, but the great changes have come from recommendations of committees and from writing groups which have suggested materials for use in the classroom to make teaching more effective and to heighten our accomplishments in mathematics.

The following partial list of writing groups and committees is not made with any order of preference in mind; rather this list is made to help the reader become acquainted with sources and origins of tides of interest in curricula. Readers who wish a more complete treatment of various programs are referred to the National Council of Teachers of Mathematics booklet "An Analysis of New Mathematics Programs."[4]

University of Illinois Committee on School Mathematics

This committee was established at the University of Illinois in 1951 to study the content and teaching of secondary school mathematics. The group introduced some new content and rearranged some of the traditional topics as well as developing some new approaches in teaching. This material is written with the structure of mathematics as the dominant feature. Social applications of mathematics are used as illustrations. Student discovery is an essential element in the teaching process.

University of Maryland Mathematics Project

This group began its work in 1957 and produced materials for 7th and 8th grade mathematics with special attention to language and mathematical structure. The topics treated are those of the usual junior high school course, but the point of view is different and the students learn much more than the traditional topics. In this program there is emphasis on the study

[4]National Council of Teachers Mathematics, *An Analysis of New Mathematics Programs* (Washington, D. C., National Council of Teachers of Mathematics, 1963).

of number systems as a tool in learning about mathematical systems in general. Also, emphasis is placed on the growth and development of mathematical concepts with strengthening of these concepts through exercises.

School Mathematics Study Group

This group was appointed by the President of the American Mathematical Society in 1958 to have as its goal the improvement of the teaching of mathematics in the schools. The study group now has its headquarters at Stanford University and, with help from the National Science Foundation, continues to produce studies and materials to increase the quality of the teaching of mathematics. There is emphasis on the structure of mathematical systems, but at the same time considerable attention is given to social applications of mathematics.

Greater Cleveland Mathematics Program

In 1959 this program originated with the Educational Research Council of Greater Cleveland which was formed to improve the quality of elementary and secondary education. The Greater Cleveland Mathematics Program "is a concept-oriented modern mathematics program in which primary emphasis has been placed on thinking, reasoning, and understanding, rather than on purely mechanical responses to standard situations."[5] Use is made of the logical structure of mathematics and discovery approaches in teaching.

The Cambridge Conference of 1963

The Cambridge Conference was initiated by two professors at the Massachusetts Institute of Technology and became an assembly of 25 mathematicians and scientists to consider what changes should be made in elementary and secondary school mathematics during the next 30 or 40 years. The findings of this conference are published in a report called *Goals for School Mathematics*.[6] This publication is often called the Cambridge Report.

The main conclusion of the report is recorded in this manner:

> A student who has worked through the full thirteen years of mathematics in Grades K to 12 should have a level of training comparable to three years of top-level college training today; that

---

[5]National Council of Teachers of Mathematics, *An Analysis of New Mathematics Programs*, op. cit., p. 10.

[6]Cambridge Conference of 1963, *Goals for School Mathematics* (New York: Houghton Mifflin, 1963).

is, we shall expect him to have the equivalent of two years of calculus, and one semester each of modern algebra and probability theory.

Irving Adler suggests that the Cambridge Report "is not a fantasy" in his discussion in the *Arithmetic Teacher*.[7]

Many of the ideas suggested in the various programs above have found their way into modern texts written by experienced writers. Indeed, it was the aim of the many writing groups to suggest new materials to be introduced by those writing texts for schools.

It should not be thought by the reader that it is only in the United States that there has been active interest in the improvement of mathematics programs. In England, F.W. Land states, ". . . tendencies all are pushing us towards a greater interest in the structure of mathematics, in logical relationships and more general considerations and away from an emphasis on computational skills and particular solutions."[8] The Organization for European Economic Cooperation (now the Organization for Economic Cooperation and Development) produced *New Thinking in School Mathematics* in 1961.[9] Other books have followed. The International Congress of Mathematicians has commissions concerned with school mathematics. It is safe to say that concern for instruction in mathematics is receiving the attention of the ministries of education in all countries. Reference 9 mentions several English publications.

Exercises

1. The "error" $\frac{2}{3} + \frac{4}{5} = \frac{6}{8}$ suggests others which you may have seen because the student did not understand the meaning of the numeral involved or because he was not aware that algorithms to perform operations in different systems of numbers might be different. Mention several other such errors which you have noted or which you may even have made yourself.
2. Using your knowledge of definitions which apply in the cases below, write correct "other names" for each of these expressions if others exist. Let us assume that we may work in any number system, including the system of complex numbers. Write also other names which might be written by students who neither understand the meaning of the notation nor have correct ideas on the type of function involved.

[7]Irving Adler, *op. cit.*, p. 179.

[8]F. W. Land, editor, *New Approaches to Mathematics Teaching* (London: Macmillan and Company, Ltd., 1963), p. 147.

[9]Organization for Economic Cooperation and Development. *New Thinking in School Mathematics* (Washington, D. C.: O. E. E. C. Mission Publications Office, 1961).

(a) $\sqrt{a^2 + b^2}$ (f) $\cos (A + B)$

(b) $\sqrt[3]{a^3 + b^3}$ (g) $\log (A + B)$

(c) $(a + b)^2$ (h) $\log A + \log B$

(d) $\sin (A + B)$ (i) $\sqrt{(-3)^2}$

(e) $\tan (A - B)$ (j) $\sqrt[2]{a^4 + b^4}$

3. In emphasizing that not all operations have the same properties, it is often interesting and instructive to have students make up operations to investigate. Let us suppose that a student by the name of *Mary* makes up this operation, we shall call it the operation of "marilation": let two natural numbers, f and s, be said to be "marilated" if the first-named number f is doubled and then added to the triple of the second-named number s. Hence

$$4\,M\,5 = 2 \cdot 4 + 3 \cdot 5 = 8 + 15 = 23$$

Also

$$5\,M\,4 = 2 \cdot 5 + 3 \cdot 4 = 10 + 12 = 22$$

Hence, marilation is *not commutative*. You use the numbers 7, 9, 15 to see if there is promise that marilation, as defined, might be *associative*. (Answer: No.) The study of such operations as marilation has a place in algebra, too.

4. Let us investigate the operation of marilation further.
   (a) Let us see if we can determine what relation must exist between f and s so that the operation of marilation will be *commutative*.
   (b) Use the numbers f, s, and t and see if there can be determined a relation which must hold for marilation to be *associative*. (Answer: f must equal s; f must equal $3t$; s may have any value.)

5. Let the operation σ be defined such that $f \sigma s$ means "to take the greater of f and s where f and s are natural numbers, and to take the common values if they are equal." Is the operation of σ as defined *commutative*? is it *associative*? Illustrate.

6. In Nos. 3 and 5 above have been defined the two operations M and σ. Study a few numerical examples to see if there is promise that M is *distributive* over σ; that σ is distributive over M. If any conjectures emerge, see if they can be *deduced* from the definitions of the operations and the properties of the natural numbers.

7. Consult Irving Adler's discussion of the Cambridge Report and note some of the good, accepted guides for teaching which he suggests may help to realize the goals as set forth in this report. Here is a challenge for us as teachers.

8. Note opinions of a partially opposing point of view in Howard Fehr's discussion of "sense and nonsense" in mathematics programs.[10]

9. In the Cambridge Report it is suggested that elementary students will compare the properties of the real numbers with those of 2×2 matrices under their respective operations. What are some differences which might be expected?

10. It is indicated in the text that other programs and points of view on curricula can be found in the National Council of Teachers of Mathematics booklet *An Analysis of New Mathematics Programs.* You might consult this booklet to learn of other programs and to learn how to find detailed information on them. References at the end of this chapter will help those who wish to consult texts in some of the programs.

4.4. Points of View in Teaching

It is not difficult to perceive that the series-of-skills approach was accompanied by much memorization through repetition and rote, though many an adventurous teacher must have developed a way to encourage student understanding. Nevertheless, the emphasis here was on skill and speed in getting the answer by the most efficient algorithm possible. Such teaching is likely to be less student centered and it often does much to instill in the learner a lasting dislike for arithmetic.[11]

We believe that there is a place for effective methodology which utilizes the student and which has sound psychological foundations under both the other two viewpoints in curriculum discussed in the previous section. Those who introduce finite arithmetic in the early grades, of course, do not use a procedure which might be used in the graduate school. Rather, the children can play games with an imaginary clock which has five hours instead of twelve; much arithmetic emerges from such activity. The idea of a Cartesian product of two sets can be introduced by finding all possible partners from a given set of boys and a given set of girls. Perhaps, as students attempt to find names for what they have created, this set of all possible partners will not be called a Cartesian product at first. The most sophisticated program urges the use of physical objects in early stages regardless of the grade level of the student. We must always remember that the young child learns through handling and that physicalization of ideas, therefore, plays an important part in his learning. Even if one finds himself using a less formal program, an attempt to show how the various arithmetic facts are related and emphasis that some parts of arithmetic exist, as they do, by agreement and some parts are logical consequences of others, will go

[10]Howard Fehr, "Sense and Nonsense in a Modern School Mathematics Program," *The Arithmetic Teacher*, 13 (February 1966), 83–91.

[11]Harrell Bassham, Michael Murphy, and Katharine Murphy, "Attitude and Achievement in Arithmetic," *The Arithmetic Teacher*, 11 (February 1964), 66–72.

far toward maintaining an atmosphere of the postulational spirit which will help students gain the feeling that mathematics is "a way of thinking" and that it can be "viewed either as an enterprise or as a body of achievements."[12] If the program is one which introduces the concept of a mathematical system very early, then effective methodology will seek to provide laboratory and inductive experience to help realize the immediate aim. Indeed, even most students in advanced courses need physical examples and models to help develop an idea.

There are several current viewpoints on methodology which demand our attention. One is that through student-centered approaches it has been found that the student can accomplish much more than was formerly thought possible and at an earlier age. Some think that our poor methodology may contribute to holding a student back; indeed, this is a serious problem when we reflect on the titanic task which our schools are attempting to perform. Another viewpoint is that we should make better use of what is known about how students learn and that there should be more basic research on how people learn mathematics. Jean Piaget,[13] for example, reports that there are four states in the development of a child's thinking and that under the age of eleven the thinking is predominantly about objects; hence, obviously, one's teaching at this level and below must take this factor into account. In fact, graduate school teaching for students whose level in a given course is "below eleven" might well consider beginning with concrete objects or examples to build up the more abstract idea. Teachers quickly perceive that the path of learning is from the concrete to the abstract and the concrete beginnings in arithmetic lend themselves well to such classroom approaches. From the concrete to the abstract is the way arithmetic has grown and it suggests the way people have learned. As teachers, we should capitalize on this natural method of advance. We might well look into history to see how ideas have developed and look upon these ideas as suggestions for classroom approaches.

4.5. Student-Centered Approaches

It is the purpose of this chapter to suggest to the teacher interested in methodology some student-centered approaches to several topics in arithmetic. It is not our intention to present the development of theoretical arithmetic in a logical formal way nor to discuss all the material usually treated in arithmetic. It is our hope, rather, that the following discussion

[12]Cassius Jackson Keyser, *The Human Worth of Rigorous Thinking* (New York: Scripta Mathematica, 1940), p. 3.

[13]Irving, Adler, *loc. cit.*, pp. 182–3.

may help the reader to gain a spirit and pattern of approach which has been found to be effective in encouraging a considerable amount of student contribution and student reflection about arithmetic. The guides suggested here will emphasize again and again how the students might be helped to discover much for themselves and how they might formulate rules after understanding rather than the often-present substitution of rules for understanding.

4.6. A Look at the Natural Number System

Ideas on determining the "how-many-ness" of a set, the difference in meaning between *number* and *numeral*, addition and multiplication of natural numbers and their inverse operations are basic to further work in arithmetic and mathematics. We include here some approaches to these topics, although it is not intended that prospective teachers will use these guides verbatim in their courses; the nature of the class and the ingenuity of the teacher will require changes in order to make them useful in given situations. The significant point is that such student-contered approaches help the student to be a part-creator of a sensible and reasonable discipline, that of mathematics. The student will eventually help to build, at his level, the realm of knowledge described by Nathan Isaacs:

> Mathematics is virtually a product of pure thought. Even though it draws its starting-points from the world of experience, it refines them into concepts well beyond the scope of that world, and then, by purely logical processes of combination, inference and construction, builds up the most elaborate thought-schemes.[14]

It must be remarked again that the manner of developing ideas illustrated here often requires a better mathematics background on the part of the teacher than other methods, but, too, such approaches help him to acquire a stronger background as he explores with students.

We shall regard as natural numbers the infinite set: 1, 2, 3, 4, 5, . . .

Assigning a "Cardinal" to a Collection

Here is a story for the student to read or a situation which might be acted out to help develop the idea of determining the plurality of a collection.

---

[14]Nathan Isaacs, "Mathematics: The Problem Subject," in *New Approaches to Mathematics Teaching*, edited by F. W. Lang (London: Macmillan and Company, Ltd., 1963), p. 25.

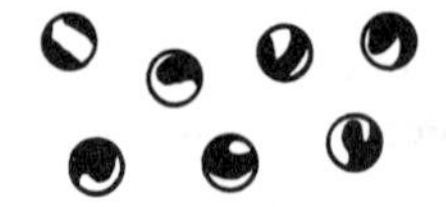

Figure 4.1

Johnny came to kindergarten one day with some marbles (Fig. 4.1). He was not sure how many marbles he had so he thought he would use the collections at school to help him with his problem.

Along the wall in his room there was a table which contained objects or pictures of objects with labels (Fig. 4.2). Johnny wondered if he had as many marbles as there are legs on a dog, so he laid his marbles on the table

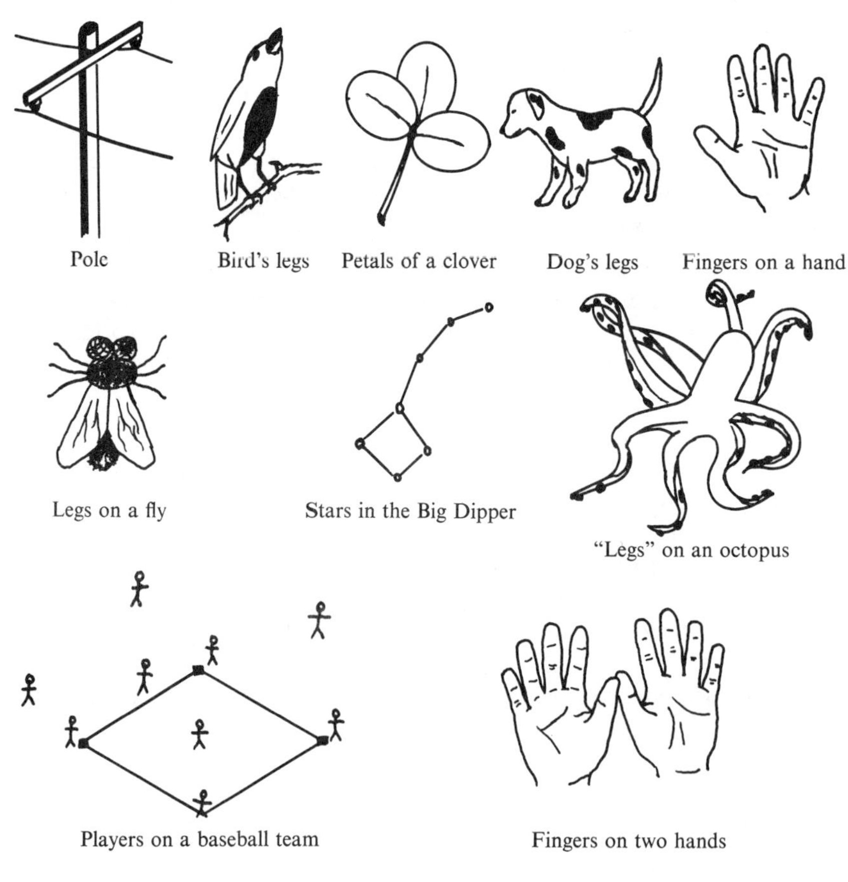

Figure 4.2

and drew connecting lines as shown in Fig. 4.3. No, there were marbles left over. Can you find the set to which his marbles will match one-to-one?

Figure 4.3

1. To which set on the table does this set of cakes match?

2. To which set on the table does this set of apples match?

3. Are there as many cakes in No. 1 as there are apples in No. 2?
4. Are there as many boys and girls in your row as there are apples in No. 2?

In using students, one might use pieces of yarn to match each student with elements of a representative set.

Exercises

1. Why is it true that Johnny might say that he has "Big Dipper" marbles? "Big Dipper" would be another name for the number which we call __________________. "Octopus" is another name for the number which we call __________________.

2. The figures or objects shown above really make up representative sets of equivalence classes of sets. Suggest five or more other sets which belong to the same equivalence class as does the set of petals of a clover leaf. The legs of a tripod are such a set.

3. The property common to all the sets in the equivalence class to which clover belongs is what we call "threeness." We give the name three to this abstraction. Three is called a *cardinal* and a first task in the elementary school is to develop the idea of a cardinal. The marks or the symbols made for these cardinals or numbers are called *numerals*. The drawing or picture of a clover might well be called a numeral if this is the mark used for three.

Suggest some numerals for the cardinals represented suggested by the sets in the figures above. For example, one might suggest merely " " for bird (which is our two). It might be interesting to contemplete what our symbols might become in several hundred years.

4. Suggest plans for experiences for very young children to help cause to emerge the idea of four, or dog.
5. Suggest some experiences and exercises which will help meaning of "less than" and "greater than" to emerge. Do these two terms represent different concepts? You might be interested in early elementary school materials in this topic.
6. Is **DOG** greater than hand? Why? How might you help a first grader understand that the size of the symbol does not indicate the size of the number? Make up some other examples which will cause the student to reflect on *number* rather than *numeral*.

From our more mature point of view we recognize that each of the sets of objects on the table in Johnny's classroom is a representative set of an equivalence class of sets and that a number is the name of the property which all the sets of a given equivalence class have in common. Too, we have seen the distinction between a number and its numeral. To suggest that much patience is required for students to develop the idea of number, or, for that matter, any abstract idea, we are reminded of Dantzig's quotation of Bertrand Russell as saying. "It must have required many ages to discover that a brace of pheasants and a couple of days were both instances of the number *two*."[15]

We should emphasize also that the marks we make as numerals for numbers might also be words: IV, 4, 1111, *four*, *vier*, *quatre*. . . are all numerals for the number four. The numeral is a concrete object but number exists in the mind; four is the abstract conception of the property common to the legs of a dog, the members of a quartette, the wheels of an auto. Number is the property of every set in an equivalence class which remains after color, shape, activity, materials, and any other distinguishing features are discarded.

[15]Tohias Dantzig, *Number, the Language of Science* (Garden City, N. Y.: Doubleday Anchor Books, Doubleday and Company, Inc., 1956), p. 3.

If we but observe children, we note that it requires a long time for a child to perceive the idea of *dog*, for example. His idea becomes refined as he goes through years of everyday experiences. Yet we expect him to feel as much at home with newly developed mathematical concepts in a matter of several minutes!

In the story above, Johnny was using the method of *matching* to determine the cardinality of his set of marbles, but we know that he could have *counted* them also. Counting can occur only after the cardinals have been ordered and after one can recite these cardinals in order. One determines the how-many-ness of a collection, then, by reciting the cardinals in order while pointing to the elements of the collection and by assigning as the how-many-ness of the set the last word recited as he points to the last element. Hence, *counting* is (a) ordering the cardinals under "less than," (b) memorizing these cardinals in order, (c) pointing to objects in the set to be counted and reciting the cardinals, (d) calling the last cardinal which is said the cardinality of the set. Determining the how-many-ness of a set by matching with representative sets and by counting are conceptually quite different.

Exercises*

1. Note the students in the classroom. Explain how one can tell how many students there are by (a) counting and by (b) matching with representative sets.
2. Eight guests are to be seated at a table and you wonder if you have enough. chairs. Suggest two ways by which this question can be answered.
3. Johnny determined the number of marbles in his collection by the process of matching. Matching one-to-one is a very important concept and it is one of the threads which go through all mathematics. Mention several concepts in mathematics which stem from one-to-one correspondence.
4. The name of our section is "A Look at the Natural Number System." If the idea of number is developed by means of equivalence classes of sets, then define a natural number.

The Sum of Two Natural Numbers

The sum of two natural numbers is developed from the unioning of representative sets which are disjoint. To help our elementary arithmetic student formulate this fundamental concept we can suggest a situation or story like this:

*These are not exercises we would have for "Johnny," rather they are for us to reflect upon and review.

1. Sandra's mother sent four cookies to school and Linda brought five (Fig. 4.4). The two girls wondered how many there were all together,

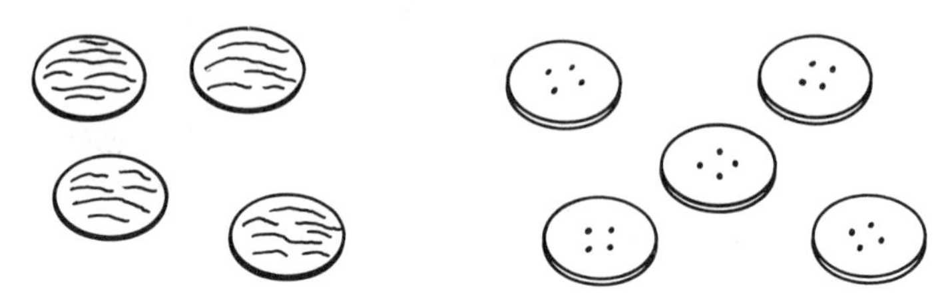

Figure 4.4

   and, more than that, they wondered how they could find out. All at once Linda remembered those sets on the table in the schoolroom. "I wonder if we couldn't use those *sets* to help us?," Linda asked. Can you suggest how Linda and Sandra might use the sets on the table?

2. With what collection on the table will the entire set of cookies match?

Experiences, activity, and discussion are requisites to developing ideas in arithmetic. Had we simply described what to do or what Linda and Sandra could have done we would not discover what the students think about the problem or how they go about solving it in their own way. When we observc how students think about a problem we are inviting many surprises. Our own ideas take on a new freshness and we as teachers gain deeper insights into the teaching of mathematics. Also, had we outlined what Linda and Sandra could do, the students in the classroom would have learned as passively as before and would have contributed little. It is up to the teacher to be quick to seize the comments which arise in class and use them to construct the idea at hand.

It may concern the reader somewhat to have Linda and Sandra say that they have "baseball team" cookies or to have a class called the Cartesian product of two sets the "all-partners-set" but the commonly used terms are also introduced at the appropriate time. Words and names devised by the student help to communicate meaning.

3. Soon Linda and Sandra learned that what they called "dog" cookies, their mothers and daddies called four cookies; "hand" cookies was called five cookies. In fact, they decided to label all their representative sets in the table by their new names: one, two, three, four, and so on. They then thought what fun it would be to add some of these numbers. In fact, they followed the suggestion of their teacher to place their sums in a table. You see this in Fig. 4.5. $4 + 5 = 9$ is already found on the chart. You fill in the other sums.

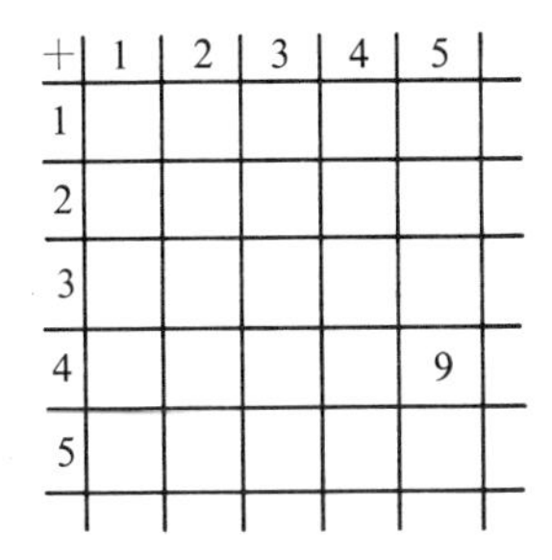

+	1	2	3	4	5
1					
2					
3					
4					9
5					

Figure 4.5

Exercises

1. Suggest questions or a situation or a classroom activity with students to have them discover that the order of addition is immaterial, that this binary operation is *commutative*. What name might students give to this property?

2. Suggest a situation or classroom activity by which students might discover that if three numbers are added, which two are added first is immaterial, or that this binary operation is *associative*. What name might students give to this property?

Using Physical Aids in Addition

The story of man's use of pebbles and marks to help him perform addition and other arithmetical processes is very interesting. As a first step in abstraction (using pebbles to stand for the real objects in sets) the teacher should give much attention to the use of physical representatives. Indeed, in the elementary school a red pebble or marker may stand for ten black ones; sticks are made into bundles of ten each to help develop the idea of "so many tens." Making use of all such devices in the elementary classroom is in keeping with the natural way in which the race has learned, and from this the teacher may gain some guides to methodology.

It is our purpose to introduce the use of the abacus as a part of what we consider the arithmetic which every mathematics teacher should know. We cannot hope to present the history of the abacus here, but the first usually mentioned is the Egyptian abacus. It consisted of ten beads on each wire with a wire for the units, one for the number of tens, one for the number of hundreds, and so on. As far as we know, the Egyptians used the abacus for computing and used (the Egyptian) numerals only for recording. The Egyptian abacus of Fig. 4.6 shows the setting for

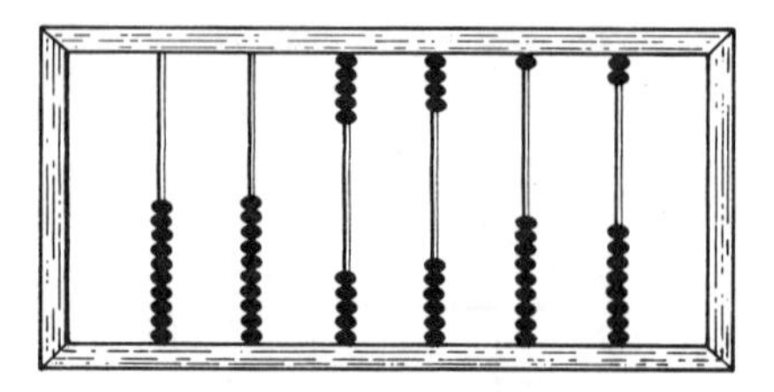

Figure 4.6

or 5412. The use of this abacus for addition will be discussed in class. Users of the abacus always began on the left when working with numbers.

Exercises

1. Students in your classes (even in the junior high school to whom you wish to introduce the abacus) may not have individual instruments. One can use lima beans on paper on which columns are indicated. Suggest other ways to make abaci.
2. Suggest a story by which you might introduce a student to the abacus.

The Persians, and, later, the Romans had abaci. Problems for computations arose in business, industry, and in scholarly discussions, but at times the extensivc calculations were done by slaves; the "calculations" worked with pebbles (the Latin *calculus* means stone, or pebble). Interesting in the development of the abacus are the divided abaci of the Chinese and Japanese. These are shown in Fig. 4.7 and demonstrations of their use might be a good class activity.

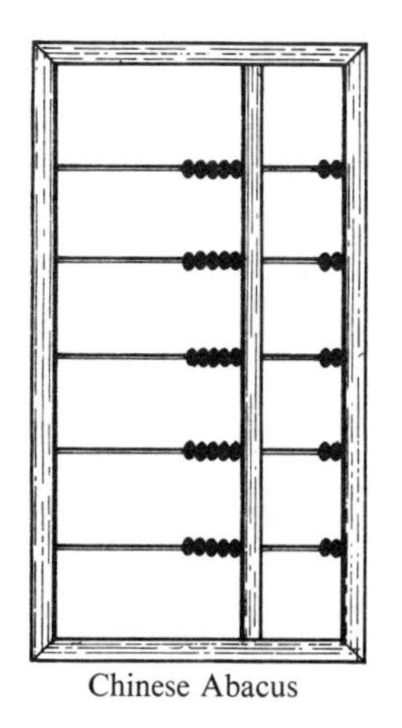

Chinese Abacus

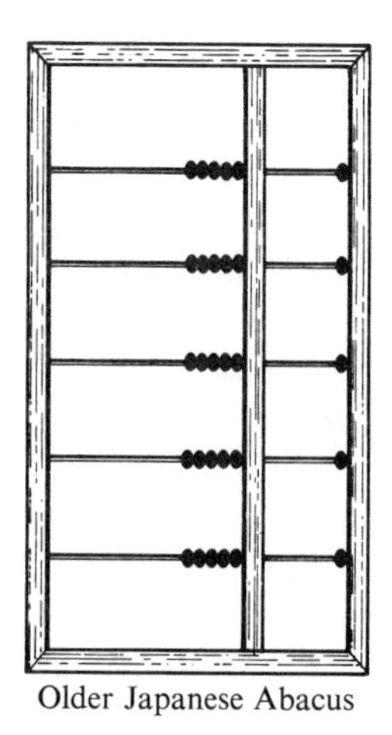

Older Japanese Abacus

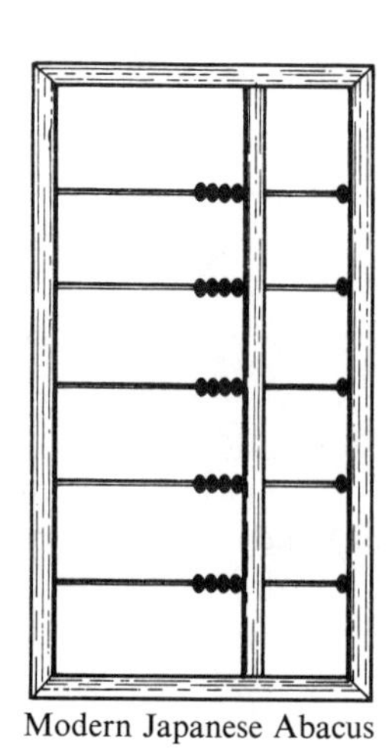

Modern Japanese Abacus

Figure 4.7

For the teacher going into the classroom it is helpful to known that along with the use of the abacus there developed some faint trace of manipulations with numerals. This is strikingly shown in the Gerbert abacus (circa 1000 A. D.) which consisted of discs labelled with 1, 2, 3, 4, ... (see Fig. 4.8).

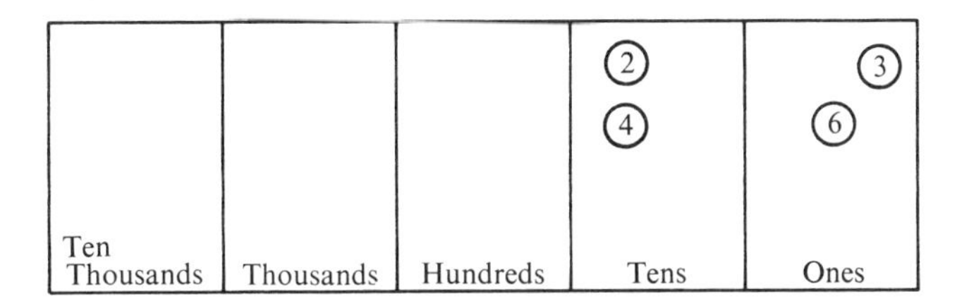

Figure 4.8

To add 23 and 46 the operator would place a 2-disc in the tens column and a 3-disc in the ones-column; also a 4-disc in the tens-column and a 6-disc in the ones-column. He would then add 3 and 6 mentally and replace these discs by a nine disc. Likewise, the 2- and 4-discs were replaced by a 6-disc. Answer: 69. This was close to not using an abacus at all! Gerbert's abacus was not popularly received and, eventually, the use of the abacus lost to those arithmeticians who were developing and encouraging the use of the algorithms on the numerals themselves. The great struggle between the abacists and algorists was won by the algorists. By the 1500's the use of the abacus in Europe had almost entirely disappeared.

One most important form of the abacus was evident in Europe during the Middle Ages; this was the Medieval Counting Board of Fig. 4.9. The board is now indicating 789. Often this diagram was painted on store

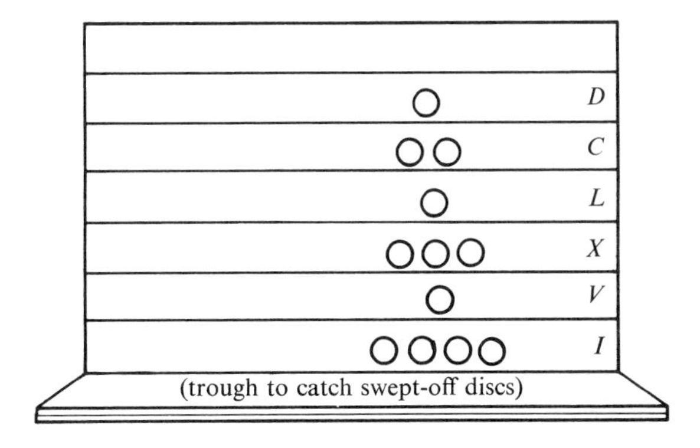

Figure 4.9

counters (the name counter in a store still persists) and some homes prided themselves in having "casting tables" with the diagram made of inlaid wood. Every store in the Middle Ages had such a device. Students may well learn to add on this kind of abacus; lima beans on a chart drawn on paper are useful.

The above very brief sketch of the use of the abacus has been presented to provide a background for several things which follow: "Tommy's method" of adding larger numbers, some historical "tidbits" which deepen insight and help to make classes more interesting, methods leading to multiplication, and others. The use of abaci has left such an imprint on our arithmetic that teachers of mathematics should be acquainted with their nature, their use, and possible absorption into usual classroom activities.

Exercises

1. Consult references listed at the end of this chapter or other sources to learn the structure of the Roman abacus; the Russian abacus.
2. You might like to look further into the history of the Gerbert abacus and into the slowly developing abacist-algorist controversy. Who were some of the abacists? some of the algorists?
3. The use of the abacus is being reemphasized in American schools. Suggest some concepts for the development of which the abacus can be helpful.

After the meaning of the base ten numeral is developed, one might well use a situation like this to help the student move gradually from forming unions in order to perform additions to the ultimate use of algorithms on the numerals themselves:

Sandra and Linda enjoyed adding so much that they added many pairs of cardinals. In fact, they nearly had the adding table of Fig. 4.5 committed to memory, though occasionally they needed to refer to it. One day Tommy came over and told the girls that he did his adding by using checkers. The girls wondered how this could be done so he explained as follows: "To add 5 and 4, I place five checkers and also four checkers in the last column. I push them together, and then I count them and I have nine." (See Fig. 4.10a). "It is easy," Tommy said. "When I have numbers like 7 and 6," Tommy explained further, "I still use sets of seven and six checkers, (see Fig. 4.10b) but as soon as I see that I have ten checkers in the last column, I pick them up and place one in the second column. Hence, the result appears as shown in my drawing (Fig. 4.10c). I call the answer 'ten and three'." Sandra and Linda liked this method very much so they added a lot of numbers this way. To add 34 and 28 they regarded 34 as three tens and four and 28 as two tens and eight. The columns appeared as in Fig. 4.11.

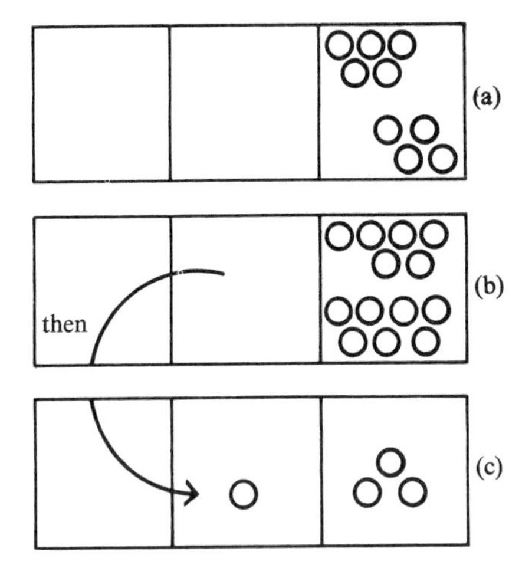

Figure 4.10

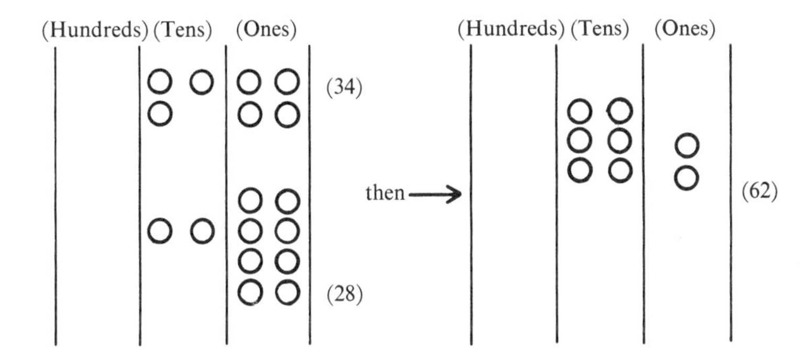

Figure 4.11

Exercises

1. Use Tommy's method to add these numbers.

(a)	(b)	(c)
4 3	6 7	1 7 3
2 9	5 8	4 6 8

2. Use Tommy's method to add these numbers.

(a)	(b)	(c)
4 7	2 4 3	6 3 4
6 9	1 7 9	2 1 6
8 8	9 8 3	7 3 8
		9 4 2

Note in the next story the introduction of something less concrete; a mere mark to stand for an object:

Georgie came to school one day and he said, "Last night I wanted to add some numbers and I couldn't find my checkers so I just made *marks* in the columns like this:

	Hundreds	Tens	Ones	
4 2 6	////	//	//////	(426)
1 3 2	/	///	//	(132)
6 7 8	//////	///////	////////	(678)

Then I used marks just like the checkers except that I just crossed them off.

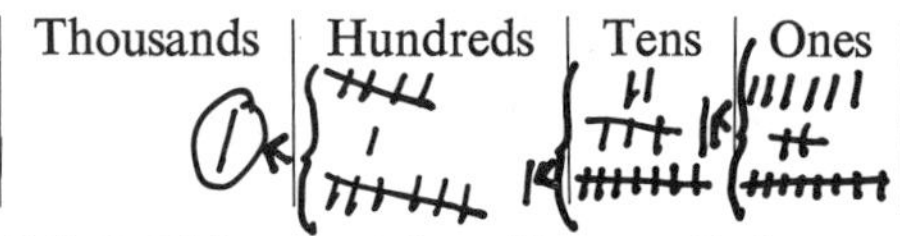

Hence, the answer is 1 2 3 6. This we might call a *pencil abacus* and students can learn much about addition from this semiphysical process. Of course, had Georgie started to add from the *left* as

Thousands	Hundreds	Tens	Ones
/ ←	~~////~~ / ~~//////~~	/ ← ~~//~~ ~~///~~ ~~///////~~	~~//////~~ ~~//~~ ~~/~~///////

with the result

Thousands	Hundreds	Tens	Ones
(1)	(//)	(///)	(//////)

he would have been adding in true abacus style. It is interesting to note that the arithmetic algorithms have changed from "left to right" only comparatively recently whereas we still work from the left in algebra.

Exercises

1. Add these by Georgie's method. You may proceed from left to right or right to left as you wish. It might be interesting to become experienced in both methods.

(a)	(b)	(c)
4 3 1	4 3 6	1 1 4 8
1 6 7	9 4 5	9 5 3

2. One sometimes sees elementary students add, let us say, 4 and 3, in this manner: marks are placed on the numerals as shown in Fig. 4.12 and then the student counts these marks! Comment on (a) the value of this method, (b) what the use of this method might show about the student, (c) what the observance of the existence of this and like methods might suggest for teachers.

Figure 4.12

3. Fingers are the handiest objects to help count and sometimes student are found using them. There are some cases on record in which students forced to quit using fingers added by using their teeth. Comment on what this might suggest to the teacher on methodology.

4. Do you think that the word "carry" describes the process well? Why? Can you think of a better word?

Multiplication of Natural Numbers

How can we lead a student into the idea of multiplication? It is in keeping with the spirit of student-centered approaches that we propose physical situations or problems and let the student think about them in his own way. Of course, we are interested in the emergence of some algorithms and some surprises may come. A suggested problem might be: Tom decided to get fifteen-cent gifts for six friends who were coming to his party. How much will the gifts cost?

Tom will no doubt write six fifteens down and add them; he may decide to do this on the abacus. The class discussion may result in Dick's suggesting that we add three thirties. Harry may suggest that the cost will be six fives or sixty and thirty. One does not know what suggestions will be made, but the teacher must utilize eventually the one which helps achieve her goal and at the same time recognize the insight gained by the use of the other suggestions. Some problems which might now follow are structured to help an algorithm emerge. In the first one, three twelves is the same as three tens plus three twos or thirty plus six. Many of these would be done in class after Harry's suggestion.

Exercises

1. Use Harry's suggestion to find the sum of

(a) three twelves	(e) four sixteens
(b) four elevens	(f) five seventeens
(c) six twenty-ones	(g) six twenty-twos
(d) seven forty-ones	(h) eight nineteens

2. Use Harry's suggestion to find the sum of

(a) four twelves. Arrange the numerals as $\begin{array}{r} 12 \\ \underline{4} \end{array}$

(b) six thirteens. Arrange the numerals as $\begin{array}{r} 13 \\ \underline{6} \end{array}$

(c) $\begin{array}{r} 17 \\ \underline{4} \end{array}$ (d) $\begin{array}{r} 23 \\ \underline{6} \end{array}$ (e) $\begin{array}{r} 34 \\ \underline{7} \end{array}$

(f) $\begin{array}{r} 43 \\ \underline{3} \end{array}$ (g) $\begin{array}{r} 43 \\ \underline{5} \end{array}$ (h) $\begin{array}{r} 63 \\ \underline{8} \end{array}$

It is not surprising that students suggest algorithms as

$$\begin{array}{r} 17 \\ \underline{4} \\ 40 \\ \underline{28} \\ 68 \end{array} \quad \text{and} \quad \begin{array}{r} 132 \\ \underline{8} \\ 800 \\ 240 \\ \underline{16} \end{array}$$

and they seem a bit more sensible, too. Of course, the student is making use of the distributive principle of multiplication and addition, but the teacher may wish to encourage students to do their algorithms well before calling attention to it.

The teacher can next plan to propose problems as 32×40 and, later, 32×45. Then the teacher can help the students reason out answers and encourage the suggestion of algorithms. Often students find it more reasonable to write

$$\begin{array}{r} 26 \\ \underline{42} \\ 800 \\ 240 \\ 40 \\ \underline{12} \\ 1092 \end{array} \quad \text{than} \quad \begin{array}{r} 26 \\ \underline{42} \\ 52 \\ \underline{104} \\ 1092 \end{array}$$

One must admit that the second shorter algorithm hides much which is done mentally and can easily become a meaningless rigmarole. The streamlining of algorithms must be gradual and, in large part, should originate from the students themselves.

Exercises

1. Multiply each of these to emphasize the use of the distributive law: 32×8 is 30×8 plus 2×8 equals $240 + 16 = 256$.

(a) 45×7 (b) 63×4 (c) 72×8
(d) 456×7 (e) 324×8 (f) 123×9

2. Do those in No. 1 again, but arrange the work as shown at the right. Note how an algorithm emerges from the way the problem was reasoned out in No. 1.

$$\begin{array}{r} 45 \\ \underline{7} \\ 280 \\ \underline{35} \\ 315 \end{array}$$

3. Multiply these using the longer algorithm demonstrated in the discussion (26×42).

(a) 27×43 (b) 65×74 (c) 325×94

4. In adding the numbers listed at the right some students think in this manner:

$$\begin{array}{r} 3 \\ 4 \\ 7 \\ \underline{+6} \end{array}$$

$$3 + 4 = 7 \quad \text{and} \quad 7 + 6 = 13 \quad \text{and} \quad 7 + 13 = 20$$

Others may say

$$3 + 4 = 7, \quad 7 + 7 = 14, \quad 14 + 6 = 20$$

Prove by the use of the associative and commutative laws for addition that

$$(3 + 4) + (7 + 6) = [(3 + 4) + 7] + 6$$

Also some will use the method of selecting the pairs of addends whose sums are 10 and adding them, as

$$3 + 7 = 10, \quad 4 + 6 = 10, \quad 10 + 10 = 20$$

You prove that

$$(3 + 7) + (4 + 6) = [(3 + 4) + 7] + 6$$

Give reasons for each step in the argument.

Some Earlier Algorithms for Addition and Multiplication

The fact that students suggest algorithms different from those which we commonly use and the awareness, by this time, that arithmetic was not always like it is now, makes it reasonable to expect that in the past centuries man used algorithms for addition and multiplication which are strange to us. The gradual movement from the use of the abacus to algorithms or numerals caused some algorithms to be patterned after the manipulation of the abacus. Even junior and senior high school students find the use of different algorithms very interesting and often there results a deeper insight into the nature of the operation itself. Students sometimes prefer the older algorithm; perhaps some of them convey an understanding which our more modern versions do not. We mention these, too, for their historical interest.

There will be some written explanation; the reader can see what is done from the illustrations.

"Scratch" Method of Addition

In Fig. 4.13, one begins at the left, adds, and scratches out each numeral as he uses it. The answer is written at the top. On adding the second column the number of hundreds is increased by one; here the 3 in 13 is

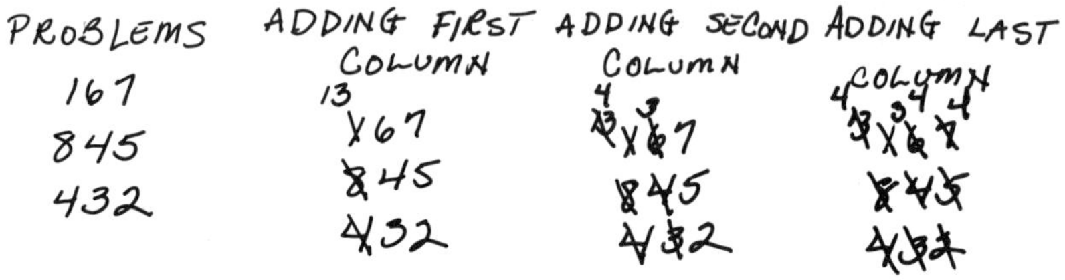

Figure 4.13

scratched out and it is replaced by 4. Finally the last column is added; the sum is 14 and the 4 is written at the top and the 3 at the top of the scond column is changed to 4. The answer is 1444.

"Scratch" Method of Multiplication

Suppose we wish to *multiply* 432 $\times$ 5. The steps are shown briefly in Fig. 4.14. One begins at the left. $5 \times 4 = 20$; scratch out the 4 and write 20. Now $5 \times 3 = 15$; scratch out the 3, write 5 above it, and change 20 hundreds to 21. Lastly, 5×2 is 10; scratch out 2, write 0 above the 2 and

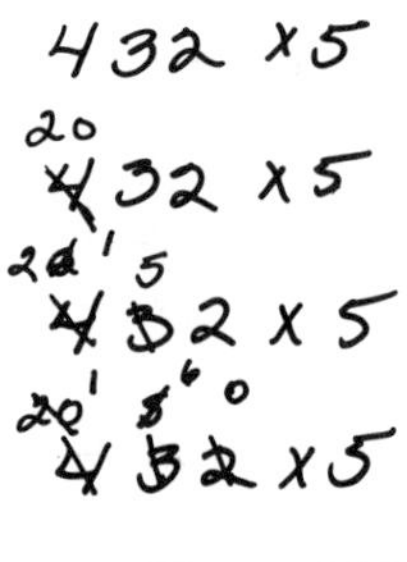

Figure 4.14

change 5 tens to 6 tens. Now all the digits are scratched out. Answer: 2160.

To multiply 432 × 56 one multiplies 432 by 5, finally scratches out the 5, since he needs it no more, and then rewrites 432 one place to the right and multiplies by 6, but adds his results to the previous product as he goes, using "scratch-out addition." This is illustrated in the successive steps of Figs. 4.15 and 4.16. Now place the numerals 432 under the original 432 but moved one place to the right and multiply and add as shown. 6 × 4 of the

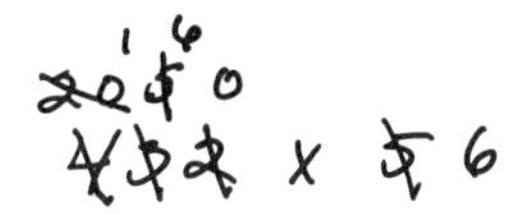

Figure 4.15

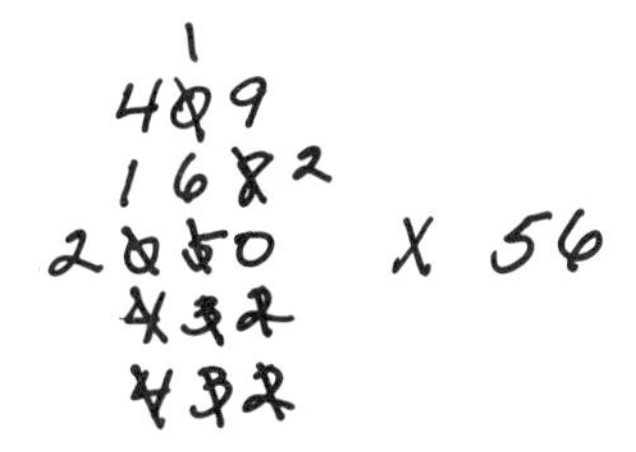

Figure 4.16

bottom 432 is 24. But 24 + 6 is 30 and the 1 in the hundreds column becomes 4. Now 6 × 3 = 18, and so on. Answer: 24192. Scratch out the 6. You will like this method once a little skill is developed.

Exercises

1. Use these problems to practice adding by the scratch method. Start at the left.

(a)	(b)	(c)
463	3165	834
217	4745	987
845	9321	642
	6785	583

   How does this algorithm show quite clearly the imprint of the abacus, especially which abacus?

2. Use the scratch method of multiplication on these problems.

   (a) 463 × 35
   (b) 493 × 76
   (c) 805 × 142
   (d) 1326 × 385

3. See if you can devise a scratch method for subtraction. Start from the left. You might try to devise one for division.
4. The Egyptians used for multiplication a method of duplation and its use continued into the Middle Ages. Find what this method is and demonstrate.
5. Consult references to learn the nature of the "lattice," or Gelosia method, of multiplication. Demonstrate.
6. The Gelosia method of multiplication led to the development of the Napier Rods (1617) for multiplication. See what these were. You might make a set for your classroom.
7. There is also a "halve and double" method for multiplication—the Russian peasant method. Learn what this is and demonstrate.
8. There are few rural areas in middle Europe in which "finger multiplication" is still done. Investigate and demonstrate this method.
9. Multiplication was also performed on the abacus. Consult texts in the history of arithmetic of other sources to see how this was done. Demonstrate.

The Operation Inverse to Addition

Early in the elementary school the student considers the "undoing" of an operation. He undoes turning on the light by turning off the light; he undoes closing the door by opening the door. Indeed, he is asked to complete exercises such as these: What is the undoing of

(a) getting up
(b) walking 2 steps toward the door
(c) opening the window
(d) closing the door
(e) finding three pennies
(f) walking 2 steps away from the door
(g) losing four apples
(h) having five pennies and receiving two more
(i) going into the swimming pool.

At the same time adding 2 cookies to a set of 5 cookies has an undoing, too. It is interesting to learn what the students call this—"unadding cookies"? taking away? Their terms may be new to the teacher, but they will give insight on how students look at the problem. Of course, the teacher will, some day, wish to introduce the word "subtract."

A physical approach to subtraction is highly imperative and through the use of the abacus algorithms may gradually emerge. The physical approach to unadding is clearly an emphasis, again, in the outward reach of mathematics. To help us, as junior and senior high school teachers, to gain a deeper insight into subtraction, let us perform these problems by the use of a pencil abacus of Fig. 4.18a. We shall do one here, although in class one should try the use of beans or a real abacus. Subtract

$$\begin{array}{r} 2\,3\,2\,6 \\ \underline{4\,1\,7} \end{array}$$

Figure 4.17

Beginning from the left of Fig. 4.17, we see we cannot take four sticks from three in the scond column, so we regroup the sticks as shown in the second figure. Now we can take away four sticks and we mark them out. We proceed for the rest of the problem in a similar manner. Hence the answer is 1909.

Exercises

1. Subtract with the use of a pencil abacus. Begin at the left or right as you prefer although the true abacist began at the left.

   (a) $\begin{array}{r} 4\,1\,3\,2 \\ \underline{1\,0\,8\,4} \end{array}$ (b) $\begin{array}{r} 4\,7\,3\,6 \\ \underline{1\,8\,8\,3} \end{array}$ (c) $\begin{array}{r} 3\,0\,6\,4 \\ \underline{1\,8\,7\,9} \end{array}$

2. Comment on the use of the word "borrowing" in subtraction. Does it accurately describe the process? Can you think of a better term?

At an appropriate time the teacher might well ask a question like this: Linda and Sandra made an adding table for us; on it we saw, for example, that 3 "added by" 2 gives 5. Can we read from the table 5 "unadded" by 2 is 3? Note that the first statement is indicated by the solid arrows in Fig. 4.18; the answer to the question can be found by the dashed arrows. Our adding table can be used for unadding of certain numbers, too! 3, for example, unadded by 7 never appears in the natural number table; such a problem (and answer) is therefore impossible among the natural numbers.

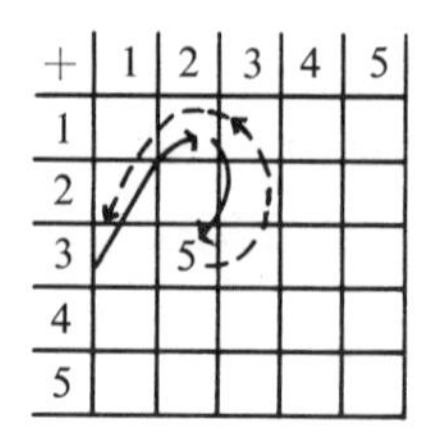

Figure 4.18

Exercises

1. From the chart above, what is

(a) 6 unadded by 2? (b) 10 unadded by 5 (c) $8\ {}^{un}+ 3$ (d) $9\ {}^{un}+ 4$

2. One can say that $9\ {}^{un}+ 4 = 5$ because $5 + 4 = 9$. Likewise $8\ {}^{un}+ 2 =$ ______________ because ______________. Indeed, we could say that $a\ {}^{un}+ b = c$ because ______________. This concept is important and useful. Even in calculus we say that $D_x^{-1}(x^2 + 3x) = x^3/3 + 3x^2/2 = k$ because________________________.

3. From the addition table in the discussion complete the following "unaddition table" (see Fig. 4.19). $7\ {}^{un}+ 2 = 5$. What shall we say to a student who asks about $4\ {}^{un}+ 5$? or $4\ {}^{un}+ 4$?

un +	1	2	3	4	5
2					
3					
4					
5					
6					
7		5			
8					
9					
10					

Figure 4.19

The above discussion describes a point of view conceptually different from taking elements away from sets. Here one sees an emphasis on the inward-looking face of mathematics. $5 + 4$ to yield 9 is *adding*; $9\ {}^{un}+ 4$

$= 5$ comes about as the operation *inverse* to adding. $9\ {}^{un}{+}\ 4 = 5$ because $5 + 4 = 9$ and there is no thought of a physical situation. The use of the word unadd emphasizes the inverse operation and we suggest that, in general, the teacher not discourage the use of such words as "unmultiply," "undifferentiate" (even in calculus), and "undistribute" for they do convey a meaning which is often more definitive than the use of conventional terms.

Exercises

1. The idea of inverse operation continues throughout all mathematics. Of course, the reader understands the operation inverse to adding but in Section 4.3 we studied an operation which we called "marilation." $f\,M\,s$ was $2f + 3s$. Find the operation which "unmarilates" and test your operation with several examples and with algebra.

2. The operation $*$ is defined as $f * s = f + s + fs$. What, therefore, is the operation inverse to $*$? Test the suggested inverse operation with several arithmetical examples and with algebra.

Subtraction Answering Several Questions; More Algorithms

When we look at subtraction from a physical point of view we see that subtraction is the answer to several different questions. Let us suppose that we can count and can add, but that we know nothing about subtraction and we are confronted with these questions:

Johnny has eight marbles and Tommy has three marbles:

(a) How many more marbles does Johnny have than Tommy?

(b) How many less marbles does Tommy have than Johnny?

(c) How many marbles does Tommy need to have as many marbles as Johnny?

(d) If Johnny loses three marbles how many does he have left?

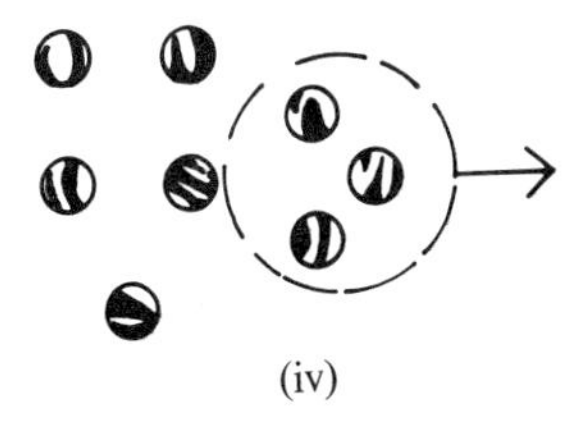

(iv)

Figure 4.20

How would we arrive at answers to these questions? You might try these on a child.

Let us look at each of these situations. The mature reader, will immediately say subtract in all cases, but the beginner must act them out either physically or mentally. Case (d) is simple, Fig. 4.20 indicates losing three, or three "walking away." There are five left. Case (c) makes use of the fundamental principle of 1-to-1 correspondence (Fig. 4.21). Three are

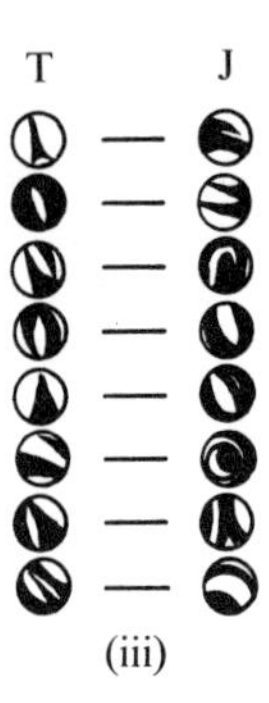

Figure 4.21

needed, but the basic ideas in (d) and (c) are fundamamentally different, "more" to many students means to add and here it means "how many marbles are required to fill the empty places." Case (a) can be diagrammed

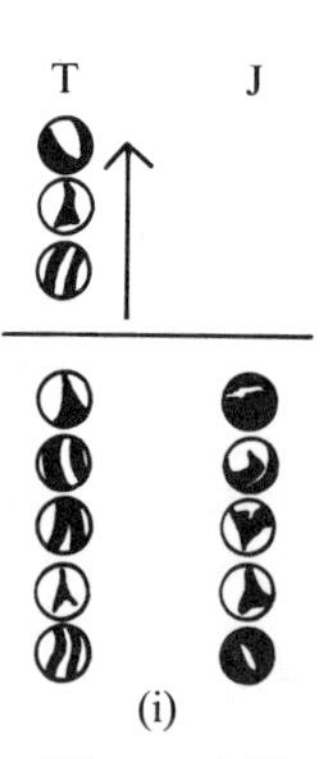

Figure 4.22

as shown in Fig. 4.22. Johnny has three more marbles than Tommy. The arrow reversed may help to indicate that Tommy has three less marbles than Johnny. The point is that all four of these questions are *linguistically and conceptually different* and time is required for the learner to perceive that all these problems are solved by subtraction. Addition is the answer to only one kind of problem; subtraction arises from at least four different situations.

Exercises

1. Joan is 32 inches tall and Mary is 36 inches tall. Phrase four different questions using these girls' heights to show four situations which give rise to subtraction.
2. You are working in a store and for a purchase worth $1.28 the customer gives you a five-dollar bill. Describe the correct way to make the change and to count it out to him. This is like which case of subtraction above?
3. Suggest several reasons why students find subtraction more difficult than addition.
4. The physical beginnings of arithmetic have left us with four situations which lead to subtraction. The inward-looking face of mathematics sees subtraction only in what way?
5. Is subtraction commutative? is it associative? is multiplication left-distributive over subtraction? is subtraction left-distributive over multiplication? is division left-distributive over subtraction? right-distributive? is addition left-distributive over subtraction? right-distributive? These are good questions for your students to consider, too.

It is not surprising that the different physical beginnings which lead to subtraction and the early use of the abacus combined to produce algorithms for subtraction which may seem strange to us. It is good for the secondary high school teacher to know that the students which come from different backgrounds may perform subtraction differently. We show here very briefly four different algorithms which are or have been in common use:

"Decomposition" Methods

Most of us, in subtraction, rewrite or decompose into a more useful form parts of the top number or the minuend when this is necessary. We do this when we borrow or when we exchange or regroup. But some people subtract by thinking "take away" and others subtract by thinking "how much must I add?" or "how much do I need?" We show these here:

Additive Decomposition

432 is decomposed
176 or rewritten as

```
 3 12 12
 4  3  2
 1  7  6
 -------
 2  5  6
```

and we ask 6 and what make 12, ? et cetera.

Take-away Decomposition

432 is decomposed
176 or rewritten as

3 12 12
4 3 2
1 7 6
2 5 6

and we say 12 less 6 is 6, et cetera.

"Equal Additions" Methods

In this method one adds 10 to 2 to make twelve; then to offset what he has done he adds 1 to 7 tens so he really subtracts ten more than the problem says. He does this because he added ten at the very start. A problem in subtraction performed by equal-additions algorithms has the outward appearance as shown. Of course, some of this is done mentally. Now the actual subtraction can be performed by take-away or additive methods.

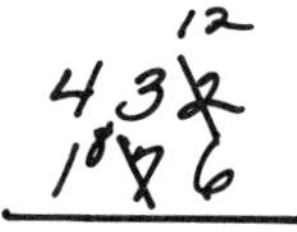

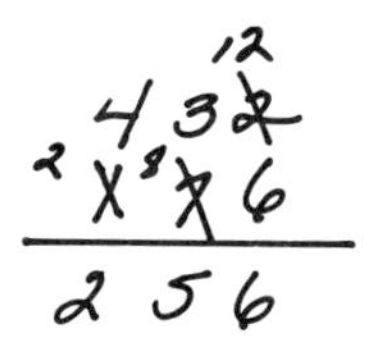

Hence we have

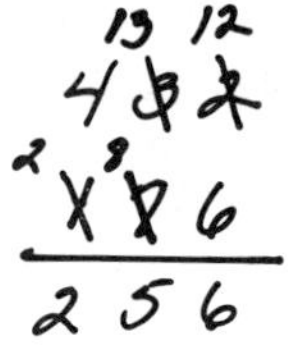

Additive Equal Additions

6 and what make 12?
8 and what make 13?
2 and what make 4? Answer: 256.

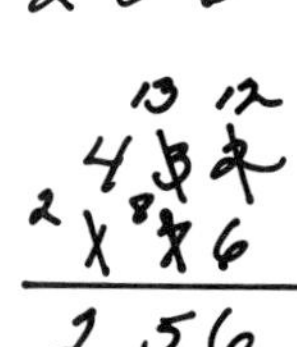

Take-away Equal Additions

12 take-away 6 is 6;
13 less 8 is 5;
4 less 2 is 2. Answer: 256.

The equal-additions method is sometimes called the borrowing-and-repaying method. How or why is this a good name?

We again emphasize the fact that the secondary and junior high school teacher will find such a background in arithmetic as described above very helpful. Not only do we become more deeply aware how man has struggled with the development of algorithms to work more quickly problems which have originated in his environment, but we begin to perceive with what care the ideas which seem simple to us must be nurtured in the minds of the student. Too, these ideas can become foci of interests on the part of the students and can lead to topics for papers and special investigations.

There is an opportunity here to have students experience that there are other ways to do things; this is realized quite easily in the university classroom for invariably there are a few students who subtract by methods other than the conventional take-away decomposition algorithm. Sometimes a student prefers one of the newly-introduced algorithms.

Exercises

1. Subtract each of these by each of the four methods outlined above.

   (a) $\begin{array}{r} 4326 \\ \underline{487} \end{array}$ (b) $\begin{array}{r} 8321 \\ \underline{1362} \end{array}$ (c) $\begin{array}{r} 2004 \\ \underline{783} \end{array}$

2. Explain in more detail than it was done in the discussion above why the equal-additions method is valid.

3. The base-ten numeral ab represents $10\,a + b$. Use the problem $\begin{array}{r} ab \\ \underline{-cd} \end{array}$ to show that the equal-additions method is valid. Assume $b < d$, $c < a$.

4. Use base-ten numerals to show that the equal-additions method of subtraction $\begin{array}{r} abc \\ \underline{-def} \end{array}$ is valid. Assume $c < f, b < e, d < a$.

5. The names which become attached to processes in arithmetic do not always describe the process correctly. What is your reaction to the appropriateness of the terms "to borrow," "decomposition method," "borrowing and repaying?" Discuss.

An Operation Inverse to Multiplication

Many times initial questions lead to concepts and algorithms which have an outreach not momentarily realized by the student. This is very much the case in division. Let us begin by a story or a situation:

(a) Harry has 15 candy bars and he wishes to give each of his five friends an equal number. How many candy bars will each friend receive. Suppose you know nothing about division—how would you figure this out? or how could a student think about this in his own way? See Fig. 4.23.

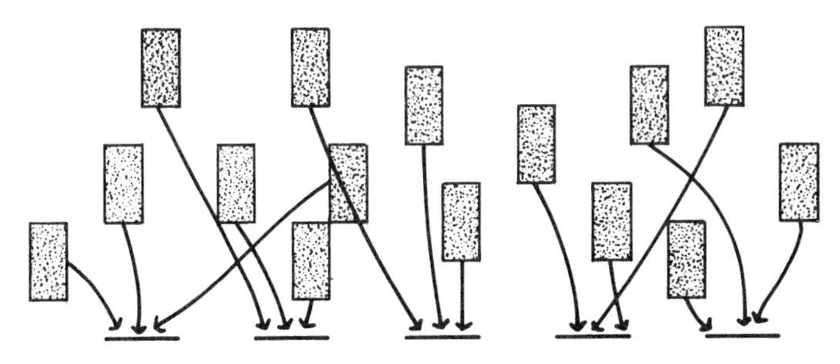

Figure 4.23

The student might make places for five groups and distribute one to each with the hope that each friend will have an equal number. There are three candy bars in each group. This process is called *partitioning* and we are said to have partitional fifteen by five.

(b) Harry has 15 candy bars and he wishes to treat each friend with five candy bars. How many friends can he so treat? Note that the answer

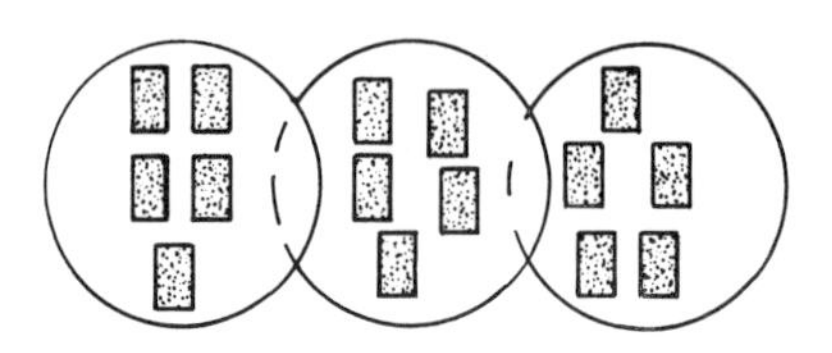

Three friends (or groups)

Figure 4.24

to this situation can be diagrammed, as in Fig. 4.24, but it would be better to ask a little boy or girl how they would do it. This process is called *quotitioning*; 15 has been quotitioned by five. Some call quotitioning measurement division.

(c) Harry has 15 candy bars and he wishes to take one-fifth of them. Is this like quotitioning? like partitioning? Some feel that this is still another different process. Without using the method of partitioning (that is, the method of forming five equal groups and taking one of the groups) one can obtain one-fifth of the set of candy bars by counting off 1 2 3 4 5, 1 2 3 4 5, 1 2 3 4 5, and then taking each one marked "one," or each marked "two," and so on.

We wish to emphasize that here are three conceptually different situations and three conceptually and linguistically different answers. Somehow the student must be led to see that there is one operation which answers all three of these questions: the operation which we call *division.* Again, multiplication is the answer to one kind of situation but division answers three different kinds. Some students find it difficult to know when to divide. All this is mentioned to help the teacher realize that division arises from situations more complicated than those of multiplication and that it may require much time for the student to learn that the operation which we call division provides an answer in all three situations.

Exercises

1. By diagram show the three different situations which give rise to division of 18 by 3. Write a statement or question for each and show the solution

by physical means. In your classes you could use lima beans, checkers, or the like. Going through the motion of these processes helps.

2. A problem asks the student to find how many feet there are in 48 inches. Is this a problem of partition or one of quotition?

3. The playground supervisor divided his 45 students into 5 teams. How many students were on each team? Is this an example of partition or one of quotition? If the playground supervisor formed baseball teams of ninc players each from his 45 students with no student playing on two teams, how many teams were formed? Is this partitioning or quotitioning?

In the development of every mathematical concept one must leave the physical or experimental beginnings (but frequently return for renourishment, reflection, and strengthening) and move in the direction of the inward look to the subject; so it is in division. This question might be asked at the appropriate time, perhaps shortly after the student learns to quotition with physical things: Is there any way by which we could figure out how many "sixes" (groups of six, as in Fig. 4.25) there are in 24 (a group of 24) without the use of beans?—and now expect student suggestions! Of course

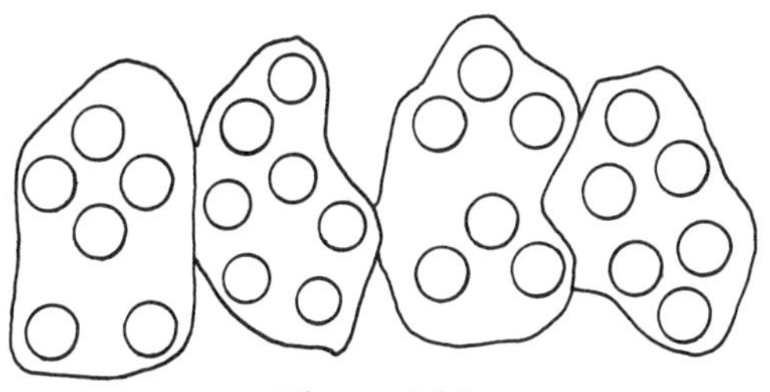

Figure 4.25

we do not know that the student will say, but he will make some kind of contribution as he thinks about this problem. It is very natural to subtract sixes and keep score on how many are subtracted.

$$\begin{array}{r} 24 \\ -\ 6 \\ \hline 18 \\ -\ 6 \\ \hline 12 \\ -\ 6 \\ \hline 6 \\ -\ 6 \\ \hline \end{array} \qquad \text{or} \qquad \begin{array}{r} 24 \\ -\ 6 \\ \hline 18 \\ -\ 6 \\ \hline 12 \\ -\ 6 \\ \hline 6 \\ -\ 6 \\ \hline \end{array}$$

and here we have some algorithm beginnings. Too, some students may volunteer, "I know that 2 sixes are 12 and we can subtract 12's twice; answer 4 sixes . . . ," and so on.

Students may at first invent something like this to see how many 14's (groups of 14) there are in 273. His first guess is 8 so he begins as shown:

```
          273
14       -112     8
         ----
          161
          126     9     next guess
          ---
           35
           14     1     next guess
          ---
           21
           14     1     next guess
          ---    --
            7    19
```

The answer is 19 with a remainder of 7. Note that no guess is wasted unless it is too large, and that the student understands exactly what he is doing. With discussion, practice, growth, this algorithm can lead to

```
        273
14      140     10
        ---
        133
        126      9
        ---     --
          7     19
```

but there is nothing wrong if the student has partial quotients which are far too small. Indeed, this method may well help to keep division within reach of the student who perceives number relations more slowly. Many texts show the algorithm (a) and this may be a development on the way to the conventional one. The streamlined version (b) which many of us use may be the

```
      9 }
     10 }  19
  14/273
     140          (a)
     ---
     133
     126
     ---
       7

      19
  14/273
      14          (b)
     ---
     133
     126
     ---
       9
```

ultimate but it may also conceal much, for when we ask "14 goes into 27 how many times?" we have changed the problem at the outset. The jump from physical things to the widely used algorithm (b) is simply too great for many students if we wish to perform division with meaning and not mere manipulation.

Exercises

1. Perform these divisions by some method other than the algorithm you use presently and assume that you are in middle elementary school:

   (a) $4632 \div 17$ (b) $1963 \div 17$ (c) $832 \div 47$

2. As we study the algorithms discussed above do they really perform quotition or partition? Which? Do you know any algorithm which performs the other operation?

3. How can it be that an algorithm for division which is really an algorithm for quotitioning can also be used to provide partitioning answers? Perhaps it will be necessary to go back to the operation of multiplication.[16]

4. When one solves the equation $4n = 12$, is this an example of quotition or partition, or do you feel that you cannot tell? Why?

5. You may be interested in consulting sources in the history of mathematics for some older algorithms for division. Demonstrate and explain why the algorithms work.

6. Although this chapter concerns arithmetic of the natural numbers and the (positive) rationals, see if it is possible to determine the quotient represented by $(x^2 + 5x + 6) \div (x + 3)$ by subtraction, in a manner like that above.

7. Try the use of the algorithm (a) to divide 34.79 by 0.14.

At some time in our growth in division we should look again at a multiplication table which the students have constructed, unless they have already called it to our attention that they can divide on a multiplication table (see Fig. 4.26). Indeed, 4×5 yields 20, but 20 "unmultiplied" by 5 yields 4. This is the same answer as 20 divided by 5, whether it be quotitioning, partitioning, or taking a unit fraction of, $20^{un} \times 5$ is 4 because $4 \times 5 = 20$; $12^{un} \times 3$ is 4 because $3 \times 4 = 12$. The inward-looking face simply sees division as inverse multiplication but, although the concept is elegant and it helps to systematize arithmetic, the idea of inverse multiplication does not bring with it the rich heritage of man's groping and struggle with ideas of numbers as does the outward-looking face. Most mathematics comes from situations in our environment and the teacher might well regard seriously this most natural method of introduction.

From the "unmultiplying" point of view, $21^{un} \times 5$ is simply an impossible problem. 21 does not appear in the "5" column. Physically, if we are

---

[16]Lyman C. Peck, and Dan Niswonger, "Measurement and Partition-Commutativity of Multiplication," *The Arithmetic Teacher*, 11 (April, 1964), 258–9.

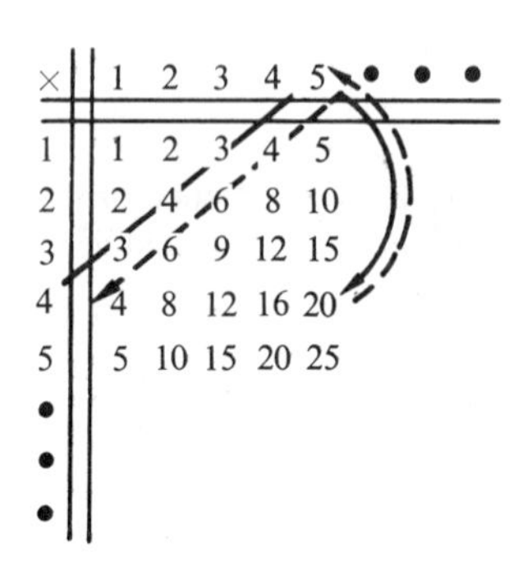

Figure 4.26

working in the domain of natural numbers, then in any of the division situations we are in difficulty.

As we close our sections on the operations on natural numbers, we might urge that every day in every arithmetic class we should do at least one problem without the use of algorithms, but solve it rather by reasoning or working in some other way.

A Discussion of Zero

Although we did not include 0 in the set of natural numbers or the counting numbers, some writers do.

First of all, zero is a new and sophisticated number, for it did not arise naturally. Alfred North Whitehead makes the interesting comment

> The point about zero is that we do not need to use it in the operations of daily life. No one goes out to buy zero fish. It is, in a way, the most civilized of all the cardinals, and its use is only forced on us by the needs of cultivated modes of thought.[17]

Secondly, any number multiplied by zero yields the product 0; and zero serves as an additive identity element since $a + 0 = a$ for every number a.

The difficulty with regard to zero occurs when one attempts to divide by zero (Fig. 4.27). We discuss the two cases $0 \div 0$ and $a \div 0$ where $a \neq 0$. The first case can be handled well by reading unmultiplication from a multiplication table as shown at the right: $0 \times 0 = 0$; $1 \times 0 = 0$; $2 \times 0 = 0$; $3 \times 2 = 6$; et cetera. Now since $3 \times 2 = 6$ then $6 \,{}^{un}\!\times 2 = 3$, also since $1 \times 0 = 0$ then $0 \,{}^{un}\!\times 0 = 1$. Likewise $0 \,{}^{un}\!\times 0 = 2$; likewise $0 \,{}^{un}\!\times 0 = 3$, and so on. Indeed $0 \,{}^{un}\!\times 0$, or $0 \div 0$, has infinitely many

[17]James R. Newman, *The World of Mathematics* (New York: Simon and Schuster, 1956), p. 442.

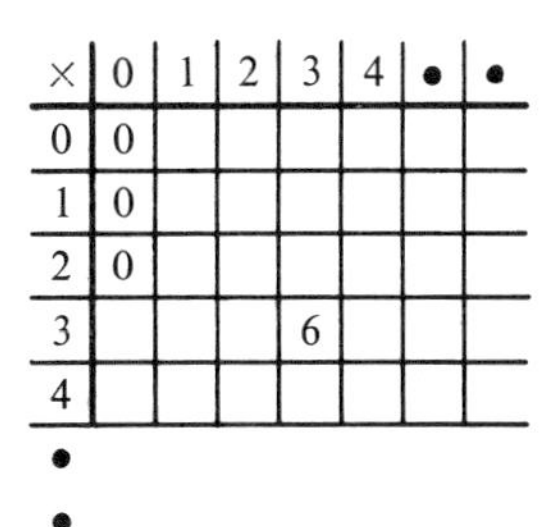

Figure 4.27

answers. The answer to $0 \div 0$ is indeterminate and if $0 \div 0$ appears in a problem we have a troublesome situation. $0 \div 0$ is undefined for its answer is not unique.

The case $a \div 0$ where $a \neq 0$, from the "table" point of view is impossible for if we attempt to unmultiply 4, say by 0, we see that 4 does not appear in the zero column. Hence, it is impossible to consider $4 \div 0$. From a more logical viewpoint, we might return to the definition of "unmultiplying." $6 \ {}^{un}\!\times 3 = 2$ because $2 \times 3 = 6$. Now let us assume that $a \ {}^{un}\!\times 0$ has an answer and let it be k where k is any number. Then $a \ {}^{un}\!\times 0 = k$ because $a = k \times 0$. But $k \times 0 = 0$: $\therefore a = 0$ which is contradictory to the assumption that $a \neq 0$. Hence, by *reductio ad absurdum* $a \ {}^{un}\!\times 0 \neq k$. That is, there exists no number which is another name for $a \ {}^{un}\!\times 0$. This is what is meant when it is said rather loosely that division by zero is impossible.

It is interesting to note that $0 \div 0$ is undefined because it is indeterminate; $a \div 0$, where $a \neq 0$, is undefined because it is impossible. Since these expressions are undefined, one never uses them to represent a number of any kind. Texts warn the student about this by saying that "division by zero is not permitted." Indeed, when such expressions occur in proofs or problems without awareness on the part of the writer, strange conclusions arise.

Students sometimes originate peculiar notions about $a \div 0$, and about $0 \div 0$. The quotient $a \div 0$ equals a, they say, "since, if you didn't divide by anything, then you still have a." This is evidence of confusion between dividing by zero and not dividing at all. We must separate the concept of zero from the idea of nothing or not doing something. Zero is a *number* which, undefined by some, belongs to the set of natural numbers, and always to the set of integers; it has its counterpart (the class $0/+1, 0/+2 \ldots$ $0/-1, 0/-2, \ldots$) in the set of rationals, and so on. That there is misunderstanding here is reinforced by hearing a radio weather forecaster say: "There is no temperature this morning—the thermometer reads zero!"

The usual reaction by the students to $0/0$ is that it is 1. It could be 1;

but it can also have the answer 2, also 3, also 4, as we see from the multiplication table. Indeed, 0/0 could be 0.

Exercises

1. That 0/0 is indeterminate is the reason that the expression $(x^2 - 4)/(x - 2)$ is not defined for $x = 2$, as we have learned before. Hence, the function $(x, x^2 - 4/2 - 2)$ has a finite discontinuity at $x = 2$ or it *has no name* at $x = 2$. What value could be assigned to $x^2 - 4/x - 2$ at $x = 2$ to make the function continuous?

2. The expression $x/x - 3$ is not defined for what value of x? Is it possible to assign it a name for this value of x? Comment on the practice of assigning the "number" ∞ to this expression for $x = 3$.

3. Consider this argument. Let $a = 4$. Now multiply each side by $a - 1$. Then we have $a^2 - a = 4a - 4$. Subtract $3a$ from each side and get $a^2 - 4a = a - 4$, and $a(a - 4) = (a - 4)$ whence $a = 1$. But it is given that $a = 4$, therefore $4 = 1$. Wherein lies the fallacy?

4. You may wish to make up problems as the above for your students. Start with "let $a = m$, multiply both sides by $a - k$..." and see if you can develop a pattern by which you can generate all problems like No. 3 above. (Must k have a particular value? What part of the procedure does m determine? Answer for k: k must equal 1.)

5. The definition of logical implication, $p \rightarrow q \leftrightarrow {}^{\sim}p \vee q$ can be illustrated by means of the accompanying truth table. Which situation do you think is illustrated by the argument in No. 3 above?

p	q	$p \rightarrow q$
T	T	T
T	F	F
F	T	T
F	F	T

6. Fallacies may be used in teaching to sharpen insight in the performance of algorithms and argument (proof) in mathematics. Also, it helps to show the student that, in some instances, extreme care must be exercised in drawing conclusions. We note, for example, that by actual computation,

$$4 - 20 = 64 - 80$$

is a correct statement. Now if we add 25 to both sides we get

$$4 - 20 + 25 = 64 - 80 + 25$$

whence

$$(2 - 5)^2 = (8 - 5)^2$$

and

$$2 - 5 = 8 - 5$$

$\therefore$

$$2 = 8.$$

There is a fallacy here but one different from that in No. 3. What is the error or fallacy? Which part of the definition of $p \rightarrow q$ does this argument illustrate?

7. You may be interested in more fallacies. Some are found in texts, placed there for students to investigate. There is available a book, *Fallacies in Mathematics* by E. A. Maxwell (England).[18] The subject of fallacies is good for special student investigation.

Tests for Divisibility

The subject of "divisibility" of one natural number by another is fascinating and useful and it is filled with many questions which junior high school students in arithmetic and algebra can well investigate. These topics lend themselves so well to experiment and to experiential approaches that, as in many other cases, one may enter this subject in a casual manner and become involved in much mathematics.

We illustrate an approach; the first step is simply to ask students to find the remainders when numbers such as these are divided by nine:

(a) 11	(e) 211	(i) 452
(b) 101	(f) 311	(j) 632
(c) 111	(g) 131	(k) 7023
(d) 1111	(h) 112	(l) 8193

Of course, the list might be longer, but it is interesting to see how many students return to class saying, "I have a quick way to find the remainder!" The teacher, "surprised," helps the students to test the conjectured algorithm on more numbers. On the next day's assignment are more problems with an invitation to "wonder why this works" or a challenge to construct a rationale for this algorithm. All of this can be done along with other review and current program work; this is an investigation concurrent with the main course.

Exercises

1. If the reader has never heard of an algorithm to be used in the digits of natural numbers to tell the remainder on dividing by nine, he should do the exercises above.

2. Study each of these numbers to determine the remainder on dividing by nine. Which are divisible by nine? (a is *divisible* by b if there exists a

---

[18]E. A. Maxwell, *Fallacies in Mathematics* (London: Cambridge University Press, 1959).

natural number k such that $a = kb$ where a and b are natural numbers.)

(a)	4326	(e)	837248
(b)	123894	(f)	6307101
(c)	630279	(g)	9304265
(d)	11432781	(h)	330014263

3. Recalling that the numeral *abcdef* (base ten) represents the number $100{,}000\,a + 10{,}000\,b + 1{,}000\,c + 10\,e + f$ and that $10 = 9 + 1$, $100 = 99 + 1$, et cetera, *test by deduction* one of the conjectures made above. Point out one or more properties which you are utilizing in your proof and which you are assuming to be valid.

4. See if you can devise a test on the digits to see if the number whose base ten numeral is *abcdef* is divisible by four. (There are several such tests; one is *abcdef* is divisible by four if $2e + f$ is.)

5. See if you can devise a test on the digits a, b, c, d, e, f to see if the number given by the base ten numeral *abcdef* is divisible by three.

6. Use the base ten numeral *abcdefghij* to see if you can devise a test for divisibility of the number by seven; by eleven.

7. When one tests a base-ten *numeral* for divisibility of the *number* by nine, the residue is the *remainder* obtained when the *number* is divided by nine. Is there a like result when one applies tests for divisibility by other numbers?

8. The test for divisibility by nine is related to the "casting out nines method" for checking addition. Learn (if you do not know) what this method is and discuss some of the assumptions made about sums of residues and residues of sums. Is this test infallible?

9. Can one use checks for addition which employ "sevens" or "elevens"? Investigate.

10. Show how the "casting out nines" method can be applied to subtraction, to multiplication, to division.

Numeration Systems

Methods or schemes by which the natural numbers one, two, three, four, . . . , are symbolized by numerals are called *numeration systems.* It is a great leap from the binary system of numerals, such as

urapun (one)	okosa okosa (four)
okosa (two)	okosa okosa urapun (five)
okosa urapun (three)	okosa okosa okosa (six)

"ras" meaning "a lot" (anything above six),

of the western tribes of the Torres Straits,[19] whose mathematical development was such that they classed all amounts above *six* as "heaps," to the decimal numeration system of today. Between the very primitive systems and those of today lie ancient systems of the Egyptians, the Babylonians, the Greeks, and many others with varying degrees of complexity. Not only do some of these systems display the rise of different bases but they use symbols for numerals rather than number words as the Torres Straits tribes did.

The entire subject of numeral systems is a very interesting one and it is rich in opportunities for student investigation which can well lead to projects, class demonstrations, written reports, and themes. Through individual or group studies of this kind it is possible to work cooperatively with areas in English, in the social sciences, and to enrich relationships between various fields of learning. Also there is opportunity here to help the student realize that the concept of *number* is independent of the numeral used to represent it and that the development and invention of numeral systems which are not cumbersome and which lead themselves to efficient use increase progress in mathematics very significantly.

It is not our purpose here to discuss the various student numeration systems of the past. These are left as possible topics for investigations if the reader wishes to pursue them; references are given at the close of the chapter. Rather, for our emphasis on methodology, we wish to suggest ways of introducing numeration systems which use the Hindu-Arabic numerals and have the same numeral *structure* as our system, but which simply have another base:

Johnny went with his parents to a country called Fivonia. The people there were about like Johnny but they had a strange way of counting; they would separate the objects in piles of five each and then write the numeral. In the case shown in Fig. 4.28 they had a pile of five and three left over and they wrote 13 for the number of objects.

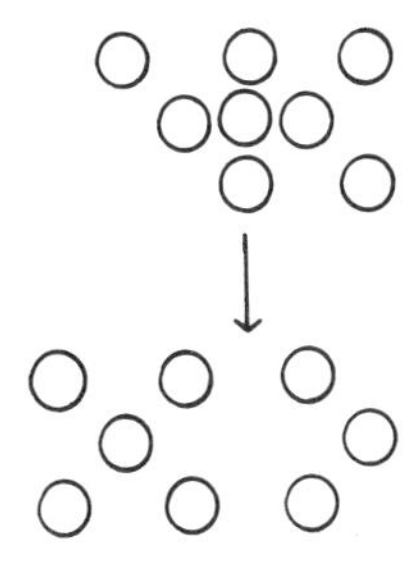

Figure 4.28

[19]Levi Conant, *The Number Concept* (New York: The Macmillan Company, 1931), p. 105.

Exercises

1. What numerals would the Fivonians use to indicate the number of objects in each of these groups?

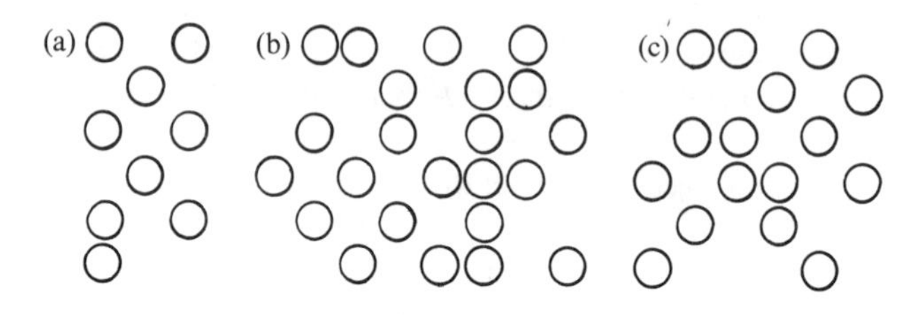

(Answer: 43)

2. When the Fivonians reached five groups of five each they wrote 100. Hence the number of each of these sets of objects would have what numeral?

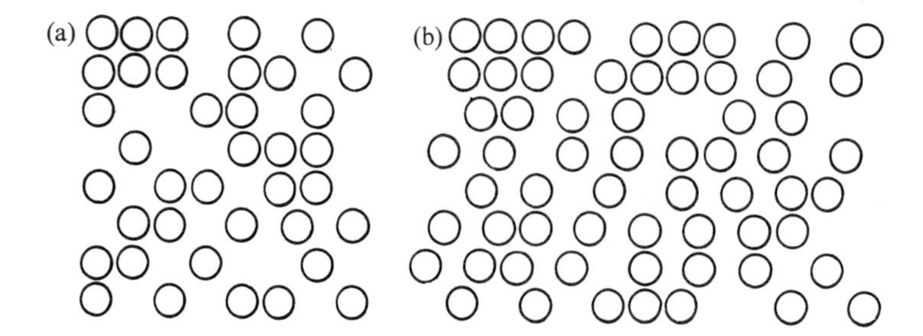

3. Each of these "Fivonian" numerals represents what number?

(a) 14 (Ans. 9) (b) 23 (c) 34 (d) 44 (Ans. 24)
(e) 103 (f) 214 (g) 434 (h) 1132

4. Had Johnny found that these people placed objects in packages by *sixes* what number would each numeral above have represented? (Answers to (a) ten; (f) forty six)

To learn to add in the Fivonian (base five) arithmetic one can use physical things: the abacus with five beads on each wire or the pencil abacus. Answers are worked out by *physical* means. The sum $22 + 14$ is the number whose numeral is 41 and after several results are arrived at in a crude manner some *hypotheses or algorithms* may be born. These are tested by more examples but finally they all seem to fit in the logical system thus far developed. Subtraction algorithms can be suggested by a similar approach. Multiplication in Fivonia can raise an interesting question for the student to handle in his own way. Seven groups of eight each must yield

$12 \times 13 = 211$ and the invention of an algorithm can come from the students themselves. Division algorithms can be developed, too.

Exercises

1. Multiply these numerals written in base five; the product should be a base five numeral also.

   (a) 2 3 × 3 = 1 2 4 (b) 2 3 × 4 (c) 2 4 × 3 (d) 1 1 4 × 4

   (e) 3 4 × 2 4 (f) 1 2 3 × 4 2 (g) 1 4 1 × 3 4 (h) 2 3 3 × 2 3

2. Subtract the numbers whose numerals are given in base five.

   (a) 4 3 1 − 1 1 (b) 4 3 1 − 1 2 (Ans. 4 1 4) (c) 4 3 1 − 4 2

   (d) 2 4 4 3 − 4 4 4 (e) 1 4 3 2 − 3 4 4 (f) 4 3 2 2 − 3 4 3

3. Multiply the numerals in No. 2 if the base were *seven*. Some answers are:

   (a) 1 0 2 (b) 1 2 5 (c) 5 4 5 4

4. See if you can develop an algorithm for *division* in base five.

   (a) 3 4 ÷ 1 2. (Ans. 2 with remainder 1 0)
   (b) 3 4 3 ÷ 2 4. (Ans. 1 2)
   (c) 1 4 4 2 ÷ 3 4. (Ans. 2 3)
   (d) 2 1 2 1 ÷ 4 2.

5. The binary system begins 1, 1 0, 1 1, 1 0 0, 1 0 1, 1 1 0, 1 1 1, 1 0 0 0, —for one, two, three, four, five, six, seven, eight, ... Continue this system to the numeral 32.

6. Perform the requested operations on these numerals in base two.

   (a) 1 1 + 1 0 + 1 1 1 (b) 1 1 1 − 1 0 (c) 1 1 0 1 − 1 0

   (d) 1 1 1 × 1 1 (e) 1 0 1 1 × 1 0 1 (f) 1 0 0 1 × 1 0 1

   (g) 1 1 1 1 1
   (h) 1 0 0 1 1 1 (Ans. 10, Remainder 1 or 10 1/11)
   (i) 1 0 1 1 1 1 0 1

7. You might like to explore the *quinimal forms* of numbers. For example 4.1 would mean four and one-fifth. What is the number represented by each of these quinimals?

   (a) 3.2 (b) 3.2 1 (Ans. 3 11/25) (c) 1 4.1 2

8. A very interesting numeral system is the duodecimal system. Let the symbols 1, 2, 3, 4, 5, 6, 7, 8, 9, D, E, 10, 11, . . . be the numerals for one, two, three, four, five, six, seven, eight, nine, ten, eleven, twelve, thirteen, ... Investigate this system for algorithms of addition, subtraction, multiplication, division, and working with duodecimals. To help in this direction see if you can perform these operations (the numerals are already duodecimal numerals):

(a) $\begin{array}{r} 4\,D\,E \\ 3\,2\,1 \\ +\,D\,E\,4 \\ \hline \end{array}$ (b) $\begin{array}{r} 8\,D\,E \\ -\,1\,E\,4 \\ \hline \end{array}$ (c) $\begin{array}{r} 1\,4\,D \\ \times\,3\,E \\ \hline \end{array}$ (d) $8\,D\,E \div 1\,D$

(e) Confirm that $1/4 = 0.3$; $1/3 = 0.4$; $1/6 = 0.2$

(f) Compute $1/5$ as a duodecimal. (Ans. $0.2444\ldots$)

(g) Compute $2/7$ as a duodecimal.

9. Consult a text on the history of arithmetic or on arithmetic for teachers and make a brief study of the numeral systems of the Egyptians; the Babylonians; the Greeks; the Romans; the Mayan civilization; the Chinese-Japanese. Here one will note some completely different structures. Each historical reference has the possibility of generating new insight into a particular concept.

Finite Arithmetics

These are also known as miniature, clock, or modular arithmetics. One could state these from definitions, but how much better it is to begin in a nonostentatious manner with "arithmetic from a clock," from a story, or some other contrived situation:

Tommy was playing a game of adding with a clock. To "add" 4 by 3 he placed the hand on 4 and then turned it through three places like the hands of a clock move. Where will the hand be now? (Answer: 1) Hence on this clock $4 + 3 =$ ______.

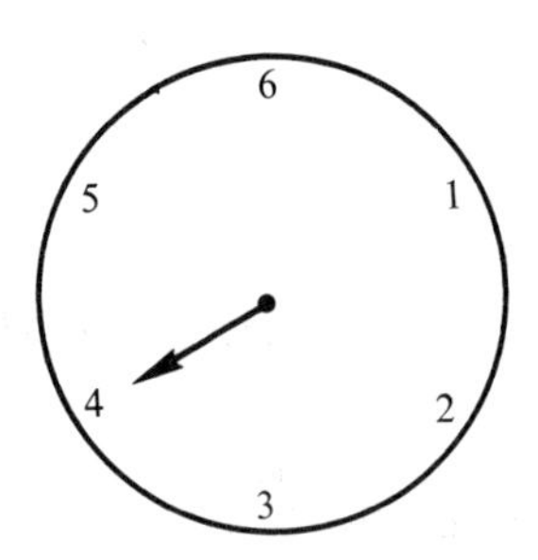

Exercises

1. (a) $2 + 5 =$ ____ (e) $3 + 5 =$ ____
   (b) $3 + 6 =$ ____ (f) $4 + 4 =$ ____
   (c) $2 + 3 =$ ____ (g) $4 + 2 =$ ____
   (d) $5 + 4 =$ ____ (h) $4 + 5 =$ ____
2. Can you find a way to add these numbers without using the clock? What method do you suggest.

Later the students and teacher can suggest summarizing the adding results in the form of a table. From this can come the idea of an "additive identity" and an "additive inverse." At some time in the discussion of finite arithmetics might come this question:

1. Is "clock addition" commutative—that is, does $4 + 3 = 3 + 4$? Does $5 + 3 = 3 + 5$?
2. How many different pairs of numbers must we study to answer completely the question of commutativity for a two-hour clock? for a three-hour clock? for a four-hour clock? for a five-hour clock? for a six-hour clock? The point to be emphasized is that in finite arithmetics we can study such properties *completely*. For a three-hour clock, we examine $0 + 0$, $0 + 1$, $0 + 2$, $1 + 1$, $1 + 2$, $2 + 2$ for commutativity and we are through.
3. Can you devise a formula which will answer the question in No. 2 for any number n of hours in the clock?

Creativity and student contribution are at work in the above situations and, if the level of the student is sufficient, one may consider attempting to investigate by mathematical induction any suggested formulae which seem to be promising generalizations.

Exercises

1. Investigate Nos. 2 and 3 above.
2. See if you can devise a formula to tell how many triples of different numbers must be used to study the property of *associativity* of addition for a n-hour clock. For the three-hour clock we could have these triples to study: $(0, 1, 2)$, $(0, 2, 1)$, $(1, 0, 2)$, $(1, 2, 0)$, $(2, 1, 0)$, $(2, 0, 1)$, where, by the triple $(1, 0, 2)$, for example, we mean that we would investigate to see if $1 + (0 + 2) = (1 + 0) + 2$.

The study of finite arithmetic has the advantage that certain questions as those of commutativity and associativity of the operations can be examined exhaustively. One can study all the cases because they are finite in number. Too, the construction of tables, as those of addition or multiplication modulo five, give practice in adding and multiplying and in determining remainders on dividing by five; perhaps practice in more skills than we had hoped, for indeed we reinforce skills while investigating other problems. Also, the study of such systems helps the student develop the meaning of the term "mathematical system." Through the study of finite or miniature arithmetic the student begins to sense the arbitrariness of mathematical systems and that the properties of the system are consequences of the way the operations are defined.

4.7. A Look at the Rationals

Adding Rationals

The study of the rationals has many opportunities for student investigation and for the creation of conjectures to be tested by more experimental work and finally by seeing if these can be made to fit into the logical system already developed. In approaches discussed here we shall assume that the student is already acquainted with the idea of a rational, and that this idea has been well nurtured by seeing and working with concrete examples of $\frac{1}{3}$ of something, $\frac{1}{2}$ of something, and so on.

It should be unnecessary to emphasize that the use of charts enhances the understanding of rationals. The student should have charts on $8\frac{1}{2}'' \times 11''$ paper along with cards to represent $\frac{1}{2}, \frac{1}{3}, \frac{1}{4}, \ldots$. The $\frac{1}{2}$-card placed beside another $\frac{1}{2}$-card fills the 1-space.

And now we are ready for questions.

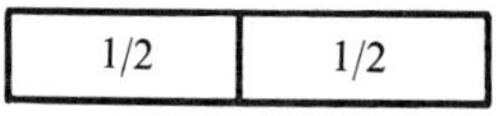

1/2 1/2

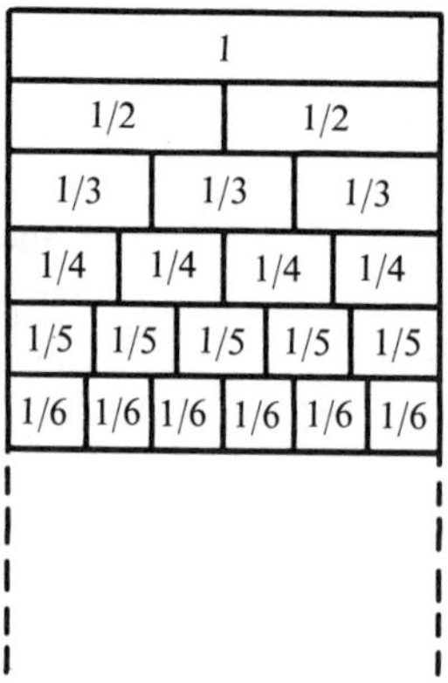

1
1/2 1/2
1/3 1/3 1/3
1/4 1/4 1/4 1/4
1/5 1/5 1/5 1/5 1/5
1/6 1/6 1/6 1/6 1/6 1/6

(a) Can anyone use his cards to show us $\frac{1}{2} + \frac{1}{3}$?

Now *every* student can do this and he experiences $\frac{1}{2} + \frac{1}{3}$ both by sight and by manipulation. Many students have never "seen" $\frac{1}{2} + \frac{1}{3}$ even with their eyes but are taught a meaningless rigmarole to add these rationals.

(b) Can you find another name for $\frac{1}{2} + \frac{1}{3}$ by the use of the cards and the charts? The student will sec that $\frac{1}{2} + \frac{1}{3}$ has $\frac{5}{6}$ for another name.

(c) Let us try $\frac{1}{3} + \frac{1}{4}$. The student finds by the use of physical devices that another name for $\frac{1}{3} + \frac{1}{4}$ is $\frac{7}{12}$.

Do other sums and invite the students to see if they can devise a method of operation on the numeral themselves to get the sum without the use of cards. What excitement there is when one suggests adding the two denominators to arrive at the numerator of the sum and multiplying them for the denominator of the sum! The teacher will err if she says, "Right!" She is using and teaching the induction process if she says only, "Perhaps! Let's try some more!" If the algorithm seems to work then this may be called Harry or Mary's conjecture and some day there may be suggested a rationale. Indeed, perhaps tomorrow someone may say, "We should have known that $\frac{1}{2} + \frac{1}{3}$ is $\frac{5}{6}$, for, from the charts $\frac{1}{2}$ has another name $\frac{3}{6}$; $\frac{1}{3}$ has the other name $\frac{2}{6}$, and $\frac{3}{6} + \frac{2}{6}$ equals $\frac{5}{6}$." This is an acceptable rationale for the elementary student of rationals. It is *not* the time to mention that he used the distributive law; detailed analysis may well destroy the spirit of discovery.

Probably someone will raise the question of two-thirds plus one-half, and rightly so. If there is no response to such problems we begin again with physical things: the cards and charts. Again there will be discoveries, conjectures, testing conjectures with more examples, and, finally, an accepted conjecture which will perhaps be supported by a rationale appropriate to the level of the student.

Students can be led easily to develop and abbreviate algorithms by themselves. There was a time when students were taught the algorithm for adding rationals which is shown. Three "goes into" 12 four times; $4 \times 2 = 8$. Likewise when we consider the $\frac{3}{4}$ we are taught to think, "4 goes into 12 three times, and 3 times 3 is nine." The 8 and 9 are added and the answer is $\frac{17}{12}$. Indeed this was taught in rote fashion and it is doubtful if even a few students really understood what was going on. Now, a student is led to consider other names for

$$\begin{array}{c|c} & 12 \\ \hline \frac{2}{3} & 8 \\ \frac{3}{4} & 9 \\ \hline & 17 \end{array}$$

$$\frac{2}{3} \quad \text{as} \quad \frac{4}{6}, \frac{6}{9}, \frac{8}{12}, \frac{10}{15}, \frac{12}{18}$$

and for

$$\frac{3}{4} \quad \text{as} \quad \frac{6}{8}, \frac{9}{12}, \frac{12}{16}, \frac{15}{20}, \frac{18}{24}$$

and he selects $\frac{8}{12}$ and $\frac{9}{12}$ as "other names" appropriate for adding. Answer: $\frac{17}{12}$. There will come a time, different for different students, at which they may not actually write other names but will discard and select mentally those names they wish to use. A form (which may lead to an algorithm) may emerge.

Exercises

1. Write five other names for each of these rationals.

(a) $\frac{1}{2}$ (b) $\frac{5}{6}$ (c) $\frac{5}{8}$ (d) $\frac{3}{16}$

2. Using other names select appropriate ones to perform addition.

(a) $\begin{array}{r} 2/3 \\ +\ 5/8 \\ \hline \end{array}$ (b) $\begin{array}{r} 1/3 \\ +\ 3/4 \\ \hline \end{array}$ (c) $\begin{array}{r} 7/3 \\ +\ 5/10 \\ \hline \end{array}$

3. What algorithm did *you* learn in the elementary school for adding rationals? Demonstrate with

$$\begin{array}{r} 5/9 \\ +\ 4/7 \\ \hline \end{array}$$

The teacher will soon observe from texts that rationals are now presented with the aid of the number ray (line), which yields insight. We suggest that the use of cards helps to make the approach concrete, and the cards and charts use the number ray idea also. Rods, bars, and rulers all can serve some purpose.

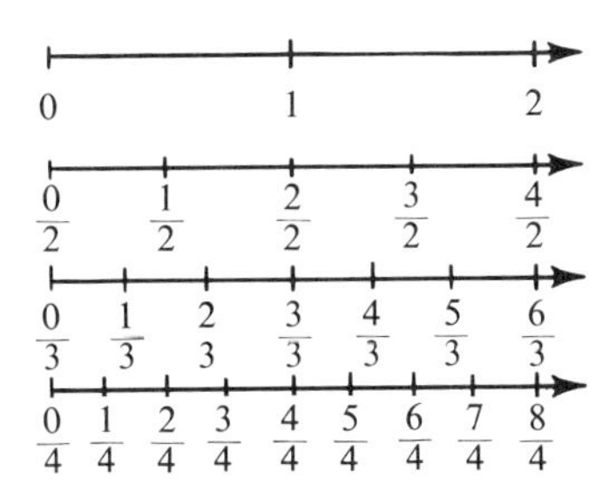

Exercises

1. Prove by the use of the postulates for a field that $1/a + 1/b = (a + b)/ab$. Perhaps this is a review of mathematics previously studied but we wish to emphasize the intellectual distance between the approach to this problem in the laboratory or concrete situations at the beginning level and the more sophisticated postulational approach.

2. Given the field postulates and the definition that $a/b = a \cdot 1/b$ prove that

$$\frac{a}{b} + \frac{c}{d} = \frac{a \cdot d + b \cdot c}{b \cdot d}$$

Multiplication of Rationals

Someday the teacher contrives this situation: Tommy has $\frac{2}{3}$ of a candy bar and he wishes to give a friend a fourth of it. How can he determine a fourth of $\frac{2}{3}$? We wonder what the student will say? This is the interesting (and instructive) facet of teaching. One may say, "It will be one half of $\frac{1}{3}$." Another may take the "length $\frac{2}{3}$" and fold it so he gets a fourth of $\frac{2}{3}$, and then hunt for a simpler name by fitting along lines on the chart. Each suggestion must be explored.

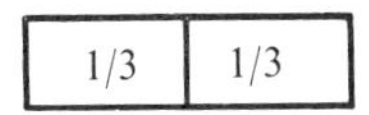

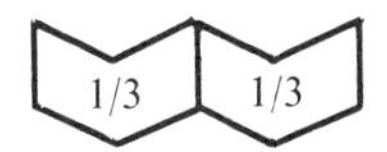

A conventional *demonstration* that $\frac{1}{4}$ of $\frac{2}{3}$ is $\frac{1}{6}$ is that which employs the use of a square or a rectangle. $\frac{1}{4}$ is shown by one type of shading whereas $\frac{2}{3}$ is shown by another type of shading. $\frac{1}{4}$ of $\frac{2}{3}$ appears in the part which contains both shadings. Indeed 2 out of the twelve small rectangles are doubly shaded, or $\frac{2}{12}$ or $\frac{1}{6}$. This example demonstrates but perhaps it lacks some of the physical "taking a fourth of" approach.

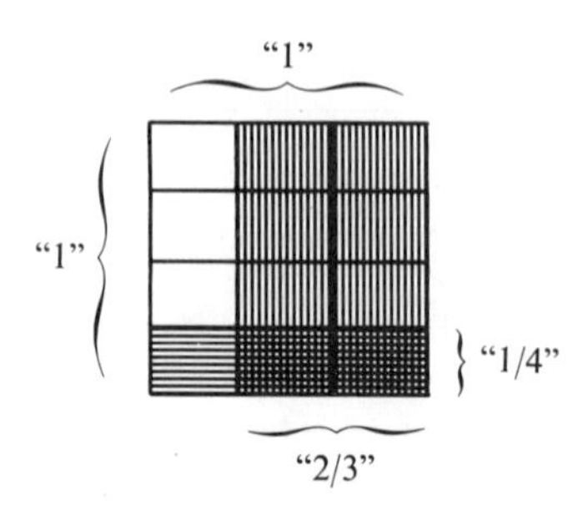

Exercises

1. By the simplest concrete approach possible show how one can determine another name for $\frac{1}{4}$ of $\frac{1}{3}$.
2. To what conjecture might this lead on taking $1/a$ of $1/b$?
3. Now use the postulates for a field to deduce that $1/a \cdot 1/b = 1/a \cdot b$.
4. The student learns by the use of concrete things that $\frac{1}{4}$ of $\frac{2}{3}$ is $\frac{1}{6}$; $\frac{1}{4}$ of $\frac{4}{5}$ is $\frac{1}{5}$; $\frac{1}{4}$ of $\frac{2}{5}$ is $\frac{1}{10}$; et cetera. Using other names, what rationale might he suggest to confirm logically that these are valid answers?
5. Using other names for $\frac{5}{6}$ (as $\frac{10}{12}, \frac{15}{18}, \frac{20}{24}, \ldots$) which one seems the best to use to reason out the answer to taking $\frac{2}{3}$ of $\frac{5}{6}$? $\frac{3}{4}$ of $\frac{7}{8}$? $\frac{3}{5}$ of $\frac{3}{4}$?
6. After working several examples the student may suggest that $\frac{3}{5}$ of $\frac{3}{4}$ can be obtained by the algorithm of multiplying the numerators for the numerator of the product and multiplying the denominators for the denominator of the product, i.e., $\frac{3}{5}$ of $\frac{3}{4} = (3 \cdot 3)/(5 \cdot 4) = \frac{9}{20}$. Explain why this algorithm actually takes care of the *rationale* of the problem as discussed in No. 5. Note that $3/5 \cdot 3/4$ and $3 \cdot 3/5 \cdot 4$ are conceptually different.
7. Using the definition that $a/b = a \cdot 1/b$ and the postulates for a field you prove that $a/b \cdot c/d = a \cdot c/b \cdot d$.

Division of Rationals

To ask a little boy, "What is $\frac{3}{4} \div \frac{1}{3}$?" makes little sense. Indeed, the more we think about this question the more baffled we become. And we

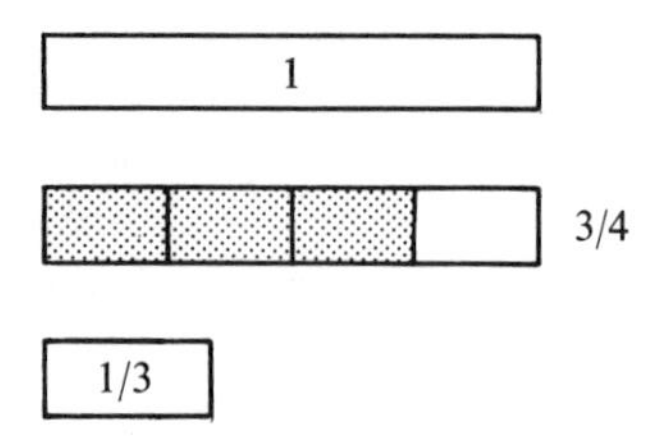

learn as teachers that it is quite possible for students to have difficulty because they cannot devise a mental picture of the question. On the other hand, "How many $\frac{1}{3}$'s are there in $\frac{3}{4}$?" has meaning; the students get out their cards and chart once more to learn how may $\frac{1}{3}$'s there are in $\frac{3}{4}$.

Most students will need the cards, although several may try to reason it out by the use of other names. Here, again, there arise conjectures to be tested by further examples.

The card $\frac{1}{3}$ "fits into" $\frac{3}{4}$ "about" $2\frac{1}{4}$ times. This can be seen physically. Hence: $\frac{3}{4} \div \frac{1}{3} = 2\frac{1}{4}$. Some student may suggest, "We do not need the cards or rods to do this: $\frac{3}{4}$ has another name, $\frac{9}{12}$; $\frac{1}{3}$ has another name, $\frac{4}{12}$; there are just as many $\frac{1}{3}$'s in $\frac{3}{4}$ as there are $\frac{4}{12}$'s in $\frac{9}{12}$; there are just as many $\frac{4}{12}$'s in $\frac{9}{12}$ as there are 4's in 9. There are $2\frac{1}{4}$ of them!" This shows insight and makes sense. Students learn to do this well and there is no rote (often meaningless to the student) as "inverting the divisor [provided they remember which rational to invert] and multiplying." *It is expected* that students may invent some algorithms and this we should encourage for efficiency of computation, but the point is that such shortcuts should, in the main, come from the students.

Exercises

1. You are an elementary student and you have just discovered or determined by cards or some other physical method that

$$\frac{3}{4} \div \frac{1}{3} = 2\frac{1}{4} \qquad \frac{5}{6} \div \frac{1}{4} = 3\frac{1}{3}$$

$$\frac{2}{3} \div \frac{1}{5} = 3\frac{1}{3} \qquad \frac{7}{8} \div \frac{1}{4} = 3\frac{1}{2}$$

What algorithm do you suggest will always give the quotient? Can you validate your conjecture by any kind of rationale.

2. Reason out by the use of other names

(a) how many $\frac{2}{3}$'s there are in $\frac{5}{6}$;
(b) how many $\frac{2}{5}$'s there are in $\frac{2}{3}$;
(c) how many $\frac{2}{5}$'s there are in $\frac{7}{8}$.

Can you suggest an algorithm which will give the quotient quickly? Can you explain why or how your algorithm "gives the answer"?

3. The "reasoning out by other names" method described in the discussion above is known as the "common denominator method." Another method, if the reader is interested, makes use of the properties of the multiplication identity element, 1.

$$\frac{2}{3} \div \frac{5}{8} = \frac{\frac{2}{3}}{\frac{5}{8}} = \frac{\frac{2}{3}}{\frac{5}{8}} \cdot 1$$

$$= \frac{\frac{2}{3}}{\frac{5}{8}} \cdot \frac{24}{24} = \frac{\frac{2}{3} \cdot 24}{\frac{5}{8} \cdot 24} = \frac{16}{15} = 1\frac{1}{15}$$

You find other names for each of these quotients by this method. Note the field approach in arithmetic; this should come as we reflect on what we have learned from physical and concrete situations.

(a) $\frac{3}{4} \div \frac{2}{3}$

(b) $\frac{5}{6} \div \frac{5}{8}$

4. There is another method which makes use of the properties of 1 along with that of the multiplicative inverse of a number; it is this:

$$\frac{2}{3} \div \frac{5}{8} = \frac{\frac{2}{3}}{\frac{5}{8}} = \frac{\frac{2}{3}}{\frac{5}{8}} \cdot 1$$

$$= \frac{\frac{2}{3}}{\frac{5}{8}} \cdot \frac{\frac{8}{5}}{\frac{8}{5}} = \frac{\frac{2}{3} \cdot \frac{8}{5}}{\frac{5}{8} \cdot \frac{8}{5}{}^{*}} = \frac{\frac{16}{15}}{1} = \frac{16}{15}$$

This method is the closest to the algorithm of inverting the divisor and multiplying and, in fact, it validates this well-known algorithm except that in the use of "inverting the divisor and multiplying" the denominator $\frac{5}{8} \cdot \frac{8}{5}$ which is marked above is left out and hence the student never knows why it is logically correct. Use this method for the problems in No. 3.

5. Since one can perform division by subtraction, students sometimes suggest that one can learn how may $\frac{1}{3}$'s there are in $\frac{3}{4}$ by subtraction. You begin subtracting $\frac{1}{3}$ from $\frac{3}{4}$ and see if you get $2\frac{1}{4}$ as the quotient. Determine by the method of subtraction these quotients:

(a) $\frac{3}{4} \div \frac{1}{6}$ (d) $\frac{2}{3} \div \frac{2}{7}$

(b) $\frac{3}{4} \div \frac{1}{5}$ (e) $\frac{7}{8} \div \frac{1}{3}$

(c) $\frac{3}{4} \div \frac{2}{5}$ (f) $\frac{9}{10} \div \frac{1}{4}$

6. Using the definition that $a/b = a \cdot 1/b$ and the postulates for a field prove that

$$\frac{\frac{a}{b}}{\frac{c}{d}} = \frac{a \cdot d}{b \cdot c}$$

An Interesting Multiplication Table

Many properties of natural numbers and properties of rationals can be gained from an ordinary multiplication table as the one shown below. Note that the table is somewhat "inverted" in that the left-hand column increases upward rather than in the usual manner for such tables. The longer students study this table the more they seem to see. It can be used to tell the areas of rectangles, it is a small "partial" table of squares, and so on. The most exciting suggestion which has come, however, is that it can be used to perform operations on the rationals—at least, it provides data by which such operations can be done quickly.

To illustrate the use of the table in adding fractions, let us suppose that the problem is

$$\frac{3}{4} + \frac{1}{6}$$

10	10	20	30	40	50	60	70	80	90	100
9	9	18	27	36	45	54	63	72	81	90
8	8	16	24	32	40	48	56	64	72	80
7	7	14	21	28	35	42	49	56	63	70
6	6	12	18	24	30	36	42	48	54	60
5	5	10	15	20	25	30	35	40	45	50
4	4	8	12	16	20	24	28	32	36	40
3	3	6	9	12	15	18	21	24	27	30
2	2	4	6	8	10	12	14	16	18	20
1	1	2	3	4	5	6	7	8	9	10
	1	2	3	4	5	6	7	8	9	10

Now from the third and fourth columns in the table $\frac{3}{4}$ has many other names ($\frac{30}{40}, \frac{27}{36}, \frac{24}{32}, \ldots$); also from the first and sixth columns $\frac{1}{6}$ has other names as $\frac{2}{12}, \frac{3}{18}, \ldots$. Now one name for $\frac{3}{4}$ is $\frac{9}{12}$ and one for $\frac{1}{6}$ is $\frac{2}{12}$.

$$\frac{3}{4} + \frac{1}{6} = \frac{9}{12} + \frac{2}{12} = \frac{11}{12}.$$

Of course, there will be a time at which the adding of $\frac{9}{12}$ and $\frac{2}{12}$ is seen to fit in the logical system by the use of the property of distributivity of multiplication over addition. Likewise the table can be used to assist in the division of rationals. Let us consider

$$\frac{3}{4} \div \frac{1}{6}.$$

Again we seek other names and find $\frac{4}{12}$ and $\frac{2}{12}$ as names which have the same denominator for $\frac{3}{4}$ and $\frac{1}{6}$, respectively. Now there are just as many $\frac{1}{6}$'s in $\frac{3}{4}$ as there are $\frac{2}{12}$'s in $\frac{9}{12}$. There are just as many $\frac{2}{12}$'s in $\frac{9}{12}$ as there are groups of 2 in 9. (Answer: $4\frac{1}{2}$) The use of the above table is found to be very helpful and, indeed, it provides good insight into the practical usefulness of the concept of other names for numbers.

The table above, though appearing to be a simple multiplication table, gives us a delightful and interesting means to build deeper understanding and to develop increased skill in performing operations among the rationals.

Exercises

With the help of the table work these problems on rationals:

1. $\frac{3}{4} + \frac{5}{6}$
2. $\frac{3}{4} \div \frac{5}{8}$
3. $\frac{2}{3} \div \frac{1}{4}$
4. $\frac{1}{4} \times \frac{2}{3}$ Here we can find "another name" for $\frac{2}{3}$ such that the numerator is 4 or a multiple of 4.
5. $\frac{2}{3} \times \frac{5}{7}$
6. $\frac{2}{3} - \frac{1}{4}$
7. $\frac{9}{10} - \frac{5}{6}$

4.8. Other Topics in Arithmetic

The limitations imposed by the writers do not permit the detailed discussion of more topics in arithmetic. Some matters commonly studied in arithmetic are included in other chapters such as square root, areas, and intuitive geometry. The spirit or manner of approach, however, which the writers have attempted to portray, is applicable to all topics. The teacher, as an engineer, contrives situations out of which the mathematics to be considered can emerge.

One topic which might be mentioned is that involving "percent." The idea of "percent" can grow naturally from rationals. A rational has many other names and that one which is called "so many hundreds" is of special interest.

$$\frac{3}{4} = \frac{75}{100}$$

Now one name is just as good as another though sometimes we may wish to use $\frac{75}{100}$ instead of its other name, $\frac{3}{4}$. Indeed, certain commodities are sold by the "hundred weight" so "75 per hundred" is not a strange idea. The Latin expression for "per hundred" is "per centum," hence $\frac{75}{100}$ is naturally called 75 per centum or 75 % and we see that percent comes from a convenient other name—so many hundred. The solution to different kinds of problems in percent might well be left to finding the truth set for certain open sentences.

$$7\% \text{ of } 60 = \square$$
$$40 \text{ is } \square\,\% \text{ of } 60$$
$$8\% \text{ of } \square = 60$$

are several such problems. Reflections on these open sentences by means of student-teacher discussion in the classroom will help algorithms emerge to give answers quickly.

4.9. Sample Study Guides

Many teachers find the use of study guides of one form or another very effective in helping to provide the kind of review and practice needed by the student. Also such guides help to invite activity by the students in exploring and carrying on investigations in mathematics at their respective levels. As a teacher gains experience there is some value in the use of the same guide sheets in succeeding years though, of course, they must be revised constantly. Also, different groups often invite different investigations and the use of study guides permits a desirable degree of flexibility.

Study guide materials may be written on the chalkboard, duplicated, and given to the student, or even dictated to the student. The fact that they are effective in encouraging much growth and active participation by the student is the important feature.

It is easier for the authors to write study guides in day-by-day situations than it is to compose several for illustration and, indeed, these which follow may fit no class whatever. Nevertheless, some samples the like of which the writers have found effective, follow.

A. *On "Percent"*

1. Review. In each of these problems write several other names for these fractions and then select appropriate ones in order to add them.

   (a) $\frac{7}{8} + \frac{5}{6}$ (b) $\frac{9}{10} + \frac{3}{4}$ (c) $3\frac{1}{6} + 2\frac{3}{5}$ (d) $2\frac{2}{3} + 1\frac{1}{4}$

2. Review. In like manner perform these divisions.

   (a) $\frac{7}{8} \div \frac{2}{3}$ (b) $\frac{1}{4} \div \frac{1}{5}$ (c) $\frac{3}{5} \div \frac{3}{4}$

3. Text (assignment from problems list or reading in the text).

4. Johnny found a job working for Mr. Smith in a candy store. Mr. Smith had the candy marked differently for different kinds of candy:

   15 ¢ for 20 pieces
   17 ¢ for 25 pieces
   7 ¢ for 4 pieces

   and so on. Johnny thought his little customers would like to compare prices, so with Mr. Smith's permission he tried to make a chart to show the cost of 100 pieces of each kind of candy.

   15 ¢ for 20 pieces would be how much for 100 pieces?
   17 ¢ for 25 pieces would be how much for 100 pieces?
   2 ¢ for 1 piece would be how much for 100 pieces?

   In No. 4 the students are beginning to think about changing 15 ¢ for 25 to ____________ for 100. Discussion in class can point out the advantage in making comparisons when each number means "so many per hundred."

B. *"Percent," Continued*

1. Review (work with rational fractions).

2. Text (work on material from the text).

3. Write other names for these expressions so they end with "per hundred."

   (a) 30 ¢ for 50 (Ans. 60 ¢ per hundred)
   (b) 15 ¢ for 25
   (c) 10 ¢ for 20
   (d) 5 out of 10 (Ans. 50 per hundred)
   (e) 12 out of 25

(f) 13 out of 25
(g) 2 out of 5

4. Write other names for these expressions so they end with "per hundred."

(a) 4 out of 200 is the same as _______ per hundred
(b) 6 out of 400 is the same as _______ per hundred
(c) 10 out of 259 is the same as _______ per hundred
(d) 15 out of 600 is the same as _______ per hundred
(e) 100 out of 300 is the same as _______ per hundred
(f) 65 out of 200 is the same as _______ per hundred
(g) 170 out of 500 is the same as _______ per hundred
(h) 43 out of 300 is the same as _______ per hundred

It is expected that soon the teacher would suggest that tradesmen long ago used Latin terminology and called five per hundred as five *per centum*. Hence, as we begin to use these two new words right away, 50 per 200 means _______ per centum.

C. *Area of a Rectangle*

1. Review work.
2. Material currently being considered.
3. Johnny heard his mother and daddy talking about placing tiles on the floor of one of their rooms. The tiles were one foot square. He thought perhaps he could help them determine how many tiles were needed. If the relative sizes are as shown above, how would you go about deciding on how many tiles are needed?

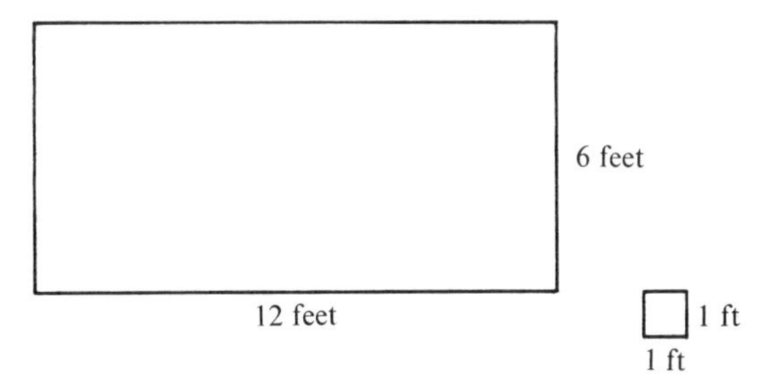

(One does not know what suggestions he will get, but they *will* reveal how the student thinks on the problem.)

D. *More on Areas of Rectangles*

1. Review exercises.
2. Current work from text or made up by the teacher.

3. From the suggestions received on the "tile problem" do we have a quick way of determining the number of tiles needed? Use one of these suggestions to determine the number of tiles necessary to cover these floors.

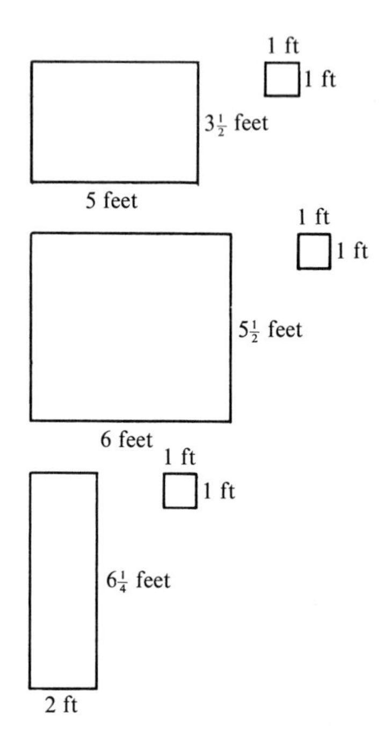

4. How could we determine the number of tiles necessary to cover the floor of the school room (we are assuming it is shaped like the floors in the preceding drawing) if the tiles are one-foot square? Perhaps you would like to do this.

5. If a rectangular floor is covered by 80 of the above tiles then the measure of its area is _______ square feet.

6. Would we need square-foot tiles to determine the number of square feet in the area? If not, what other method could be used.

Exercises

1. Write a series of guide sheets which will help start and nurture the idea of percent gain.

2. Write a series of guide sheets which will help the student to construct a method or formula to determine the area measure of a triangle.

3. Do the same for the area measure of a trapezoid.
4. Select another topic commonly studied in arithmetic and plan a series of study guides from which the topic emerges through exercises done by the student.

As one might expect, there are many, many suggestions for further reading in the teaching of arithmetic since this subject has always been present in school programs. The bibliography lists some books which contain sections on the teaching of arithmetic and some of the more recent articles on phases of arithmetic teaching. One who wishes to pursue this study further should not overlook entire books on the teaching of arithmetic and teacher guides which have been prepared to accompany modern arithmetic programs.

Bibliography

Adler, Irving, "Mathematics for the Low Achiever," *NEA Journal*, (February, 1965), pp. 28-30.

Butler, Charles H., and F. Lynwood Wren, *The Teaching of Secondary Mathematics* (New York: McGraw-Hill Book Company, 1960), Chapter 11.

Bacon, Marjorie, "Operation Bottle Caps," *The Arithmetic Teacher*, XII (October, 1965), pp. 466-8.

Beaumont, Andrie, "How an Abacus Can Help You Teach Mathematics," *Grade Teacher*, (April, 1965), p. 24.

Chirko, Thomas, "More on Venn Diagrams, G. C. D. and L. C. M.," *The Arithmetic Teacher*, XIII (November, 1966), pp. 552-5.

Cochran, B. S., "Children Use Signed Numbers," *The Arithmetic Teacher*, XIII (November, 1966), pp. 587-8.

Crist, R. L., "Use of a Programmed Multiplication Text Under Group-Paced and Individual-Paced Conditions," *Audio-Visual Communications Review*, XIV (Winter, 1966), pp. 507-13.

Cunningham, G. C., "Making a Counting Abacus," *The Arithmetic Teacher*, XIV (February, 1967), pp. 132-4.

Cunningham, George, "So You're Going to Teach Modern Mathematics," *The Instructor*, LXXV (October, 1965), pp. 36-8.

Fremont, H., "Pipe Cleaners and Loops; Discovering How to Add and Subtract Directed Numbers," *The Arithmetic Teacher*, XIII (November, 1966), pp. 568-72.

Geyser, G. W. P., "On the Teaching of Decimal Fractions," *The Arithmetic Teacher*, XIII (December, 1966), pp. 644-6.

Gillcrist, William A., Jr., "A Study of Grading in Arithmetic," *Education*, LXXXVI (November, 1965), pp. 177-81.

Glennon, V. J., "Is Your Mathematics Program on an Even Keel?" *The Instructor*, LXXVI (February, 1967), pp. 70-1.

Hamilton, E. W., "Manipulative Devices," *The Arithmetic Teacher*, XIII (October, 1966), pp. 461-7.

Humphrey, J. H., "Mathematics Motor Activity Story," *The Arithmetic Teacher*, XIV (January, 1967), pp. 14-6.

Johnson, Donovan A., and Gerald R. Rising, *Guidelines for Teaching Mathematics* (Belmont, California: Wadsworth Publishing Company, 1967).

Kennedy, Joseph, "Multiplication Tables and Dominoes," *The Arithmetic Teacher*, X (May, 1963), p. 283.

Mauro, C., "Developing an Understanding of Inverse Operations," *The Arithmetic Teacher*, XIII (November, 1966), pp. 556-63.

May, L., "Card Games Make Difficult Concepts Come Alive," *The Grade Teacher*, LXXXIV (October, 1966), pp. 92-4.

May, L. J., "Search for Patterns Helps Children Think for Themselves," *The Grade Teacher*, LXXXV (September, 1967), pp. 113-4.

Metzner, S., "Classroom Tested Learning—Games for Use in Urban Elementary Education," *Journal of Education*, CIL (December, 1966), pp. 3-48.

Morton, R. L., "Fractional Numbers with a Sum of One," *The Arithmetic Teacher*, XIII (December, 1966), pp. 658-61.

National Council of Teachers of Mathematics, *Instruction in Arithmetic* (Twenty-Fifth Yearbook). Washington: The National Council of Teachers of Mathematics, 1960. *Topics in Mathematics for Elementary School Teachers* (Twenty-Ninth Yearbook). Washington: The National Council of Teachers of Mathematics, 1964.

O'Donnell, J. R., "Number, Numeral, and Plato," *The Arithmetic Teacher*, XIII (May, 1966), pp. 401-2.

Olberg, R., "Visual Aid for Multiplication and Division of Fractions," *The Arithmetic Teacher*, XIV (January, 1967), pp. 44-6.

Phillips, J. M., "Jo Phillips Discusses Counting," *The Instructor*, LXXVI (August, 1966), pp. 108-9.

Ross, Ramon, "Diagnosis and Correction of Arithmetic Underachievement," *The Arithmetic Teacher*, X (January, 1963), pp. 22-6.

Salzer, R. L., "Discovering What Discovery Means," *The Arithmetic Teacher*, XIII (December, 1966), pp. 656-7.

Sganga, F. T., "Bee on a Point, a Line, and a Plane," *The Arithmetic Teacher*, XIII (November, 1966), pp. 549-52.

Sganga, F. T., "World Formulas in Math," *The Instructor*, LXXVI (June, 1967), pp. 118-ff.

Snow, John T., "An Experiment with Remedial Students," *The Arithmetic Teacher*, IX (April, 1962), p. 215.

Strehler, Allen F., "What is New About the New Math?" *Saturday Review*, LXXXIV (March 21, 1964), pp. 68-9.

Struthers, Joseph A., "The Challenge of Mathematics in the Elementary School," *The Arithmetic Teacher*, XIII (January, 1966), p. 48.

Townsend, Myrtle M., "Mathematics in the Elementary School—Problems or Processes," *Childhood Education*, XLII (February, 1966), p. 6.

Van Engen, Henry, "The Reform Movement in Arithmetic and the Verbal Problem," *The Arithmetic Teacher*, X (January, 1963), pp. 3-6.

Wehrman, K., "Math Gets Exciting When They Make Their own Discoveries," *Grade Teacher*, LXXXIV (January, 1967), pp. 88-91.

Welsing, W. C., "Teaching Division by the Reciprocal Method; a Programmed Instruction Unit," *Journal of Business Education,* XLII (January, 1967), pp. 149-51.

Welson, J. W., "What Skills Build Problem-Solving Power?" *The Instructor*, LXXVI (February, 1967), pp. 79-80.

Williams, J. D., "Effecting Educational Change; Some Notes on the Reform of Primary School Mathematics Teaching," *Educational Research*, VIII (June, 1966), pp. 191-5.

Winthrop, H., "Arithmetical Brain-Teasers for the Young," *The Arithmetic Teacher*, XIV (January, 1967), pp. 42-3.

5

Teaching Algebra

Algebra is a low form of cunning.

William James[1]

Modern mathematics does not replace classical mathematics. It generalizes it, supplements it, unifies it. But classical mathematics... are as important as they ever were.

Irving Adler[2]

5.1. Algebra from Different Points of View

As it was suggested with reference to arithmetic, algebra can also be considered from several points of view and each of these contributes insight into both algebra and mathematics.

One point of view is that algebra is a *collection of skills* and methods to solve equations and to find the unknowns. Algebra is a set of algorithms for working problems and solving for x. True, algebra taught from this point of view may provide for creativity and the devising of methods, or, sometimes, "tricks" to solve problems. There is little attempt, however, to demonstrate that algebra has a logical structure and that there is a certain logical cohesiveness which unifies the entire discipline; formal deduction is used very sparingly. A course in algebra given as a set of skills grows merely

[1]James R. Newman (editor), *The World of Mathematics* (New York: Simon and Schuster, 1956), p. 26.

[2]Irving Adler, "The Changes Taking Place in Mathematics," *The Mathematics Teacher*, LV (October, 1962), 449.

by increasing the manipulative ability of students so that they are more skillful in assigning letters to unknown quantities and acting on them as if they were numbers. The more adept the algebra student is at these skills, the better student he is. Many students have done well in algebra only to meet disappointment and frustrations when they meet "proof" for the first time. The set-of-skills and devising-methods approach to algebra may have suggested to William James his oft-quoted statement that algebra is "a low form of cunning."

A proper use of creativity and invention on the part of students leads to the second point of view of algebra. In the *logical* point of view, the student begins with some postulates—which may be suggested by observations on the natural numbers or from some other more logically primitive source, such as the postulates of Peano—and ultimately develops the ordered real number field, or perhaps the complex number system. Of course, there is emphasis on manipulation, but such an approach helps the student to build a good logical structure of algebra in which every step taken is defended by deduction from what has been developed before. The student becomes acquainted with deduction in algebra and he perceives that he is working in a mathematical system which includes arithmetic as a special case. Algebra, at one stroke, "has a way of making assertions that are valid for many numbers or even for all numbers in the real number system."[3] The second quotation at the beginning of this chapter was written in this spirit.

There are strengths and dangers in both points of view. The empirical point of view may well give rise to exciting adventures and challenging conjectures to be tested, but if they are never considered more deeply and woven into a logical fabric, the students will remain as "manipulators"—with all the accompanying dangers of forgetting and of having special methods for each particular type of problem. On the other hand, student-and-teacher examination of ideas can help to build a logical system such that students will recognize themselves as part-creators of the structure, providing deep and abiding experiences in the algebra which they have helped to build. The real teacher of mathematics is one who can help turn scattered findings into a logical system. The real teacher is characterized by Fawcett when he says,

> Only those with a creative imagination can see the mighty oak in a tiny acorn, and it is the exciting responsibility of the mathematics teacher to develop that kind of creative insight as he guides his students to recognize the broad generalization which serves as a continuing thread on which the beads of experience can be hung.

---

[3] *Ibid.*

> . . .(The teacher) guides them as they examine their experience, looking for the acorn that becomes the mighty oak.[4]

On the other hand, there is a danger present in the strictly logical system approach to algebra. The closely-adhered-to postulational approach may be devoid of life for many students if it becomes a series of deductions in the hands of a teacher who cannot communicate the beauty of mathematics in the language of students, or who does not understand how students learn. The strength of the postulational approach to algebra lies in the insight which it gives the student to the nature of proof and to the nature of mathematics. It introduces and emphasizes ideas which will grow as the student studies more mathematics.

Differing points of view in algebra, however, do not exclude each other. Indeed, the skillful teacher may at one time have one emphasis and at another time a different emphasis; another example of the two faces of mathematics. Some teachers may wish only to postulate much, but have short sequences of theorems such as those on odd numbers, on even numbers, on divisibility, on solving equations, et cetera, and leave the construction of the real number system to later courses. Under any plan there is opportunity for originality, for careful examination of conjectures, and for the use of deduction in algebra.

The series-of-skills character of algebra which was filled with rules and with examples to be followed by the student is gradually giving way to programs which encourage more attention to proof and to logical structure. Indeed, it is held by some that algebra is a better discipline in which to teach deduction than is geometry, for, in the latter, the student has a tendency to rely on what he *sees* in the figure rather than on what is given. Fallacies in earlier work in geometry were made by learned mathematicians through this deceptive procedure.

5.2. A Look into this Chapter

The purpose of this chapter is to discuss the teaching of various topics in algebra with emphasis on student-centered approaches. It is not intended that it will include a substantial development of any subject usually found in algebra. It is surely not the purpose to include here a thorough and complete construction of the real number system, nor is it meant to suggest that the order of topics discussed here should be the order in which they should be studied in your algebra courses.

It it hoped, however, that the discussion in this chapter will illustrate

---

[4]Harold P. Fawcett, "Reflections of a Retiring Teacher of Mathematics," *The Mathematics Teacher*, LVII (November, 1964), 455.

a meaningful and effective kind of methodology which is immediately applicable and which may encourage teachers to attempt like approaches to other topics.

The techniques introduced here are not meant to be used blindly. It is, rather, the nature of the approach being discussed which is really important. Nor is it meant to imply that texts in algebra may not have good approaches for students; the teacher can learn much by studying the methodology suggested in his own or in other texts. The study of these books as well as periodicals on teaching mathematics helps much.

In brief, it is our aim to present several topics in a manner which has drawn from students exciting, interesting, and fruitful suggestions.

5.3. A Beginning on Equations

Regardless of the point of view concerning the development of algebra in the school program, whether this be largely empirical or postulational, much emphasis is given eventually to solving equations. The setting up and solving of equations is one of the outward facets of algebra and a very useful one in mathematics and in related fields. Indeed, we recall that the Egyptians struggled with skills of this kind for several centuries in the use of the method of false position.

We shall illustrate two approaches: (a) a semiphysical or semilaboratory empirical approach and (b) one which cultivates more the inward-looking approach.

Pictorial Beginnings

Teachers who are interested in situation approaches can devise and engineer many different kinds of experience out of which the ideas of equation solving can develop. We present this situation in which Fred finds himself as an example:

> Fred and Tom were talking about their smaller brothers. Tommie said to Fred, "My little brother is so tall that if you would double his height and add 4 you would get seven feet. Can you tell me how tall my brother is?"
> Now this was quite a problem for Fred and he thought about it for awhile. Finally, he drew a figure like that in Fig. 5.1.

How can you use this figure to tell how tall Tommie's brother is? Of course another student might have another suggestion. One does not know what the reaction of any one student will be, but elementary students perceive that one can "sweep off" four feet from the top of each column so the

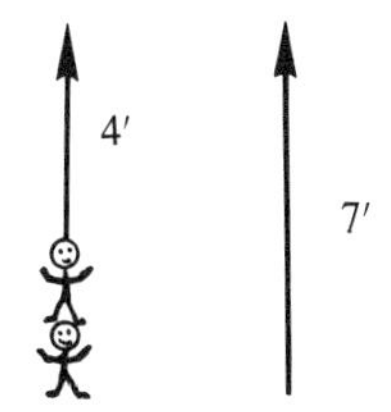

Figure 5.1

resulting figure appears as in Fig. 5.2. Now if the brother "used twice" was 3 feet tall then the brother "used once" was 1-1/2 feet tall!

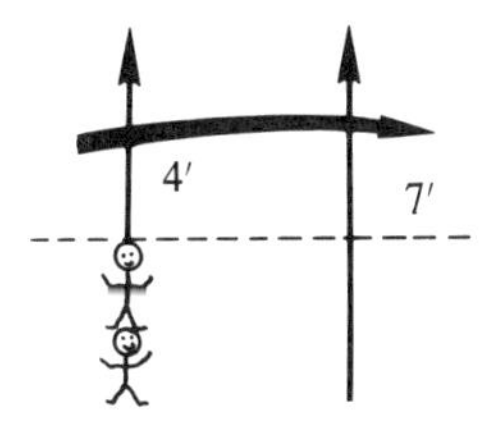

Figure 5.2

> One can use sacks and beans to solve a problem such as:
> Two equal sacks containing beans plus four more beans had a total of ten beans. How many were there in each sack? (see Fig. 5.3.)

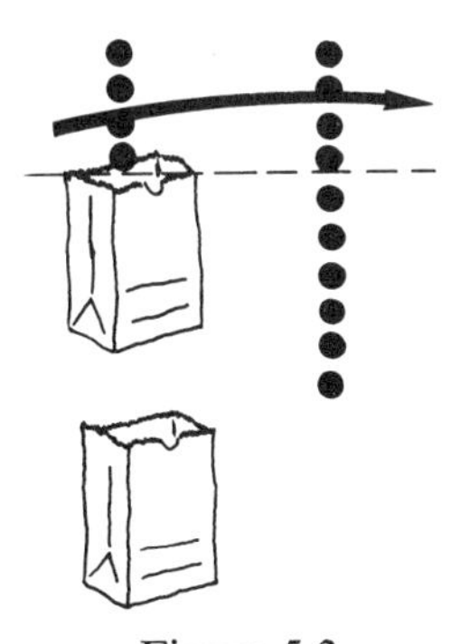

Figure 5.3

How close we are to the subtraction axiom when we sweep off equal amounts from the tops of each column!

After several such situations, depending on the level of the student, the teacher may gradually help the student to perceive that open sentences can be used to express the messages conveyed by diagrams such as those above. Rather than "How many beans are there in each sack?" we simply now ask, "What number makes the given open sentence true?" Indeed, much of the above work in algebra is now being done in the elementary grades.

Use pictorial approaches to solve these problems:

Exercises

1. My brother is so tall that if you would triple his height and then add 6 you would get 19 feet. What is his height?
2. My brother is so tall that if you multiply his height by four and then subtract 3 you would get 17. What is his height?
3. My brother is so tall that four times his height plus 10 equals three times his height plus 13. What is his height?
4. Six times the height of my brother plus 3 feet is the same as four times his height plus 15 feet. How tall is my brother?
5. John had some boxes each of which contained the same weight of candy. Four of these boxes taken together with a 2-pound weight made the scale read 12 pounds. What is the weight of each box?
6. Five times a number decreased by two is the same as twice the number increased by ten. Form a "picture approach" to help select the number which makes the above statement true.
7. Twice a certain number increased by ten equals five times the number increased by one. Form a "picture approach" to find what value of x makes the statement true.

The use of pictorial diagrams in this very elementary approach, the sweeping off of equal amounts from each column, and the partitioning of the amounts in one of the columns give rise intuitively to "axioms" which the students might help to postulate. Eventually, of course, we wish students to write equations in horizontal form and we should suggest that this is the way mathematicians write them. Is it possible to guide our students from the vertical pictorial diagrams to something horizontal? Yes. For problem 5 in the exercises above, one student indicated

$$\square + \square + \square + \square + 2 \quad \text{balances} \quad 12$$

this might be graphically represented as in Fig. 5.4.

$$\square + \square + \square + \square \quad \text{balances} \quad 10$$

Each box contains 2-1/2 pounds.

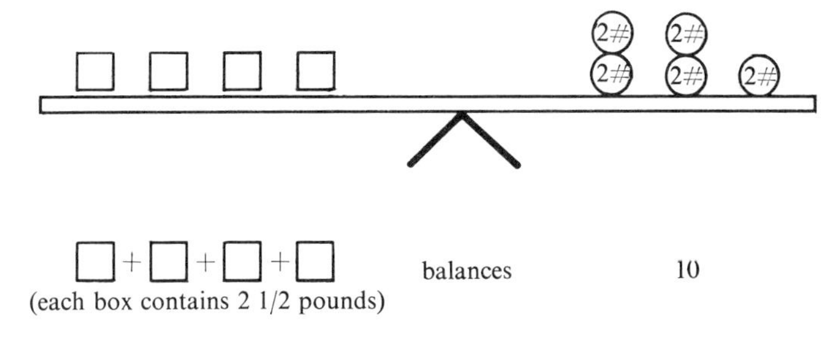

Figure 5.4

Much exploratory work at the beginning levels—and this may begin in the elementary school—makes the methods of solving equations much more understandable. Rules and axioms may well be the outgrowth of student experience and come from student suggestions.

Exercises

1. In the previous exercises the "situation approach" which involved Fred and Tom's brothers was mentioned. Make up a story of a situation which will encourage the use of pictorial diagrams or figures to solve the equation $3x + 4 = 19$.
2. Do the same for $5x + 10 = 2x + 19$.
3. Suggest a pictorial approach to arrive at the solution to $x/4 + 3 = 5$.
4. Suggest a pictorial approach to solve $3x/4 + 3 = 9$.
5. In the late elementary scool or in the early junior high school, pictorial beginnings are found to be very effective. Yet we must provide for transition to symbolic writing of equations, as $3a + 4 = 13$. Devise a list of exercises or a study guide sheet which will help to do this. You may wish to begin with a story.
6. How might you try to relate a pictorial-diagram solution to the more formal equation solving? Plan a study guide which you think will help do this.
7. Make up four problems each of which requires for its solution one of the four axioms of equation solving. Each problem should employ a different axiom. Now solve each equation pictorially.
8. A student has the equation $5x - 6 = 3x + 8$. He hesitates in knowing what to do. One teacher says, "Show him what to do." Another says, "Let him use pictorial diagrams." Which is your reaction, or do you have another suggestion?
9. Pictorial beginnings perhaps illustrate the outward face of mathematics. How?

Open-Sentence Approaches

Another approach to solving equations in algebra, which can also be used with elementary students, might be called the "open sentence" approach. In each case, now, we are simply asking the question, "What number makes this statement true?" Of course, we can use boxes, □, circles, ○, or just blanks, ___. The general plan of this approach is that after the student has found the correct answers by guess-and-check, then reflection on the open sentences along with the truth sets will help more mechanical methods of solution to emerge.

Assuming that the students know what the term "truth set" means, or what it means for a number to satisfy an open sentence, we can plan a series of exercises to help students experience growth in methods of solving equations.

Exercises

1. Find the truth set of each of these open sentences.

(a) $\square + 4 = 6$
(b) $\square + 3 = 7$
(c) $\square - 4 = 8$
(d) $\square + 5 = 8$
(e) $2\_\_ = 6$
(f) $2\bigcirc = 10$
(g) $\frac{\bigcirc}{3} = 4$
(h) $\square - 3 = 2$
(i) $4\bigcirc = 20$
(j) $\frac{\bigcirc}{4} = 7$
(k) $\bigcirc + 7 = 16$
(l) $\bigcirc + 3\ 1/2 = 7$
(m) $\square - 4\ 1/2 = 3\ 1/2$
(n) $\bigcirc + 2 = 8$
(o) $2 + \bigcirc = 8$
(p) $\square - a = b$
(q) $\square + r = s$
(r) $\square - m = n$
(s) $m\square = n$
(t) $\frac{\square}{m} = n$

2. Have you found any quick methods to determine the solutions to the open sentences above? What?

3. Find the solutions for these open sentences.

(a) $3\square - 4 = 14$
(b) $3\square - 4 = 8$
(c) $3\square + 2 = 11$
(d) $4\square - 1 = 19$
(e) $3\bigcirc - 4 = 16$
(f) $6\_\_ - 2 = 22$
(g) $7\square - 3 = 18$
(h) $6\square + 2 = 38$

4. What quick methods can you suggest to find the solution set for each of the above open sentences?

5. Find solution sets for each of these open sentences.

(a) $7\square + 3 = 3\square + 27$
(b) $8\square - 2 = 2\square + 28$

What mechanical or algorithmic methods applied to these open sentences produce the solution?

5.4. A Field Approach to Equations

The use of the postulates for an ordered field* to solve equations can be introduced through student experiences and student contribution, too. Of course, the various stages in development of equation solving described in the previous section, which leaned to experiential and laboratory approaches, can be guided into and summarized by the use of the field axioms. One method, which uses field approaches in solving equations, is the algorithm which directs us to "undo" what the equation indicates has already been done. Let us illustrate.

> Tommy was confronted with problems (open sentences) like $3x + 5 = 20$. Tommy said, "This is the way I do it: the problem says that three times some number added by 5 makes 20. Hence, 'unadding' by five, I know that three times the number must be fifteen. Now three times the number is 15 so the number 15 'untimesed' by five is three. The answer is three." Well, the other students became interested in Tommy's method and they tried some problems.

Note that the idea of undoing what an operation did has arisen quite naturally, but if the teacher wishes to hold strictly to the field postulates, it will be necessary to lead the student to see, for example, that $(3x + 5)$ unadded by 5 is another name for $(3x + 5)$ added by $^{-}5$ where $^{-}5$ is the *additive inverse* of 5. Indeed, we see the idea of "adding the additive inverse of five" to be a part of the inward look of mathematics for, surely, "adding an additive inverse" is not physically related very closely to the four concrete situations which give rise to subtraction.**

Exercises

1. Use Tommy's "method of undoing" to find the solution to each of these equations.

(a) $4x = 12$ (d) $ax + b = c$ (g) $\frac{2a}{3} + 4 = 12$

(b) $3a - 6 = 12$ (e) $\frac{x}{4} = 12$ (h) $\frac{3a}{5} - 2 = 7$

(c) $2x + 7 = 10$ (f) $\frac{a}{3} - 7 = 9$

---

*The postulates for a field are found in books on algebra or in books describing mathematical systems.

**See Ch. 4.

2. What experiences might you plan to convince a student that unadding by five is the same as addition by the additive inverse of five?

At this moment even our inward look needs careful investigation and rethinking for in creating the additive inverse of 5, the *natural number five* has been replaced by a kind of number which can be associated with another number called its additive inverse. Of course, this new kind of number is the *integer*. To distinguish between the natural number five and the positive integer five, for example, it is found instructive by some to use the numerals 5 and $^{+}5$, respectively. Also the use of the raised sign as a part of the symbol 5 emphasizes the difference in concept between *adding five* and *adding positive five*; the first is written $\cdots + 5$ and the second $\cdots + {}^{+}5$. As students mature this method of writing may be dropped to conform with conventional styles.

To emphasize the use of the field axioms in determining the truth set for the open sentence $3x + 5 = 11$, for example, we can record our reasoning in the manner which follows. First of all, however, the sentence $3x + 5 = 11$ might be rewritten as $({}^{+}3/{}^{+}1)x + {}^{+}5/{}^{+}1 = {}^{+}11/{}^{+}1$ so that the coefficients belong to a system which, under addition and multiplication, make up a field. For convenience we shall simply write the open sentence as ${}^{+}3x + {}^{+}5 = {}^{+}11$. Hence we have

$$
\begin{array}{rll}
{}^{+}3x + {}^{+}5 &= {}^{+}11 & \text{(given)}\\
({}^{+}3x + {}^{+}5) + {}^{-}5 &= {}^{+}11 + {}^{-}5 & \text{(logical identity)}\\
{}^{+}3x + ({}^{+}5 + {}^{-}5) &= {}^{+}11 + {}^{-}5 & \text{(associativity of addition)}\\
{}^{+}3x + 0 &= {}^{+}11 + {}^{-}5 & \text{(additive inverse property)}\\
{}^{+}3x &= {}^{+}11 + {}^{-}5 & \text{(additive identity property)}\\
&= ({}^{+}6 + {}^{+}5) + {}^{-}5 & \text{(another name for } {}^{+}11)\\
&= {}^{+}6 + ({}^{+}5 + {}^{-}5) & \text{(associative law)}\\
&= {}^{+}6 + 0 & \text{(additive inverse property)}\\
{}^{+}3x &= {}^{+}6 & \text{(additive identity property)}\\
{}^{+}3^{-1}({}^{+}3x) &= {}^{+}3^{-1}({}^{+}3 \cdot {}^{+}2) & \text{(logical identity)}\\
({}^{+}3^{-1} \cdot {}^{+}3)x &= ({}^{+}3^{-1} \cdot {}^{+}3){}^{+}2 & \text{(associative law for multiplication)}\\
{}^{+}1x &= {}^{+}2 & \text{(multiplicative inverse property)}\\
x &= {}^{+}2 & \text{(multiplicative identity property)}
\end{array}
$$

The above solution really demonstrates that if x is the solution for ${}^{+}3x + {}^{+}5 = {}^{+}11$, then x can have no value other than ${}^{+}2$. It is still necessary to show that the truth set is $\{2\}$; that is, we need yet to prove that the element

2 makes $^{+}3x + {}^{+}5 = {}^{+}11$ a valid statement. To do this we merely note that

$$^{+}3 \cdot (^{+}2) + {}^{+}5 = {}^{+}6 + {}^{+}5 = {}^{+}11$$

Hence the solving of equations as ordinarily done constitutes only half the necessary process. On arriving at the open sentence $x = {}^{+}2$ we have eliminated all the candidates for the truth set for $^{+}3x + {}^{+}5 = {}^{+}11$ except $^{+}2$, but at this stage we do not know *if* $^{+}2$ is a solution. To answer the latter question requires further testing. The conventional method of solving equations tells clearly what elements are not in the truth set, and those which remain as possibilities must be examined one by one.

The above solution and discussion surely emphasize the logical structure and the postulational basis of algebra, as opposed, let us say, to pictorial methods and other laboratory-type approaches. Indeed, the student experiences the inward-looking face of mathematics at work.

In a manner similar to that above, determine the truth set for each of these open sentences. Emphasize the use of the field postulates in eliminating candidates and then examine the one remaining to see if it is a solution.

Exercises

1. $^{+}7x + {}^{+}4 = {}^{+}25$
2. $^{+}6x - {}^{+}3 = {}^{+}15$ (First reword this open sentence to read $^{+}6x + {}^{-}3 = {}^{+}15$)
3. $^{+}5x + {}^{+}4 = {}^{+}25$
4. $x + {}^{+}a = {}^{+}b$
5. $^{+}2 + \sqrt{x + {}^{-}2} = 0$
6. $\sqrt{^{+}2x + {}^{-}4} + {}^{+}2 = 0$

We note in the example above that if x is in the truth set for $^{+}3x + {}^{+}5 = {}^{+}11$ then it is in the truth set for $x = {}^{+}2$. Now $^{+}2$ is in the truth set for $x = {}^{+}2$ and there are no others. We can learn if $^{+}2$ is in the truth set for $^{+}3x + {}^{+}5 = {}^{+}11$ by seeing if $^{+}3(^{+}2) + {}^{+}5 = {}^{+}11$ makes a valid statement *or* we can begin with the sentence $x = {}^{+}2$ and attempt to retrace our steps, with logical support, to construct the sentence $^{+}3x + {}^{+}5 = {}^{+}11$. The retracing of steps can be done for $x = {}^{+}2 \to 3x = {}^{+}6 \to {}^{+}3x + {}^{+}5 = {}^{+}6 + {}^{+}5 = {}^{+}11$ and we have employed no more than the field postulates and theorems. The useful point of the above procedure is that the field properties have enabled us to change the more complicated open sentence $^{+}3x + {}^{+}5 = {}^{+}11$ into one whose truth set can be read at a glance, namely $x = {}^{+}2$.

In Exercise 5 above if x is in the truth set of $^{+}2 + \sqrt{x + {}^{-}2} = 0$ then it is in the truth set of $x = {}^{+}6$ but the element ${}^{+}6$ in the truth set of $x = {}^{+}6$ is not in the truth set of ${}^{+}2 + \sqrt{x + {}^{-}2} = 0$. In fact, the truth set for the radical equation is empty. The raised-sign notation is used in this exercise but we shall now use it only when we wish to emphasize the use of the properties of a field.

Exercises

1. Problems such as these are found useful to develop the idea of having two open sentences with the same truth set. See if these pairs of open sentences have the same truth sets.

(a) ${}^{+}3x = {}^{+}12$ and ${}^{+}3x + {}^{+}3 = {}^{+}15$

(b) $x + {}^{+}5 = {}^{+}13$ and $x = {}^{+}8$

(c) $x = {}^{+}4$ and ${}^{+}2x = {}^{+}8$

(d) ${}^{+}2x = {}^{+}8$ and ${}^{+}3x = {}^{+}24$

(e) ${}^{+}3x - {}^{+}1 = {}^{+}5$ and $x = {}^{+}1\text{-}1/3$

(f) ${}^{+}2x + {}^{+}6 = {}^{+}12$ and $x = {}^{+}3$

(g) ${}^{+}6x - {}^{+}4 = {}^{+}20$ and $x = {}^{+}4$

(h) ${}^{+}6x = {}^{+}42$ and ${}^{+}3x = {}^{+}21$

(i) ${}^{+}6x = {}^{+}30$ and ${}^{+}6x + {}^{+}4 = {}^{+}34$

(j) ${}^{+}3x - {}^{+}8 = {}^{+}28$ and ${}^{+}3x + {}^{+}4 = {}^{+}40$

(k) ${}^{+}5x = {}^{+}2x + {}^{+}9$ and ${}^{+}5x^2 = x({}^{+}2x + {}^{+}9)$

(l) $x = {}^{+}3$ and $x + {}^{+}6 = {}^{+}9$

(m) $x - {}^{+}3 = {}^{+}7$ and $x = {}^{+}10$

2. Let us consider the open sentences

(a) $x - 4 = 8$ and (b) $x = 12$

Which open sentence is simpler? Which suggests the truth set immediately? What can be done to both sides of (a) to yield (b)?

3. Note Exercises 1(b), 1(f), 1(l), 1(m). Each part contains two open sentences. What can be done to each side of each equation to produce the simplest open sentence with the same truth set.

4. What operations or processes do you suggest can be applied to both members of open sentences to make simpler open sentences? One might think of open sentences as $4a = 8$, $a+4=8$, $a-4 = 8$, $a/4=8$. For example, it might be Charles who suggests that if the "opposite of a" be added to a and to b in $x + a = b$ then one obtains a simpler open sentence with the same truth set. This might then be called "Charles' conjecture" or "Charles' process." Looking into the rationale or the logical substantiation may come immediately or later; perhaps there will be "Charles' theorem."

5. What descriptive term might we apply to open sentences which have the same truth set? (Mathematicians call them *equivalent* open sentences. Students often understand better the definitions and terms which they formulate themselves; later they may be told what term others use.)

6. "Charles' conjecture" mentioned in Exercise 4 is a special case of the conjecture that

$$\text{if } ax + b \text{ has the truth set } \{e\}$$

$$\text{then } ax + (b + m) = d + m \text{ has the truth set } \{e\}$$

See if you can prove this by the use of the field properties. This is an addition property of equations.

7. Now try to prove that

$$\text{if } ax + (b + m) = d + m \text{ has the truth set } \{e\}$$

$$\text{then } ax + b = d \text{ has the truth set } \{e\}$$

If this can be done then we will have proved along with the work in Exercise 6 that the two open sentences are *equivalent.*

8. Prove that if $ax = b$ has the truth set $\{e\}$, then $(ma)x = mb$ has the truth set $\{e\}$ where $m \neq 0$. This theorem we can call the multiplicative property of equations.

9. Prove that if $(ma)x = mb$ has the truth set $\{e\}$ then $ax = b$ has the truth set $\{e\}$ where $m \neq 0$. If we can deduce the suggested conclusions in both this and the preceding exercise, what relation can be used to describe these two open sentences?

10. Using the properties of a field, prove that $ax/m = b/m$ and $ax = b$ $(m \neq 0)$ are equivalent statements.

11. Use the properties of a field to prove that if $\{c\}$ is the truth set for $ax = b$, then the set $\{c\}$ is *unique.*

We have observed that some open sentences yield truth sets more readily than others and we have provided logical defense for many of the algorithms suggested to produce simpler open sentences from more complicated ones. Gradually we have moved from the experimental approach to the more sophisticated inward-looking face of mathematics—the facet which examines the logical structure of the subject. We must help the student realize that the manipulations he performs on equations simply product equivalent open sentences from which he can recognize at once candidates for the truth set of the original equation. Furthermore, although one often closes the solution to an equation with a statement as $x = 5$, the student is not through until he writes the answer as a truth set. That is, 5 is the answer rather than $x = 5$, for $x = 5$ is still an open sentence.

There are still other properties of a field which should be emphasized in the solving of equations; one of these is the property of multiplication being distributive over addition. When one solves $4a + 3a = 21$ by writing $7a = 21$, he is really thinking $4a + 3a = (4 + 3)a$, in which case he is using the *distributive law* that $ba + ca = (b + c)a$. $4 + 3$ is then rewritten as 7 and we have $7a = 21$. The expression $4a + 3b$ cannot receive a simpler name for the distributive law does not apply, since the terms have no factor in common.

The advantage of the use of the distributive law in explaining that some sums can be expressed by simpler names and some cannot is that the student develops more understanding and knowledgeable insight into the manipulation involved. This basic understanding becomes a guide in the same and similar situations so the student is not dependent on memory of tricks of various kinds. The understanding produced through the application of the distributive law stands far elevated in intellectual accomplishments over the explanation heard in one classroom:

> You can add apples to apples; therefore, $7a + 3a = 10a$. You can't add apples to bananas; therefore, you can do no more with $7a + 3b$.

How this teacher would defend the idea that although they cannot be added, "apples and bananas" can be multiplied, as $7a \cdot 3b = 21ab$, we do not know. The distributive law provides an inward look even to the simple rule still found in many texts that "similar" terms can be added, but "dissimilar" terms cannot be. Compare the firmness in argument that $4a + 3a = 7a$ because $4a + 3a = (4 + 3)a = 7a$ on the strength of the distributive law relative to the validity of this statement based on any other rules you have learned.

To emphasize the use of the distributive principle, students might well be asked to arrange steps in some of their solutions for equations like this:

$$4x + 3x + 9 = 30$$

$$(4 + 3)x + 9 = 30$$

$$7x + 9 = 30$$

$$7x = 21$$

$$x = 3$$

$\{3\}$ is the truth set, or, 3 is the solution.

Exercises

1. Write steps which lead to the solutions of these equations emphasizing the use of the distributive law:

(a) $3a + 6a + 2 = 20$

(b) $\frac{a}{3} + \frac{4a}{3} = 20$

(c) $\frac{5}{2}a + \frac{1}{3}a + 4 = 38$

2. In a similar manner solve these equations:

(a) $7x - 3x + 4 = 2x + 8$

(b) $8x + 4 + \frac{3x}{4} = \frac{x}{8} + 73$

The field approach helps us to recognize and to use the logical structure on which algebra is based. The study of algebra through the use of the field postulates, which themselves might have been arrived at inductively, is far superior to looking at the subject as an accumulation of algorithms which are used "because they yield the answer."

The soundness and clarity which the field approach gives can be summarized by the somewhat detailed solution of this equation:

$$^{+}13x + {}^{-}7 = {}^{+}4x + {}^{+}11 \tag{1}$$

Now any number x which belongs to the truth set for (1) will also be a member of the truth set for

$$(^{+}13x + {}^{-}7) + {}^{+}7 = (^{+}4x + {}^{+}11) + {}^{+}7 \tag{2}$$

by the addition theorem on equations (see Problem 6 in the preceding group of exercises). It will also be a member of the truth set for

$$^{+}13x + (^{-}7 + {}^{+}7) = {}^{+}4x + (^{+}11 + {}^{+}7) \tag{3}$$

by the associative property of addition, and in the truth set for

$$^{+}13x + 0 = {}^{+}4x + {}^{+}18 \tag{4}$$

by the property of additive inverses on the left, and since $^{+}18$ is another name for $^{+}11 + {}^{+}7$ on the right, and in the truth set for

$$^{+}13x = {}^{+}4x + {}^{+}18 \tag{5}$$

by the property of the additive identity element, and in the truth set for

$$^{+}13x + (^{-}4x) = (^{+}4x + {}^{+}18) + {}^{-}4x \tag{6}$$

by the addition theorem on equations, and in the truth set for

$$(^{+}9x + {}^{+}4x) + {}^{-}4x = {}^{+}4x + (^{+}18 + {}^{-}4x) \tag{7}$$

by logical identity or "another name for" on the left and by the associative property of addition on the right, and in the truth set for

$$^{+}9x + (^{+}4x + {}^{-}4x) = {}^{+}4x + (^{-}4x + {}^{+}18) \tag{8}$$

by the associativity property of addition, and in the truth set for

$$^{+}9x + (^{+}4 + {}^{-}4)x = {}^{+}4x + ({}^{-}4x + {}^{+}18) \qquad (9)$$

by the property of right-hand distributivity of multiplication over division, and in the truth set for

$$^{+}9x + 0 = (^{+}4x + {}^{-}4x) + {}^{+}18 \qquad (10)$$

by the property of additive inverses and the multiplication property of the additive identity element on the left and the associative property of addition on the right, and in the truth set for

$$^{+}9x = (^{+}4 + {}^{-}4)x + {}^{+}18 \qquad (11)$$

by the property of the additive identity element on the left and by the property of multiplication being right-distributive over addition on the right side, and in the truth set for

$$^{+}9x = 0x + {}^{+}18 \qquad (12)$$

by the property of additive inverses, and in the truth set for

$$^{+}9x = 0 + {}^{+}18 \qquad (13)$$

by the multiplication property of the additive identity element, and in the truth set for

$$^{+}9x = {}^{+}18 \qquad (14)$$

by the property of the additive identity element, and in the truth set for

$$^{+}9x = {}^{+}9 \cdot {}^{+}2 \qquad (15)$$

since $^{+}18$ has the "other name" $^{+}9 \cdot {}^{+}2$, and in the truth set for

$$\frac{^{+}1}{9}(^{+}9x) = \frac{^{+}1}{9}(^{+}9 \cdot {}^{+}2) \qquad (16)$$

by the multiplication property proved previously in this section and in the truth set for

$$\left(\frac{^{+}1}{9} \cdot {}^{+}9\right)x = \left(\frac{^{+}1}{9} \cdot {}^{+}9\right) \cdot {}^{+}2 \qquad (17)$$

by the associative property of multiplication, and in the truth set for

$$^{+}1 \cdot x = {}^{+}1 \cdot {}^{+}2 \qquad (18)$$

by the property of multiplicative inverses, and in the truth set for

$$x = {}^{+}2 \qquad (19)$$

by the property of the multiplicative identity element. Hence any number in the truth set for (1) is in the truth set for (19). We should now try to reverse the steps and hence show that any element in the truth set for (19) is also in the truth set for (1). If we are successful we will have shown that the open sentences

$$^{+}13x + {}^{-}7 = {}^{+}4x + {}^{+}11 \qquad (1)$$

and

$$x = {}^{+}2 \qquad (19)$$

are equivalent. This is an exercise for the reader.

Exercises

1. Show by steps which use the properties of a field that any element in the truth set of

$$x = {}^{+}2 \qquad (19)$$

is also in the truth set for

$$^{+}13x + {}^{-}7 = {}^{+}4x + {}^{+}11 \qquad (1)$$

2. In each case show that these open sentences are equivalent by transforming the first one given to the second and then the second to the first.

(a) $x + ({}^{-}3) = {}^{+}8;\ x = {}^{+}11$

(b) $^{+}4a + {}^{+}3 = {}^{+}26;\ a = {}^{+}6$

(c) $^{+}5a + {}^{+}7 = {}^{+}2a + {}^{+}16;\ a = {}^{+}3$

3. Find the truth set for each of these open sentences in this manner: use field properties and properties of operating on equations to arrive at the simplest open sentence. Then, regarding this as merely scratch work, retrace all steps to show that the set which makes the simplest open sentence true makes also the given sentence true.

(a) $13x + {}^{-}5 = {}^{+}6x + {}^{+}30$

(b) $^{+}3/4x + {}^{+}10 = {}^{+}19$

(c) $(x + {}^{+}1)(x + {}^{+}5) = x(x + {}^{+}3) + {}^{+}35$

5.5. Postulational Approaches Elsewhere in Algebra

To solve equations one makes use of the postulates for a field but not all processes and operations in algebra require the entire set of postulates. The logical substantiation of the conjecture that $(ab)^2 = a^2b^2$, for example, requires merely the definition of equality, and the commutative and associative laws of multiplication. Indeed, the test by deduction that $(ab)^2 = a^2b^2$ and like expressions are good topics by which to help students begin to become aware of and skilled in deductive thinking in algebra.

$$(ab)^2 = (ab)(ab) \quad \text{(definition)}$$

but

$$(ab)(ab) = a[b(ab)] \quad \text{(multiplication is associative)}$$

and

$$a[b(ab)] = a[b(ba)] \quad \text{(multiplication is commutative)}$$

and

$$a[b(ba)] = a[(bb)a] \quad \text{(multiplication is associative)}$$

and

$$a[(bb)a] = a(b^2a) \quad \text{(definition)}$$

and

$$a(b^2a) = a(ab^2) \quad \text{(multiplication is commutative)}$$

and

$$a(ab^2) = (aa)b^2 \quad \text{(multiplication is associative)}$$

and

$$(aa)b^2 = a^2b^2 \quad \text{(definition)}$$

and hence

$$(ab)^2 = a^2b^2 \quad \text{by the } \textit{transitive property}$$

of the equals relation. Often this proof is written more briefly, but then there is danger that the role of the transitive law and perhaps other basic properties are not fully realized by the student.

Explorations and investigations will stimulate the challenge to test, by deduction, the conjecture that $(ab)^3 = a^3b^3$, and soon the student begins to sense the beginning of a logical thread which continues throughout algebra. For those interested in the use of the mathematical induction principle the investigation of the conjecture that $(ab)^n = a^nb^n$ for *every* natural number n is, indeed, a crowning achievement. This is just one example of sequences of investigations and possible resulting theorems which can be carried on in algebra.

It is interesting to the prospective teacher to examine high school texts in algebra to see how various writers argue, for example, that $(ab)^3 = a^3b^3$. Sometimes many of the logical steps are glossed over and the student does not gain a clear-cut picture of the logical support involved. Indeed, this lack of careful argument in some cases emphasizes again the important role of the teacher.

Exercises

1. A student makes the conjecture that $(ab)^3$ has the other name a^3b^3. Test this conjecture by deduction.
2. Test by deduction the conjecture that $(ab)^4 = a^4b^4$.

3. See if the conjecture that $(ab)^n = a^n b^n$ can be made the logical consequence of the postulate of mathematical induction and other beginning postulates and definitions in algebra.

4. Examine several texts, some of recent origin as well as older ones, to investigate their argument that $(ab)^3 = a^3 b^3$, and similar relations.

5. Look through texts in algebra for high schools and select another topic which you think could be used well to emphasize logical structure in algebra. Outline or develop this sequence of theorems and proofs.

5.6. Beginnings on Word Problems

The beginning experiences which students have with word problems are highly significant in that they can set the emotional stage which determines how such problems are approached in the future. An attitude of distaste against such problems is regrettable for it is through word problems that we see mathematics at work; we see the outward-looking face of mathematics serving man in many ways with a continuing emphasis on the variety of problems in the solution of which mathematics is essential. The aim of this section is to suggest some approaches to word problems.

At some time in your teaching of algebra you will wish your students to be able to solve problems like this one:

> Henry has 40 grams of a 7% solution of sugar. How many grams of sugar must be added to make it an 11% solution of sugar?

Work this problem. Did you have difficulty? We hope not, but if you did, something about the meaning of "percent composition" was probably overlooked. A teacher interested in the *gradual* strengthening of this students would not start with problems of this complexity; rather, he would plan simpler experiences to build up correct concepts on percent composition. To help proper ideas and methods to emerge in the mind of the student, rather than merely compelling him to work problems by rule and rote, is the purpose of the genuine teacher.

Skill in working word problems comes largely from being able to extract from the given situation significant facts and to express them in mathematical language. But to do this the student must know something of the natural setting of the problem. He must know more than mathematics (the field approach to solving equations); rather, he must sense that his problem is imbedded in a network of principles which are applicable to the particular problem. If it concerns percent strength of a solution of sugar, then he must know that the percent concentration equals the weight of the sugar divided by the weight of the sugar and water. If sugar is added, both the numerator and the denominator are increased. If the problem concerns coins, then he must be aware that different coins have different value and he must know how to use this knowledge to state in mathematical language relations among the coins as expressed in the problem.

It requires skill to work word problems, but some of this skill is the ability to think in terms of the discipline in which the problem occurs. Hence, the teacher must often feel responsible for providing experiences to nurture the student's thinking in terms of the subject matter of the problem.

To help the student learn to use facts and expressions relative to a given kind of problem it is often effective to proceed through carefully planned exercises to develop these ideas gradually. This was discussed briefly in Section 3.5. If we consider the percent concentration of sugar solution again, we might "several days ago," have begun some simple experiences like these:

1. Tom had a chemistry set and one bottle contained 20 grams of sodium chloride solution which was marked 4%. How many grams of salt were in this solution?
2. John's father bought 1000lbs. of fertilizer marked "3% nitrogen." How many pounds of nitrogen are present in the fertilizer which he bought?

Next can come problems like these:

1. Johnnie dissolved 15 grams of sugar in 200 grams of water. How many grams of solution are these? The resulting solution is what percent sugar? What percent of the solution is water?
2. Now Johnnie placed 5 grams more sugar in the solution. How many grams of sugar are there in the solution now? How many grams total solution are these? What is the percent strength of the solution?

And we can have problems like these:

1. Henry had 300 grams of solution marked sodium nitrate 8%. How many grams of sodium nitrate were present in the solution?
2. Henry's little brother accidentally spilled 50 grams of water into the solution? What is the percent concentration now?
3. Next Henry wished the solution to be a little stronger so he added 10 grams of sodium nitrate. What is the percent concentration of sodium nitrate now?

Not until sufficient groundwork is laid so the student understands thoroughly the meaning of percent composition and related computation should the teacher consider "algebra" problems.

We are now ready for a next step in helping algebra methods emerge from experiences of students, and we can introduce it in this way:

Henry has 50 grams of an 8% solution of salt. How many grams

> of salt are in the solution now? How many grams of salt must be added to make the concentration 12% salt? You might guess at the answer and see if you are correct.

Had the student guessed 6 grams (and some students would regard six as a very good guess—why?) then he could have checked his guess by asking, "Does

$$\frac{50(.08) + 6}{50 + 6} = 0.12?\text{"}$$

The answer is no, for the left-hand member is approximately equivalent to 0.18. Now the student might guess 5 grams and he writes

$$\frac{50(.08) + 5}{50 + 5};$$

He checks again. The teacher can carefully encircle the guessed answers and the following form comes to light each time:

$$\frac{50(.08) + \quad}{50 + \quad} = 0.12$$

The position which the unknown takes arises from the process of arithmetical checking. In determining if a guessed answer is correct, the student *goes through the very steps in thinking necessary to set up the equation.* Hence it is seen that although arithmetic emerges from the use of concrete objects, algebra emerges from arithmetic. The climax to this approach comes when the student writes

$$\frac{50(.08) + g}{50 + g} = 0.12$$

and solves for g by methods which he has learned. Of course, with practice and growth the student will gain facility in setting up equations immediately for kinds of problems which he understands. Again, the arithmetical checking not only strengthens fundamental ideas in the subject matter of the problem and helps students to review arithmetical computation skills, but it serves also as a guide to the formulation of open sentences whose solution we seek in order to answer the original question. This approach is wellcalled the "guess-and-check" method.*

We will not be satisfied until we have helped our students work problems like this:

> John has 50 grams of a 12% solution of NaCl. How many grams of this solution must be removed and replaced with pure NaCl to have a 15% solution?

---

*The writers developed more insight into possibilities of this method after work with Dr. Nathan Lazar, The Ohio State University.

and others which involve percent composition. Note in the following problems how there is a shift from arithmetical work to an emergence of an algebra method.

Exercises

1. Henry had 50 grams of a 12% solution of NaCl. He took out 5 grams of the solution. What weight of salt was removed?
2. Henry had 50 grams of a 12% solution of salt. He removed 10 grams of the solution and then replaced it with 5 grams of pure salt. The solution is what percent salt now?
3. George had 120 grams of a 15% solution of sugar. He removed 50 grams of the solution and replaced it with 5 grams of water. The solution is what percent sugar now?
4. John had 150 grams of a 20% solution of hypo. He added 50 grams of an 8% solution of hypo. What is the concentration of the resulting solution with reference to hypo?
5. Henry had 200 grams of a 12% solution of H_2SO_4. How many grams of solution must he remove and replace with an 8% solution of H_2SO_4 to have a resulting concentration of 9% Guess at an answer and check by arithmetic?
6. Try to get a closer guess for the answer to the problem above. Write your work to encourage the emergence of an algebra method.
7. Work Exercise 5 by "algebra"
8. John had 120 grams of a 15% solution of $NaNO_3$. He removed 40 grams of the solution. He then added 15 grams of pure $NaNO_3$. Now how many grams of 10% $NaNO_3$ solution must be added to cause the solution to become a 25% solution of $NaNO_3$? Guess the answer. Check by using arithmetic methods.
9. See if you can guess an answer which checks more closely. Check in a manner to show how an algebra solution can emerge.
10. Work Exercise 8 by methods of algebra.

It is not the writers' practice to give all such problems on one sheet. Rather, from day to day there is increasing complexity introduced so the algebra methods for percent composition emerge in keeping with the program of topics set by others if such timing is necessary. With such nurturing experiences the student is better able to work percent composition problems when he comes to them in the text if a rigid topical outline is followed. One can be preparing the students for new topics and be strengthening old ones simultaneously. This was discussed in Chapter 3.

Under actual operating classroom conditions with such approaches an assignment might someday appear as this:

1. *Review*. Solve these equations:

   (a) $\frac{3x + 2}{4} = 7$ (b) $\frac{.08x + 4}{64} = 0.12$, et cetera.

2. *Review*. Desirable and necessary items as determined by need.
3. Problems made up by the teachers or assigned from the text on topics being studied currently.
4. Johnny's chemistry set had a bottle marked "60 grams of hypo; 3% solution." How many grams of (actual) hypo are in the bottle?
5. John's father bought 3000 pounds of fertilizer marked 5% nitrogen, 10% phosphorous, 3% potash. How many pounds of each are present in the amount purchased?

On the next student guide sheet there would be the usual review exercises plus problems from topics being studied at the present time, but there would be a slight variation on the arithmetic problems being used to lead to algebra problems concerning percent composition. We might now have

.
.
.

4. Henry had 50 grams of a 5% solution of salt. He added 25 grams of water. What is the concentration now?
5. George had 80 grams of a 10% solution of acid. He added 20 grams more of acid. What is the percent concentration now?

.
.
.

We are trylng to emphasize that the growth we have been describing is developed over a period of several days. True, a few students can progress rapidly; others need more time. The small steps in growth described above have much in common with programmed materials.

It is not only a challenge to teachers, but it is most helpful to students who plan to be teachers, to develop sets of problems to lead students from simple arithmetic beginnings to the emergence of methods of algebra. We shall attempt this in the next group of exercises.

Exercises

1. Suppose you wish your students to develop algebraic methods to work a problem like this:

> I am thinking of three consecutive odd integers such that four times the first plus twice the second plus three times the third is 79. What are the numbers?

Construct a set of seven or eight arithmetic problems which will lead to an algebraic method of solution of the above. Let one or two of these problems be guess-and-check in nature. The first problem might be to name the four consecutive integers beginning with 2. The meaning of the terms "consecutive integer" and "consecutive odd integers" must be included in these experiences.

2. Write several arithmetical problems to review the use of the expression $d = rt$. Now suggest more difficult problems so that through a guessing-and-checking approach an algebra method will emerge to work problems of this kind:

> John and George are 200 miles apart and they are in autos headed toward each other. John drives at an average speed of 42 miles per hour while George's average is 38 miles per hour. If they both start at the same time, in how many hours will they meet?

3. Devise several problems, in an arithmetic setting, which increase in complexity but which, over a period of several days, will help the student gain an "algebra" approach to problems such as

> A grocer had two kinds of grass seed: one which sold at 80¢ per pound and one which sold at $1.20 per pound. How much of each kind should be taken to form 200 pounds of a mixture to sell at 95¢ per pound?

4. Suggest several arithmetic problems which will strengthen the student in the concepts needed to work this problem:

> At what time after four o'clock are the two hands of a clock together?

5. Let us suppose that your class needs practice on "tea, coffee, and grass seed" problems. A grocer has seed which sells for q cents per pound and some which sells for r cents per pound. He wishes to make up p pounds of a mixture to sell for s cents per pound where $q < s < r$. Determine the solution and then tell what relations you must keep in mind if you wish the problems to come out with an answer which is a counting number. Let the unknown x be the number of pounds of the seed which sells for q cents. Your students should be encouraged at all times to work out such problems in general terms; this illustrates the power of algebra and the effect of each component of the problem on the answer.

6. With the use of the expressions developed in Exercise 5, make up to problems for your students on mixtures of tea.

7. Treat Problem 2 in a similar manner; determine how the distance and speeds must be related so the answer comes out even.

8. Make up two problems from your expressions developed in Exercise 7 so the problems will come out to an even number.

5.7. More "Beginnings" on Word Problems

The last section emphasized the effective use of simple arithmetic problems to nurture the understanding of principles involved in various kinds of word problems and to provide settings out of which the algebra equation could emerge. In the school programs, as the year progresses, the customary practice has long been simply to give the student more difficult problems to be solved with one unknown. During the latter part of the school year, simultaneous equations are then introduced and the student often begins to solve word problems with a new ease and confidence, for the new method of attack helps him to set up the equations more easily.

Some teachers find it effective at the very outset to have the students use as many letters as there are unknowns in the problem. Indeed, most problems have several unknowns. Of course, such methods introduced early require a reorganization of the algebra course, but this approach to word problems exhibits and uses a more ready flow from English sentences to mathematical sentences. Actually, the transition is quite natural for there is a close parallel between mathematical sentences and English sentences. Note these statements:

A collection of quarters and dimes totals 19 coins, which says:

the (number of) quarters + the (number of) dimes is 19,

which can be written as

$$q + d = 19$$

Is it any wonder that the algebra student takes on a new outlook when he perceives the similarity of the English language statement and the algebraic statement!

If we state further that the value of the collection is \$1.65, it is natural to write the relation about the value in cents as

$$25q + 10d = 165$$

We hence have the two equations

$$q + d = 19$$

$$25q + 10d = 165$$

which the student may well solve by noting that from $q + d = 19$ we can write another name for q, namely $19 - d$. Students can discover that they could have done this with one unknown, but many pupils find it more natural to think in terms of two unknowns first.

Consecutive-number problems lend themselves readily to this natural language flow from words to mathematical symbols.

> There are three consecutive integers such that three times the first one plus seven times the second one plus twice the third makes a total of 71. What are the numbers?

Using the letters f, s, t for the first, second and third numbers, we have from the above description of the three integers,

$$3f + 7s + 2t = 71 \qquad (1)$$

Also, the second integer is one more than the first, so

$$s = f + 1;$$

likewise

$$t = f + 2 = s + 1$$

To solve this we use the "other names" for s and t and replace s and t by these in (1) to get

$$3f + 7(f + 1) + 2(f + 2) = 71$$

which the student can readily solve. Here there is emphasis on the use of "$=$" as "another name for" and the student perceives the real power of this concept and the meaning of *substitution.*

Of course, for most students, the skills in translating from English descriptive phrases and sentences to mathematical formulations need to be fostered by carefully planned exercises. The student needs to be able to translate "four more than twice the first number," for example, readily into the expression $2f + 4$. The expression $2f + 4$ is called an *open phrase* and it is the open phrase which is really the building block of open sentences, or of equations. The statement of relations among open phrases is an open sentence and it is the ability to write these open phrases and subsequent open sentences which is the key to successful solving of word problems. Effective practice to develop this skill is a must for most students. Practice by the use of short drill exercises, quizzes, oral exercises, "warming up" activity at the beginning of class, all help to strengthen the student in speaking, writing, and thinking in terms of open phrases. This is to be encouraged. Some typical exercises might well be as follow.

Write open phrases for each of these expressions:

1. twice a number
2. twice a number plus 4
3. three times a number increased by 6
4. twice a number less 4
5. four more than a number
6. seven more than three times a number
7. the sum of three consecutive numbers

8. four more than the second consecutive number
9. the sum of three consecutive odd numbers
10. twice a number decreased by three
11. twice the square of a number increased by four
12. three times the number of dimes decreased by five
13. ten more than Charles' age
14. six less than twice John's height
15. eleven times three less than George's age
16. a number which is ten less than n
17. a number which is n less than ten
18. the difference between 6 and one-half of any number.

Indeed, frequent 5- or 10-minute written or oral drills on expressions just as simple as these will pay big dividends. Then there should be much practice merely on setting up open sentences. Students usually have more skill in solving equations than they do composing them. Some simpler practice items might appear as these:

Write open sentences for these expressions:

1. the second number is four more than the first
2. the second number is three times the first
3. the third number is four more than three times the first
4. the sum of Helen's age and John's age is three more than twice George's age
5. twice the length plus three times the width is 10
6. the number of grams of salt added to 100 grams of a 15% solution makes it a 20% solution.

Of course, this list could be made much longer, but we emphasize that many short quickly written exercises as these strengthen building blocks for problem solving.

Another example to show the more natural flow from English to mathematical expressions by the use of several unknowns appears in "rectangle problems."

> The length of a rectangle is four more than twice the width. What are the dimensions of the rectangle if the perimeter is 38?

If the letters ℓ and w are used, we have these relations written directly from the English statements:

$$\ell = 2w + 4$$

$$2\ell + 2w = 38$$

We could use the other name for ℓ in the second equation and have

$$2(2w + 4) + 2w = 38$$

If the student eventually perceives that he could have used $2w + 4$ for the length immediately, then, truly, he has changed this to a one-unknown problem, but this one-unknown problem has emerged as a result of looking at the problem in a more natural way. Not infrequently students have difficulty and find it awkward to express all the unknowns in a given problem in terms of *one* letter. The use of several unknowns facilitates more ready translation from English to mathematics; also the student experiences for a longer time the use of "another name for" as well as other topics related to equations in several unknowns.

The manner in which students have learned some of the fundamental processes in arithmetic may be reflected in how they manufacture equations to be solved. For example, the problem

> Five times a given number increased by seven is four more than twice the number

may be looked upon as

> Five times a given number increased by seven
> has the same name as
> four more than twice the number

which, in symbols, becomes

$$5n + 7 = 2n + 4$$

Now the student who writes the equation

$$5n + 7 - 2n = 4$$

may like to think of the given data in the problem as describing a "take away" or "how many are left" approach to subtraction. One who writes

$$5n + 7 = 2n + 4$$

may be employing the "how many more are needed" approach. The equation

$$5n + 7 - 4 = 2n$$

may come from a how-many-more-than approach and four is subtracted, therefore, from the larger to make the two expressions equal. *It is often difficult to perceive how a student thinks; to give him maximum understanding the teacher should look at the same problem from as many different points of view as possible.*

Again, the ability to translate English expressions into mathematical language along with skill in sensing the physical or subject-matter principles underlying the word problem is usually in much need of development and practice. Teachers must be patient here and they may well consider student-centered and inductive approaches seriously. Indeed, even the university student has difficulty in this part of mathematics study if he suddenly finds himself in a new field. Witness, for example, one's immediate reaction to this problem:

> At the moment a tank contains 600 gallons of solution containing 60 lbs. of $Na_2S_2O_3$. A pipe lets water in at the rate of 10 gallons per minute. Simultaneously, another pipe lets in 30 gallons per minute of a solution containing 0.1 lb. per gallon of $Na_2S_2O_3$. A third pipe lets out 40 gallons of the solution per minute. The solution is stirred constantly. Set up a differential equation whose solution will yield the number of pounds of $Na_2S_2O_3$ in the tank at any time t.

Try to express what is happening by letting n be the number of pounds of $Na_2S_2O_3$ in the tank at any time t; set up an expression for dn/dt. Some teachers sense the need of instructive and inductively planned simple problems which will climax in a problem like the one given above.

Exercises

1. Set up the differential equation for the problem given above.
2. Write several exercises so that as one goes from one problem to the next, he will be able to set up the differential equation requested in No. 1.
3. Write the open sentences revealed by this problem and then solve. Use four unknowns.

   > The number of women at a mixed gathering was 1 less than 1/4 of the number present. The number of men is 8 more than 1/6 of the group. The number of children was 29 more than the number of men. What was the total number in the mixed gathering?

4. Consider this problem: Mr. Jones has $10.00 invested, part at 6% and the remaining at 4% per annum. What sum is invested at each rate? Write several problems in arithmetic which will cause the student to think in the manner which this problem requires and from which the method of solution may emerge.
5. At what time between 2:00 and 2:15 are the minute hand and the hour hand on a clock together? Write several arithmetic problems which introduce and emphasize principles involved in working the problem.
6. Study a section of word problems in a high school algebra text and make a list of open phrases which need to be mastered before the student can have much success relative to the given problems.

5.8. Working with Integers

Through the entire elementary school experience the student has become acquainted with the number line extended in both directions and he learns to do elementary problems involving the integers. It remains for the teacher in the formal course in algebra to capitalize on his previous study and to emphasize in a manner suitable to the class some of the underlying logical structure of algebra.

There are several avenues to the introduction of integers and in each of these methods of introduction one can use student-centered approaches to much advantage. The student can be led to the integers by an ordered-pair-of-natural-number approach, by a vector approach leading to ordered pairs, or simply by postulating the existence of "signed numbers" as names for points on the integer number line. We shall illustrate two of these approaches to the definition and addition of integers.

Ordered-Pair-of-Natural-Number Approach

This approach is used in school texts; it is a way to emphasize that the set of integers is erected on the set of natural numbers. Too, this approach is one which rests more firmly on the inward look in mathematics. One may well begin our discussion like this:

> Tommy took a job running a small produce shop for Mr. Smith. Each day he had to pay out money for the produce and each day he took in money from sales. Just after he started, Mr. Smith was called away for a week or more so Tommy invented a system of bookkeeping by himself:
> If he takes in \$5 and pays out \$3, he summarizes his business by writing (5, 3): if he takes in \$4 and pays out \$7, he writes down (4, 7).

Exercises

1. Describe the day's business if Tommy writes (2, 5); (7, 6); (6, 6).
2. Are Tommy's pairs of numbers ordered pairs? Why?
3. Would you say that the days for which the entries are (7, 3) and (12, 8) represent the same net business? Suggest four or five other entries which describe the same net business.
4. Do the ordered pairs (3, 1) and (8, 7) describe the same net business? Why?
5. All the ordered pairs which represent the same net business form a class or a set of ordered pairs. What might such a class be called?

6. The class or ordered pairs (3, 7) (2, 6) (1, 5) (10, 14) (12, 16) (19, 13)... might well have what *name*?

Such exercises as these above lead the student gradually to the concept of an equivalence class of ordered pairs of natural numbers where $(a, b) \sim (c, d)$ if and only if $a + d = b + c$. The name of each such equivalence class is an integer; the name of the equivalence class above may be designated as $^{-}4$. The introduction of technical terms as "equivalence class" and "integers" should come *after* the student has the idea; ideas *first*, names afterward. How often we do just the reverse, and sometimes so quickly that the student is still attempting to understand the definition while we are expecting him to *use* it.

We might now introduce the situation or story which has Tommy interested in finding the total net business for two days. We continue with:

Exercises

1. Tommy's record for Monday was (7, 9) and for Tuesday it was (13, 3). What ordered pair will represent the "net business" for two days?
2. Describe by ordered pairs the net business if on separate days the entries are
   (a) (3, 2) and (8, 10)
   (b) (14, 3) and (10, 3)
   (c) (2, 2) and (8, 8)
   (d) (8, 8) and (16, 3).
3. How do you propose to *define* the "addition" of the ordered pairs (a, b) and (c, d)?
4. Does $(3, 4) + (9, 17)$ have the same sum as $(9, 17) + (3, 4)$? If so, what property of addition of ordered pairs of natural numbers does this suggest?

Unless the above approach leads the student to integers as we know them and as he will use them, our approach has been made in vain. There comes a time at which the teacher should ask, "Do these ordered pairs of natural numbers have physical meaning," or, better, often the student asks. We consider more exercises for the student:

Exercises

1. When Tommy made the entry (3, 5) did he gain or lose money that day and how much?
2. Describe the "gain or loss" state of affairs for each of these entries.
   (a) (1, 7) (b) (6, 2) (c) (10, 3) (d) (4, 4).

3. Interpret the gain-or-loss situation for each day's business and then tell the net gain or loss. Is the same net gain or loss indicated in the sum of the ordered pairs?

   (a) (1, 6) + (2, 9). [The sum is (3, 15) which means that there is a net loss of 12. Indeed, that is the sum of the losses on the two separate days.]

   (b) (1, 6) + (9, 3) (d) (10, 7) + (3, 8)

   (c) (10, 3) + (11, 4) (e) (4, 2) + (7, 11)

4. If the day's business is (5, 1) then Tommy would make a gain of 4 dollars. If the day's business were (1, 5) then Tommy would say that he lost 4 dollars. What symbol or notation do you suggest to distinguish between $4 gained and $4 lost? (The student might suggest 4*g* and 4*l* or 4^g and 4^l or +4 or −4 or other notations. Whatever is suggested should be used to help fix ideas and principles in the students' own language and knowledge —changes, if necessary, may come later.)

5. What single numeral as suggested in Exercise 4 will identify the integer to which each of these representatives belong? The answer to (c) is 8g, meaning a gain of 8.

   (a) (4, 7) (e) (1, 1) (i) (6, 3)

   (b) (11, 3) (f) (3, 8) (j) (6, 4)

   (c) (12, 15) (g) (2, 7) (k) (4, 2)

   (d) (15, 12) (h) (1, 6) (l) (18, 16)

All of the above came from the simple story of Tommy's work with Mr. Smith and the situation which arose when Tommy was left to devise his own summary system. Very quickly students will gain a feel for adding positive and negative integers (these the teacher and students will formally define), but if any question arises there is ready recourse to using an ordered pair to represent an integer. Note that in this approach there is no appeal to motion nor to directed line segments.

The teacher who wishes to pursue this ordered-pair approach further may consult books on modern algebra, but he will need to invent his own classroom situations from which such an approach can emerge.

Exercises

1. Try to devise another situation which will give rise to ordered pairs of natural numbers and to equivalence classes of these ordered pairs and hence to integers as is presented here in Tom's method of keeping his records.

2. Let us now consider *multiplication* of ordered pairs and hence the multi-

plication of integers. Of course, three of the integers (2, 5) added, as (2, 5) + (2, 5) + (2, 5), will yield (6, 15). But *three* belongs to the equivalence class which has representatives such as (4, 1), (5, 2), (6, 3), (7, 4), et cetera, and the product (7, 4) × (2, 5) must belong to the equivalence class one of whose representatives is (6, 15). See if you can devise a situation or story from which will emerge multiplication of ordered pairs.

3. Multiply these integers given their ordered-pair representatives. Do the results agree with what you might expect?

   (a) (3, 6) × (7, 12) (d) (4, 8) × (7, 3)

   (b) (8, 4) × (1, 1) (e) (1, 1) × (2, 2)

   (c) (3, 2) × (2, 3) (f) (7, 6) × (1, 8)

4. Suggest some observations from the above list of exercises which might become hypotheses and possible investigations for attempts at proof.

5. Set up a short sequence of definitions and theorems and see if the hypotheses suggested in Exercise 4 might be admitted as a part of the logical structure. One of the significant products in the classroom is the scrutiny and discussion of attempts at logical organization by the student.

6. In certain algebra groups the teacher and students may wish to continue the ordered-pair approach to integers and show that the entire system under addition and multiplication has the properties of an *integral domain*. What, in your opinion, is the place of such a program in the school curriculum? Support your contention. Which facet of mathematics does this approach emphasize?

A Number-Line Approach to Integers

In the modern school program students will already be well acquainted with the integer number line extended in both directions. Since the elementary school uses the number line to help give concreteness to arithmetic, it is only natural for students to explore scales on the other side of zero and, indeed, this is being done. We need only to look at thermometers, east and west house numbers, and other such number lines, to sense the existence of "opposites" in a ready and natural manner.

The number-line approach develops in the student a much more accurate point of view of positive and negative numbers than that heard from students sometime ago when the negative numbers were through of as numbers which represented "less than nothing." Of course, the numbers may be given other names such as 2R, 2L, 3U, 6D (meaning right, left, up, down) and so on, and the student in working with names he has invented will develop a more satisfying insight than if names are used which he does not understand.

Exercise

Suggest eight or ten pairs of opposites which in situations interesting to the student can later be abstracted into positive-and-negative number concepts.

Addition of Integers

Addition is an important operation in the set of integers and it should be emphasized that we cannot expect integer addition to be the same as natural number addition. The number line is a concrete aid for the student to formulate a definition for addition of integers and it gives assistance in determining sums once a definition has been made.

What shall we mean by addition of two integers? The only kind of addition we have thus far is that among the natural numbers. Do we wish 2R "+" 3R to mean 5R? Indeed the student should at first be asked such simple questions as those that follow without mention of the term "addition."

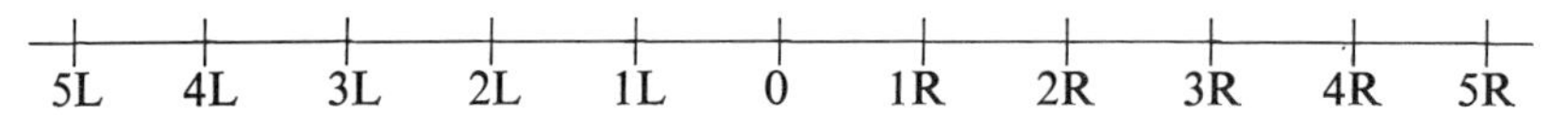

(a) John is at 1R (which means he is located at one to the right) and he moves to the right 2 units. Where is he now?
(b) John is at 1R and then makes a change of 2R. Where is he now?
(c) Harry was at 0, then changed his position by 3R and then by 2R. Where is he now?
(d) Helen was at 2R and then changed her position by 5L. What is her position now?
(e) Jeanne had a position in the scale and then made two changes: 5L and 2R. This is the same as what single change?

In situations like the above, the integers are combined with ease and with no fear; of course, the student should have exercises in several such settings.

As teachers, however, we see that these R- and L-numbers have two interpretations. When we start at 3R and change by 2R, then the first integer indicates a *position* and the second represents a *change*—hence, we have a position "acted upon" by change. 3R followed by 2R might mean also a change of 3R followed by a change of 2R; the result is a *combined change* of 5R. Students should experience both these interpretations. Although the teacher can best do this as he becomes better acquainted with the needs of his students, we suggest some sample exercises in addition to those in the preceding list.

1. Tommy's team takes the ball from 5N through a change of 4N. Where is the ball now?

2. A throw from 7N through 10S leaves the ball at what position?
3. A throw of 6N followed by one of 7S has the same effect as what kind of throw?
4. A movement of 5R followed by one of 3R produces what change in position?
5. A bird flying 3N, then flying 4N, has made what change in position?
6. A bird starts at 3N and then flies 4N. What is the new position?
7. What shall we call the operation of combining 3N with 4N to yield 1S? Why do you think your name is a good one?

Note especially Exercise 4 above. Some students may say, “I would call this adding because a change of 5R followed by a change of 3R is a change of 8R” or “because a position of 5R affected by a change of 3R produces a new position of 8R”—“and all of this is like adding 5 and 3.” Indeed, these students already have an idea of isomorphism of two sets of elements under respective operations of addition. The teacher should, of course, ask if the student is willing to call “3R followed by 5L” addition also. Such an approach encourages students to help erect their own mathematical structure and they sense that mathematics is *created.*

Exercises

1. As a teacher what will you do if the students suggest a name for the operation other than “addition?”

2. Make up a situation which will give practice in the “position followed by a change” approach. Write some exercises.

3. Develop a situation which will emphasize the “change followed by a change” approach. Write several exercises.

4. Using “degrees rise” and “degrees drop,” write interpretations to these problems:

(a) $4r$ “+” $3d$ (a rise of 4° followed by a drop of 3°)

(b) $7d$ “+” $2r$

(c) $8d$ “+” $2d$

(d) 8^{-} “+” 2^{-}

(e) 8^{-} “+” 2^{-}

(f) $^{-}7$ “+” $^{+}3$

(g) $\begin{array}{r} ^{-}11 \\ \text{“+”}\ \underline{^{+}3} \end{array}$ (h) $\begin{array}{r} ^{+}10 \\ \text{“+”}\ \underline{^{-}6} \end{array}$ (i) $\begin{array}{r} ^{+}11 \\ \text{“+”}\ \underline{^{+}6} \end{array}$

(j) $\begin{array}{r} ^{+}3 \\ \text{“+”}\ \underline{^{-}10} \end{array}$ (k) $\begin{array}{r} ^{+}2 \\ \text{“+”}\ \underline{^{-}9} \end{array}$ (l) $\begin{array}{r} ^{-}2 \\ \text{“+”}\ \underline{^{-}8} \end{array}$

5. When problems are written in a vertical pattern, one often simply verbally directs the student to add and then writes:

$$\text{(a)}\ \begin{array}{r} {}^{+}4 \\ \underline{{}^{-}3} \end{array} \qquad \text{(b)}\ \begin{array}{r} {}^{+}6 \\ \underline{{}^{+}7} \end{array} \qquad \text{(c)}\ \begin{array}{r} {}^{-}8 \\ \underline{{}^{+}3} \end{array} \qquad \text{(d)}\ \begin{array}{r} {}^{-}8 \\ \underline{{}^{-}3} \end{array}$$

and so on. Yet some texts write the same problems in this manner:

$$\text{(a)}\ \begin{array}{r} 4 \\ \underline{{}^{-}3} \end{array} \qquad \text{(b)}\ \begin{array}{r} 6 \\ \underline{7} \end{array} \qquad \text{(c)}\ \begin{array}{r} {}^{-}8 \\ \underline{3} \end{array} \qquad \text{(d)}\ \begin{array}{r} {}^{-}8 \\ \underline{{}^{-}3} \end{array}$$

after suggesting "omit the plus sign," when the number is positive. What can you say in favor of such an agreement? against such a convention?

Subtraction of Integers

Subtraction of integers comes readily if the students have been taught additive methods, although here we are assuming that the definition of subtraction for integers is the same as that for natural numbers. If we are defining subtraction among the integers such that "differences" are obtained by the additive methods of subtraction used in arithmetic, then this choice should be clearly indicated and the teacher and students should agree that the definition is a tentative one. Does this proposed definition give answers which are also suggested by other points of view, as "subtraction in the number-line" or "subtraction" arrived at from ordered-pair approaches? Here is an opportunity to concern ourselves with consistency of agreements.

Exercises

1. Write answers to these subtraction problems if we assume that the additive method of arithmetic yields the difference.

$$\text{(a)}\ \begin{array}{r} {}^{+}4 \\ (-)\ \underline{{}^{+}3} \end{array} \qquad \text{(b)}\ \begin{array}{r} {}^{+}4 \\ (-)\ \underline{{}^{+}7} \end{array} \qquad \text{(c)}\ \begin{array}{r} {}^{+}4 \\ (-)\ \underline{{}^{-}8} \end{array} \qquad \text{(d)}\ \begin{array}{r} {}^{+}3 \\ (-)\ \underline{{}^{-}6} \end{array}$$

$$\text{(e)}\ \begin{array}{r} {}^{-}7 \\ (-)\ \underline{{}^{-}6} \end{array} \qquad \text{(f)}\ \begin{array}{r} {}^{-}10 \\ (-)\ \underline{{}^{-}13} \end{array} \qquad \text{(g)}\ \begin{array}{r} {}^{+}4 \\ (-)\ \underline{{}^{+}6} \end{array} \qquad \text{(h)}\ \begin{array}{r} {}^{-}10 \\ (-)\ \underline{{}^{+}11} \end{array}$$

2. Some students are helped by the use of the number line. Do the problems in Exercise 1 again and make some conjectures on "which way to count" as one subtracts a negative integer and as one subtracts a positive integer. Can the subtraction of natural numbers be accomplished in this manner if the set of natural numbers is considered to match 1:1 with the positive integers?

3. Through the use of the integer number line the subtraction of a negative integer may be defined as the process of moving to the right and the subtraction of a positive integer may be defined as the process of moving to the left. Explain how the system of integers under this definition of subtraction is an extension of the system of natural numbers under natural number subtraction. The use of the words *isomorphic* and *subsystem*, per-

haps introduced in other courses you have had, are necessary to explain the meaning of *extension of a mathematical system.* The thought of this definition should be emphasized.

The reader notes that nothing has been said about a certain rule which, at one time, all algebra students memorized with such blindness that many never did understand why it was done. We refer to the rule that "to subtract two numbers, change the sign of the bottom one and add." Much deadening of learning in algebra resulted in the emphasis of such rules—never "why" but just "do." Now if such a suggestion comes from students, then the class might well test this as any other hypothesis and accept it as a method which seems to work. It is "*as if* we replaced the bottom integer by its opposite and then added." To speak of changing the sign of a number introduces as much error as removing the horizontal bar from the symbol 5; the "+" part of $^{+}5$ is as much a part of the positive integer 5 as the bar is of the natural number 5. Indeed, all this is evidence that we must speak and write with care in algebra and in all mathematics.

Subtraction is sometimes done horizontally and the student should practice often with problems as

$^{+}4 - (^{-}3) =$ ______; $^{-}8 - (^{+}7) =$ ______

$^{-}4 - (^{-}3) =$ ______; $^{+}4 + {}^{-}3 - (^{-}6 + {}^{-}4) =$ ______

Exercises

1. Plan a series of exercises which will help the student through the transition from problems as $^{+}5 - (^{-}3 + {}^{-}6 + {}^{+}4)$ to the more standard form $5 - (-3 - 6 + 4)$.
2. Examine a text in algebra to see how it introduces work on "removing parentheses." Examine the logic behind the explanation.

Multiplication of Integers

The multiplication of integers in the number-line approach sets forth some interesting problems. First of all the class must discuss what it chooses to mean by, let us say, $^{+}4$ times $^{+}5$. Perhaps this is the time to reexamine the meaning of the multiplication of natural numbers. 4 times 5 means $5 + 5 + 5 + 5$ and we can have a physical interpretation of this definition, such as four packages of beans containing five each. But the numbers $^{+}4$ and $^{+}5$ are either *position* on the number line or they represent *changes*. What is the *product* of two positions on a line? or two changes? or a position by a change?; it seems that we are in difficulty. A discussion of this situation would be advisable and also would be a good situation in which to emphasize the arbitrary quality of definition. One may point out that a definition must be consistent with the other parts of a discipline and must permit the subject matter to reduce, if necessary, or to relate otherwise to results found by other means. For example, if the definition of area for a portion of a surface does not yield, $A = \pi r^2$, for a very special surface (the sphere) then we would question the definition.

In the foregoing discussion we defined the addition of two integers in such manner that $^+4 + {}^+3$, for example, has the sum $^+7$. The set of positive integers under integer addition is isomorphic to the set of natural numbers under natural number addition. Do we wish multiplication of positive integers to have the same property? If so, we must so define the multiplication of positive integers.

Exercises

1. Argue why it seems reasonable to define $(^+a)(^+b)$ to be $^+(a \cdot b)$ on the basis of our experience with natural numbers; a is the natural number corresponding to the integer ^+a. What, in detail, does this definition mean?

2. Some texts provide a rationale like this for the multiplication of two positive integers, say $^+3$ by $^+5$; $^+3 \cdot {}^+5$ means $3 \cdot {}^+5$ which means $^+5 + {}^+5 + {}^+5 = {}^+15$. Comment on the mathematical soundness of this approach.

3. Try to devise a physical interpretation, if any, to explain why $^+3 \cdot {}^+5$ yields $^+15$?

4. Plan a series of numerical exercises for the algebra student to help him sense that there is a distributive law of multiplication of positive integers over the addition of positive integers. Of course, $a \cdot (b + c) = a \cdot b + a \cdot c$, where a, b, and c are positive integers, requires either postulation or proof.

5. Now do we wish the distributive law in No. 4 to be valid for the addition of positive and negative integers—as $^+4 \cdot (^+3 + {}^-2)$? If so, we must prove or postulate the statement that $a \cdot (b + c) = a \cdot b + a \cdot c$ where a, b are positive integers and c is a negative integer. If this property be accepted, then using the existence of additive inverses and the additive identity element, prove that the product of a positive integer by a negative integer is a negative; i.e., $^+a \cdot (^-b) = {}^-(^+a \cdot {}^+b)$ where ^+a is a positive integer and ^-b is a negative integer.

6. Hence, why is $(^+a) \cdot (^-b)$ also equal to $^-(^+a \cdot {}^+b)$?

7. If we postulate or prove that the multiplication of positive integers is commutative then show from the result in No. 5 that

$$^+a \cdot (^-b) = {}^-b \cdot {}^+a$$

or that the multiplication of a positive integer by a negative integer is commutative.

8. Plan a series of exercises which will help the student prove that the product of two negative integers is a positive integer.

9. In the foregoing exercises we have made some agreements and we have also made some deductions. Outline the logical organization we have created.

Some Intuitive Approaches to Multiplication

An approach to the multiplication of integers sometimes suggested in the junior high school is the use of velocities of automobiles, the number line, and time in a manner like this:

Let us identify time progressing as *positive* time; distance to the right of the observer as distance *positive*; moving to the right at, say 10 mph as *positive* velocity. Time "hours ago" will be considered as *negative*; distance to the left of the observer will be regarded as *negative*; motion to the left as *negative*.

Hence an auto moving to the right at 10 mph 3 hours from now will be 30 miles to the right, hence a "positive number times a positive number yields a positive number."

An auto moving to the right at 10 mph was where 3 hours ago? It was at a position 30 miles to the left of the observer, hence a "positive number times a negative number produces a negative number."

The rigorous mathematician may not like the argument above (it is readily agreed that the foregoing development is not the ultimate in mathematical discourse), yet "proof" to the student is what makes the mathematics convincing at the moment in his present intellectual stage. Proof that addition of natural numbers is commutative is far different for the second-grader from that for the upperclassman, yet there are features in common.

Exercises

1. Devise an intuitive argument, as that above, to convince young students that the product of a negative and a positive number yields a negative number.
2. Use the above conventions and argue that a negative number times a negative number is a positive number.
3. Devise another story or situation to argue that the product of two negative integers is a positive integer.

Another intuitive approach to the consideration of the product of two negative numbers makes use of the recognition of a pattern which seems to emerge in the following table:

$$\begin{array}{l} \cdot\ \cdot\qquad\cdot \\ \cdot\ \cdot\qquad\cdot \\ {}^{-}4\cdot{}^{+}3 = {}^{-}12 \\ {}^{-}4\cdot{}^{+}2 = {}^{-}8 \\ {}^{-}4\cdot{}^{+}1 = {}^{-}4 \\ {}^{-}4\cdot\ 0 = 0 \\ {}^{-}4\cdot{}^{-}1 = ? \\ {}^{-}4\cdot{}^{-}2 = ? \\ {}^{-}4\cdot{}^{-}3 = ? \\ \cdot\ \cdot\qquad\cdot \\ \cdot\ \cdot\qquad\cdot \\ \cdot\ \cdot\qquad\cdot \end{array}$$

We arrive at ${}^{-}4\cdot{}^{+}3 = {}^{-}12$ and like products by other means and list them as shown above. Note the last column; what products are predicted for

$${}^{-}4\cdot{}^{-}1,\qquad {}^{-}4\cdot{}^{-}2,\qquad {}^{-}4\cdot{}^{-}3\ldots ?$$

In the discussion of the multiplication of integers we must move from the intuitive and mathematics-in-situations approaches to the more inward (postulational) approach in order to place work with integers on an acceptable foundation. That is, the student should formulate assumptions and definitions concerning the integers. This may be an important and significant place for the student to become aware of the postulational nature of mathematics. The postulational character of mathematics is more readily sensed in a study of the set of integers than it is merely in the study of the positive numbers, for not all the integers have simple and natural physical representatives and hence more is left to deciding what kind of arithmetic we would like the integers to have. The teacher should capitalize on the newness of the integers at the early high school level with suitable concrete introductions to generate hypotheses or conjectures to be studied in a logical setting.

Exercises

1. Devise a pattern argument similar to that above to make it plausible that the product of a positive integer by a negative integer is a negative integer.
2. Algebra texts in use in schools may contain approaches to the multiplication of negative integers in way not mentioned here. Consult a text on this topic, study the argument found there in, and comment on the approach. What does it have in common with approaches mentioned here or others

with which you are acquainted? (Does the text use an ordered-pair-of-natural-number approach? Vectors? Directed distances? Integers as indicators of changes? Integers to denote positions? Is there mentioned or unmentioned isomorphism of nonnegative integers with the whole numbers?)

5.9. Logarithms

The introduction of logarithms early in courses in algebra not only helps to show another example of mathematics in the service of man but it provides also opportunity for mathematical investigations and growth. The writers have found it interesting not to mention the term "logarithm" at the beginning, but later at some convenient time to emphasize such problems as

$$2^2 \cdot 2^3 = 2^5 \qquad 2^4 \cdot 2^3 = 2^7 \qquad 3^2 \cdot 3^4 = 3^6$$

and to encourage the feeling that this is a quick way to multiply; perhaps we could construct something rather powerful from this idea of multiplication.

The student is off to an unpretentious but exciting exploration if he receives as part of an assignment the construction of a table of powers of two, at which time the teacher should have available a large wall chart containing the same powers. The teacher might ask and discuss, "What is $8 \cdot 4$?" as he points on the table to 8, 4, and 32. The use of bigger numbers will cause some students to employ paper-and-pencil techniques to arrive at answers, but one or more students may begin to report answers with extreme promptness. The teacher might now ask, "How can you do this so fast?" and the students who explain will have another opportunity to gain poise and practice in expressing themselves as they explain their methods. As the reader knows, the students who perform the multiplications quickly have *observed* or *discovered* that they can do this by adding exponents.

$2^1 = 2$	$2^9 = 512$	$2^{17} = 131072$
$2^2 = 4$	$2^{10} = 1024$	$2^{18} = 262144$
$2^3 = 8$	$2^{11} = 2048$	$2^{19} = 524288$
$2^4 = 16$	$2^{12} = 4096$	$2^{20} = 1048576$
$2^5 = 32$	$2^{13} = 8192$	$2^{21} = 2097152$
$2^6 = 64$	$2^{14} = 16384$	$2^{22} = 4194304$
$2^7 = 128$	$2^{15} = 32768$	$2^{23} = 8388608$
$2^8 = 256$	$2^{16} = 65536$	$2^{24} = 16777216$

Exercises

1. Compute these products by the use of the table you have constructed.

(a) 8192×8 (c) $8 \times 32 \times 64 \times 16$

(b) $256 \times 8 \times 4096$ (d) $1024 \times 32 \times 8$

2. Compute these quotients by the use of the table.

(a) $4096 \div 16$

(b) $16384 \div 1024$

(c) $(256 \times 512 \times 2048) \div 4096$

(d) $(4096 \times 256 \times 1024) \div 32768$

(e) $\dfrac{4096 \times 512 \times 2048}{3278 \times 4}$

(f) $\dfrac{256 \times 512 \times 4096 \times 131072}{8192 \times 16384}$

3. Indeed the table above enables us to perform multiplications and divisions quickly, but are there any disadvantages to the table?

During the next class hour one will wish to gain more practice in the use of the table, but there should be much discussion of the disadvantages also. Students will probably suggest that the table contains only those numbers which are powers of two and that one could not multiply, say, 83×251. Hence the question which can come up for consideration is, "Can we construct a table of powers of two which will contain as many consecutive natural numbers as we wish?" Our algebra class now has a real problem which the teacher may emphasize. The kind of table we wish to construct is illustrated by the partial table shown.

To begin to "complete" our table the student usually says, "It will be easy to find the power of 2 which gives 3! Since $2 = 2^1$ and $4 = 2^2$ then $3 = 2^{1.5}$." In like manner, students suggest that $6 = 2^{2.5}$, $5 = 2^{2.25}$, and so on. Perhaps part of the assignment for the next day will be to use this method to complete the table of powers of 2 to the power for 32. To repress or stifle this suggestion by teacher edict, by saying that is wrong because the function $y = 2^x$, or $\{(x, 2^x)\}$ is an exponential function rather than a linear function, would dampen and/or extinguish the spirit of experimentation and exploration we wish to encourage. Rather, we should

$2 = 2^1$	$9 = 2^?$	$15 = 2^?$
$3 = 2^?$	$10 = 2^?$	$16 = 2^4$
$4 = 2^2$	$11 = 2^?$	$17 = 2^?$
$5 = 2^?$	$12 = 2^?$	$18 = 2^?$
$6 = 2^?$	$13 = 2^?$	$19 = 2^?$
$7 = 2^?$	$14 = 2^?$	$20 = 2^?$
$8 = 2^3$	.	.
.	.	.
.	.	.

permit the student to continue in his quest for a better table in his own way. A moment's reflection all convince us that the strength he gains in handling decimals and fractions and the feeling of confidence he has as he manipulates the numbers in this table are rewarding enough. Moreover, he is actually performing *interpolation* at his own invention; this is precisely what we do when we use tables. Perhaps he does not even know that he is assuming that the numbers vary linearly as the exponents.

Students usually develop a table as shown. The exercises below will help us to study the table.

$$2 = 2^{1.00} \qquad 5 = 2^{2.25} \qquad 7 = 2^{2.75}$$

$$3 = 2^{1.50} \qquad 6 = 2^{2.50} \qquad 8 = 2^{2.30}$$

$$4 = 2^{2.00}$$

Exercises

1. Complete the table of powers of 2 to the power which gives 32, using the method suggested above.
2. Check your table for consistency and correctness. Does the table show that $4 \times 3 = 12$? that $5 \times 3 = 15$? that $6 \times 3 = 18$? that $7 \times 3 = 21$? that $7 \times 4 = 28$?
3. Explain why our interpolation gives erroneous results. In what kind(s) of functions does interpolation give correct results?
4. In our continuing work in algebra we might now ask for other suggestions to express numbers as powers of 2. Imagine yourself a high-school freshman, do you have another suggestion?

In the approach above the student has seen his hypothesis fail and he seeks another solution to his problem, exactly as man has groped and learned throughout history. The student has strengthened some arithmetic skills and he has learned something about interpolation. Sometimes the teacher must make suggestions too, and he may rightfully do this, since he is often a co-worker in the investigation. The teacher may mention "I wonder if it would help to make a *graph* of powers of 2 relative to the exponents?" The class has begun another adventure.

With high school freshmen, the teacher often finds it necessary to help much with the construction of the graph, for the use of centimeter graph paper may be a new experience. But now the student is making a graph with a purpose and such matters as desirable scales, making accurate reading, and so on, are skills which receive attention. The partial graph obtained appears roughly as that shown in Fig. 5.5. In practice, the student attempts to read the desired exponents from the centimeter graph paper and one who

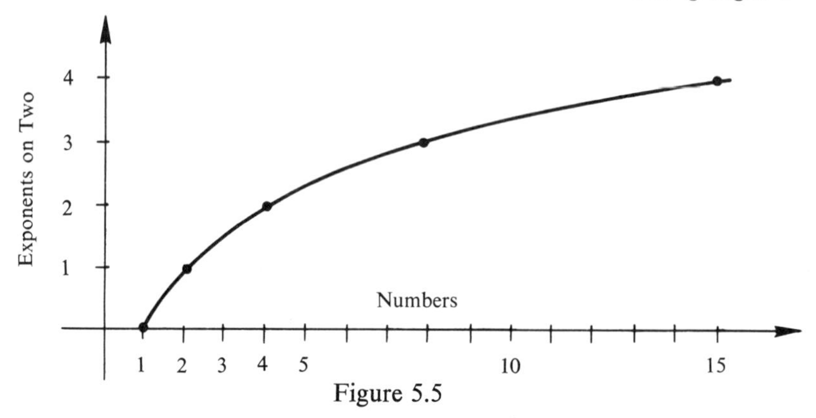

Figure 5.5

works carefully can become quite successful in constructing small tables. It is interesting to note also that the student understands precisely what these numbers are meant to do.

Finally, the teacher can furnish to the student from other sources powers of 10 which yield the *prime* numbers from 1 to 100 and ask the students to complete the table. For example, from data as these: $10^{.301} = 2$, $10^{.477} = 3$, $10^{.699} = 5$, $10^{.845} = 7$, $10^{1.041} = 11$, and $10^{1.114} = 13$, students can construct a table.

Number	Exponents in Powers of 10
1	0.000
2	0.301
3	0.477
4	—
5	0.699
6	—
7	0.845
8	—
9	—
10	—
11	1.041
12	—
13	1.114
.	.
.	.

The student computes the exponents of the powers of 10 for the composite numbers and, in doing so, not only strengthens his arithmetic skills, but gains a deeper insight into the role of the primes in our arithmetic system.

The above approach acquaints the student with some mathematical ideas with no mention of any of the terms associated with logarithms (such as the mantissa and the characteristic). As students gain practice in computing products, quotients, powers, and roots of problems (all of whose answers are less than 100), the teacher might ask, "What would be a good name for these useful numbers?" The usual answer is, "Let us call them 'exponents' or 'multiplying numbers'." In the experience of the writers, no student has ever suggested the name "logarithm"; the teacher introduces the word. Here we have a good example of developing a concept before giving it a name—regarded by most as one facet of good teaching.

Exercises

1. So that the reader also experiences the fact that the prime numbers are the builders of the others, compute the exponents for powers of some of the composite numbers.
2. Prepare a graph on centimeter or millimeter paper to read the exponents of the powers of 2 accurately. You might prepare a transparency of your graph to use on an overhead projector.
3. Select an algebraic idea or concept which you think you could develop *before* you give it a name. Outline your approach.

5.10. A Summary

It was indicated at the beginning of this chapter than our discussion would not include all the topics ordinarily studied in secondary school algebra. Rather the purpose of the chapter is to emphasize ways to introduce the numbers of algebra and means by which equations and methods of solution might emerge in a natural way. A few topics have been discussed to illustrate how student-centered or discovery approaches might be used.

Indeed, after considering the materials of this chapter, the teacher still has many opportunities for experimenting with approaches and for student-teacher investigation of topics. There is no end to the challenge of starting with "things experiential and experimental" to use if they fit into the logical (inward-looking) structure of the subject.

Some of the rules for multiplication of special kinds of numbers can come under the scrutiny of the algebra student. Can we defend in logical manner the old rule that numbers as 73×77 can be multiplied quickly writing 21 and then placing the product of 8×7 in front, to get 5621? This rule works if the sum of the two right-hand digits is 10 and if the two left-hand digits are alike. In the process one multiplies the two right-hand digits and writes down the answer; he then increases *one* of the left-hand digits by one, multiplies them, and places these digits in front of the two just written. Hence 84×86 equals 9×8 placed in front of 24 or 7224. There are many such rules which can become the subject of algebraic investigation and proof. The secondary school algebra student enjoys learning how the binomial expansion $(a + b)^2 = a^2 + 2ab + b^2 = a^2 + b(2a + b)$ is related to the well-known process of taking square root. After seeing this relation one can use the expansion for $(a + b)^3 = a^3 + 3a^2b + 3ab^2 + b^3 = a^3 + b(3a^2 + 3ab + b^2)$ as a guide for taking cube root—all this at the appropriate level of the student. The binomial theorem can be applied to probability and to heredity in classes in biology, both of which are fruitful subjects for directed investigation. The use of ordered pairs of numbers to introduce vectors can be done in algebra. One of the many exciting paths which might be pursued could be to begin with the division of $x^2 + 6x + 5$ by $x + 1$ by *subtraction* and end with the algorithm we now use. Algebra abounds with inviting problems to help the student perceive that rather than being a "bunch of unrelated tricks" algebra can become a strong logical structure, every procedure of which is defended by reason and proof.

Exercises

1. Plan a classroom approach for introducing the "taking of square root by division." This means that if the student estimates the square root of 128 to be 11.2 he then divides 128 by 11.2 to find the quotient. Of course, he is pleased if the quotient equals the divisor. If not, he takes the average of the two as a second estimate. Let the student suggest this, however.

2. Using the second estimate from No. 1, make a third estimate.

3. Show that if the first estimate for the square root of $a^2 + b$ be taken as a then the "quotient" is $a + b/a$ and hence the second estimate is

$$a + \frac{b}{2a}$$

How does this compare with the first two terms of $(a^2 + b)^{1/2}$?

4. Review the algorithm for taking square root which appears as that shown. The problem illustrates the extraction of the square root of 1225. Explain how this is related to the binomial expansion

$$(a + b)^2 = a^2 + 2ab + b^2 = a^2 + b(2a + b)$$

or, how the binomial expansion can be seen to be a guide in constructing the algorithm.

$$\begin{array}{rrr} & 12^{1}25 & \underline{35} \\ & \underline{9} & \\ 65 & 325 & \\ & \underline{325} & \end{array}$$

5. Now by the use of the binomial expansion for

$$(a + b)^3 = a^3 + 3a^2b + 3ab^2 + b^3 = a^3 + b(3a^2 + 3ab + b^2)$$

as a guide, take the cube root of 15,625. See if you can devise a set of directions for other to follow.

6. Plan a *series* of problems or experiences so your students in advanced algebra will see the relation between the algorithm which they learned to take square root (if they learned the traditional one) and the expansion of $(a + b)^2$.

7. Prove that numbers such as 62 and 68 can be multiplied as described in the text, that is, by writing 7×6 or 42 in front of 2×8 or 16 to make 4216.

5.11 Some Sample Study Guides

Indeed, the form of the study guide is not important. Rather, the *plan* of student-centered approaches (the plan of constant reviewing, often in new settings, the strengthening of those skills which will be used later, the discussion of current material, and, finally, the invitation to explore and to manufacture ideas which will become the center of attention later) is the methodology which the writers have found most effective. Study guides may have the form shown below, yet any teacher may develop a style which he finds workable and which results in the most effective kind of learning.

Multiplication of Binomials

1. *Review.* Write the names of single terms for each of these products.

   (a) $3x^2 \cdot 2x^4 = 6x^6$ (f) $^-4a \cdot {}^-2a =$ ________
   (b) $3x^2y \cdot 2xy^3 =$ ____ (g) $4^2 \cdot 4^2 =$ ________
   (c) $(x^3 \cdot x^2)x =$ ____ (h) $4^2 \cdot 4 =$ ________
   (d) $4a^2b \cdot 3ab^3 =$ ____ (i) $(4^2 \cdot 3^2) \cdot (4^3 \cdot 3) =$ ____
   (e) $^-4a \cdot 3a^2b =$ ____

2. Write at sight the products of these binomials:

   (a) $(x + 2)(x - 3)$ (d) $(x - 4)(x - 5)$ (g) $(x + 4)(x - 4)$
   (b) $(a - 4)(a + 2)$ (e) $(x - 3)(x + 3)$ (h) $(x + a)(x - a)$
   (c) $(a + 3)(a - 5)$ (f) $(x - 6)(x - 7)$ (i) $(y + 3)(y - 2)$

3. *Text* (problems)

4. Can you suggest a way to multiply, at sight, expressions as $(2x + 5)(x + 3)$? Show your method. Try it on

   (a) $(3x + 7)(x + 5)$ (b) $(3x + 5)(2x + 1)$ (c) $(2x + 7)(x + 3)$

5. In Exercise 2(a) we see that the product $(x + 2)(x - 3)$ can be written as $x^2 - x - 6$. See if you can devise a way to learn that the expression $x^2 - x - 6$ can be considered as arising as the product of $(x + 2)$ and $(x - 3)$. Try to determine the pair of binomials which has each of these expressions as its product:

(a) $x^2 - x - 20$ (e) $x^2 + 7x + 12$ (i) $x^2 - 8x - 9$

(b) $x^2 + x - 6$ (f) $x^2 - 7x + 12$ (j) $x^2 - x - 72$

(c) $x^2 + 2x - 48$ (g) $x^2 - x - 12$ (k) $x^2 + x - 72$

(d) $x^2 - 2x - 15$ (h) $x^2 + x - 12$ (l) $x^2 - 6x + 5$

6. Would it be possible for the expression $x^2 - x - 30$ to be the product of two pairs of binomials? If so, what? How is this different from products in arithmetic? How the same?

7. Henry found two dice and started tossing them. Sometimes a 4 and a 3 appeared, sometimes a 5 and a 2. List all the pairs which could appear whose sum would be 7.

Multiplication of Binomials (cont.)

1. *Review.* Write at sight the products of these binomials:

(a) $(x - 3)(x - 4)$ (g) $(a - 3b)(a - b)$

(b) $(a + 2)(a - 5)$ (h) $(m + r)(m - 2r)$

(c) $(x + 3b)(x - 5b)$ (i) $(13 - 2)(13 + 2)$

(d) $(x + 3c)(x - 5c)$ (j) $15 \cdot 17 = (16 - 1)(16 + 1)$

(e) $(x + a)(x - a)$ (k) $18 \cdot 20$

(f) $(a - 3)(a - 1)$ (m) $17 \cdot 21$

2. *Review.* Of what pairs of binomials is each of these the product?

(a) $x^2 - 3x + 2$ (d) $x^2 - 8x - 20$ (g) $a^2 - ab - 6b^2$

(b) $x^2 - x - 2$ (e) $x^2 - x - 12$ (h) $a^2 + 6ab - 40b^2$

(c) $x^2 + 4x + 3$ (f) $a^2 - 3a - 18$ (i) $a^2 - 49b^2$

3. Write at sight these products

(a) $(2x + 7)(x - 3)$ (f) $(x - 3y)(2x + 7y)$

(b) $(x + 4)(x - 6)$ (g) $(a - b)(a - 3b)$

(c) $(2x - 3)(x + 5)$ (h) $(3a + 2b)(a - b)$

(d) $(x - 6)(x + 7)$ (i) $(3x - 4y)(3x + 4y)$

(e) $(3x - 5)(2x + 9)$ (j) $(3x + 7)(2x + 8)$

4. *Text* (selected problems)

5. Suggest how you might attack the problem of finding the pair of binomials whose product is $6x^2 + 29x + 35$; whose product is $2x^2 + 13x + 21$; whose product is $2x^2 + x - 3$.

6. *Review.* A certain collection contained 20 coins: dimes, nickels, and quarters. There were two less dimes than nickels and four more quarters than nickels. The value was \$3.50. How many of each coin were there in the collection?

7. The product $(x + 3)(x + 2)$ has the equivalent name $x^2 + 5x + 6$. Show how this equivalence is a consequence of the distributive property of multiplication over addition. Are there any other properties of multiplication necessary to prove that $(x + 3)(x + 2) = x^2 + 5x + 6$? Write the argument in steps and give a reason for each step.

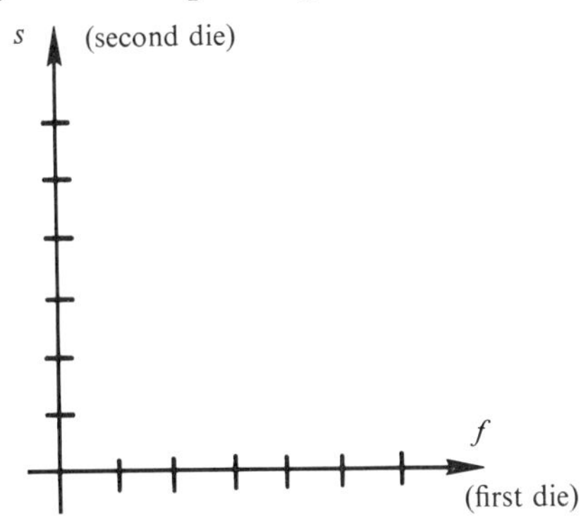

8. In Exercise 7 in the preceding study guide we were asked to list all the dice throws which would result in a sum of 7. Construct a graph of these pairs. What algebraic statement in f and s summarizes these possibilities?

9. Graph all the pairs such that the sum is *less* than 7.

Quadratic Equations

1. *Review.* Show the use of the associative and commutative properties of multiplication in arguing that $(ab)^2 = a^2b^2$. State reason for each step.

2. *Review.* Write pairs of binomials which yield these products:

(a) $x^2 - 7x + 12$	(e) $2x^2 - x - 15$
(b) $x^2 - 6x + 9$	(f) $6x^2 + 17x - 14$
(c) $a^2 - 8a + 16$	(g) $a^2 - a - 132$
(d) $b^2 - 36$	(h) $6a^2 - 13a - 63$

3. *Review.* Use the method of factoring to solve these equations:

(a) $a^2 - 6a + 8 = 0$	(c) $x^2 + x = 12$
(b) $x^2 - 3x = -2$	(d) $x^2 = x + 6$

(e) $a^2 - 8a + 16 = 0$ (h) $3a^2 - 48 = 0$

(f) $a^3 - 7a^2 + 10a = 0$ (i) $a^3 = a^2 + 30a$

(g) $a^3 - 16a = 0$

4. *Text* (selected problems)

5. Suggest numbers in the blanks so that these expressions will be the square of a binomial.

(a) $x^2 + 4x +$ ______ (e) $x^2 - 10x + 3 +$ ____

(b) $x^2 - 6x +$ ______ (f) $a^2 - 12a - 3 +$ ____

(c) $x^2 + 8x +$ ______ (g) $a^2 + a +$ ________

(d) $x^2 + 8x + 4 +$ ____ (h) $a^2 - \frac{1}{2}a +$ ______

6. A ladder is set at an angle of 60°. Tom moves up the ladder 8 feet. How far *up* from the ground has he gone? How could you either compute this directly or arrive at an approximation?

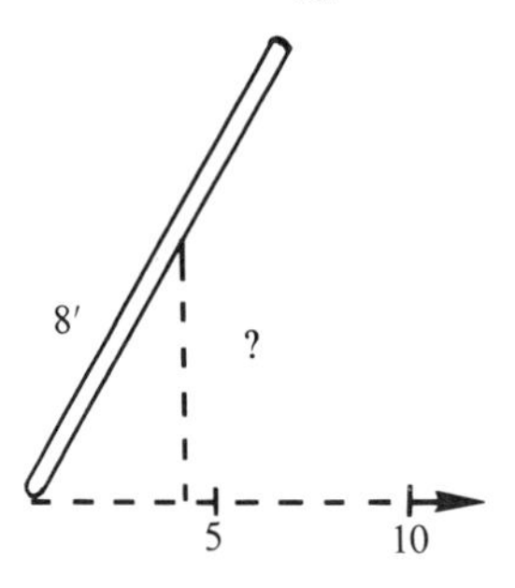

The question in Exercise 6 sets the stage for the emergence of trigonometric ratios; which may be looked upon as *functions* later.

Bibliography

Ballew, H., "Discovery Learning and Critical Thinking in Algebra," *The High School Journal*, L (February, 1967), 261-70.

Barnett, I. A., "Introducing Number Theory in High School Algebra and Geometry," *The Mathematics Teacher*, LVIII (January, 1965), 14-23.

Bhushan, V., "Social Need and Learning in Algebra with Programmed Instruction," *Journal of Experimental Education*, XXXV (Fall, 1966), 94-6.

Butler, Charles H., and F. Lynwood Wren, *The Teaching of Secondary Mathematics* (New York: McGraw-Hill Book Company, Inc. 1960), Chapters 13, 14, 15.

Callanan, C., "Scientific Notation," *The Mathematics Teacher*, LX (March, 1967), 252-6.

Coxford, A., "Classroom Inquiry into the Conic Sections," *The Mathematics Teacher*, LX (April, 1967), 315-22.

Davis, David R., *The Teaching of Mathematics* (Cambridge, Mass: Addison-Wesley Press, Inc., 1951), Chapters 9, 11.

Foster, B. L., "Euclid's Algorithm Revisited," *The Mathematics Teacher*, LX (April, 1967), 358.

Friedland, A., "Abstract Algebra in High School: A Curriculum with Applications to the Traditional Course," *High Points*, XLVIII (April, 1966), 19-30.

Glicksman, A. M., "Vectors in Algebra and Geometry," *The Mathematics Teacher*, LVIII (April, 1965), 327-32.

Gurau, P. K., "Individualizing Mathematics Instruction," *School, Science and Mathematics*, LXVII (January, 1967), 11-26.

Horton, G. W., "Boolean Switchboard," *The Mathematics Teacher*, LVIII (March, 1965), 211-20.

Johnson, Donovan A., and Gerald R. Rising, *Guidelines for Teaching Mathematics* (Belmont. California: Wadsworth Publishing Company, Inc. 1967).

Langer, S. K., "Algebra and the Development of Reason," *The Mathematics Teacher*, LVIX (February, 1966), 158-66.

Leonard, W. A., "Preparing Solidly for Algebra, Using Prime Numbers," *The Arithmetic Teacher*, XI (October, 1964), 418-20.

Lloyd, D. B., "Some Considerations in the Teaching of Modern Elementary Algebra," *School, Science and Mathematics*, LXVII (October, 1967), 600-2.

Mallory, C., "Intuitive Approach to $x^{\circ} \equiv 1$," *The Mathematics Teacher*, LX (January, 1967), 41.

Marks, J. L., "Using the Analytic Method to Encourage Discovery," *The Mathematics Teacher*, LX (March, 1967), 241-5.

McCreery, L., "Applications, Discovery and $y = mx + b$," *School Science and Mathematics*, LXIV (December, 1964), 799-800.

McDonald, I. A., "Abstract Algebra from Axiomatic Geometry," *The Mathematics Teacher*, LIX (February, 1966), 98-106.

Moser, J. M., "Geometric Approach to the Algebra of Solutions of Pairs of Equations," *School Science and Mathematics*, LXVII (March, 1967), 217-20.

National Council of Teachers of Mathematics, *The Teaching of Algebra*, Seventh Yearbook (New York: Teachers College, Columbia University, Bureau of Publications, 1932).

National Council of Teachers of Mathematics, *Enrichment Mathematics for High School*, Twenty-Eighth Yearbook (Washington: National Council Teachers of Mathematics, 1963).

O'Donnell, J. R., "Don't Shy Away from the Zero Exponent," *The Arithmetic Teacher*, XIV (April, 1967), 299 ff.

Read, C. B., "Next Three Terms of a Sequence," *School, Science and Mathematics*, LXVII (June, 1967), 518-22.

Snyder, H. D., "Impromptu Discovery Lesson in Algebra," *The Mathematics Teacher*, LVII (October, 1964), 415-6.

Wright, F. W., "Motivating Students with Projects and Teaching Aids," *The Mathematics Teacher*, LVIII (January, 1965), 47-8.

6

The Teaching of Geometry

6.1. Beginnings of Geometry

Geometry is said to have had its origin in Egypt, for each year the overflowing Nile River washed out boundaries and the necessity of relocating them developed methods which later became classified as *geometry*. Indeed, the word "geometry" has the Greek origin, or *γεω-μετρικο'σ* "earth measure." Most certainly, however, more elementary rudiments of geometry must have existed in the mind of man long before the time of the Egyptians. The sense of something being round, the idea of an angle, the recognition of parallel lines, the concepts of between, inside and outside, and other ideas must have preceded even the contributions of the Egyptians.

The earliest record of mathematical knowledge is found on the Egyptian scroll which was compiled in 1650. B. C. by the scribe Ahmes. This scroll is 18 feet long, 13 inches wide, and contains solutions to 85 problems to serve as guides for working like problems. This document was purchased from the Egyptian government by Henry A. Rhind in 1858 and now found in the British Museum; hence the Ahmes papyrus is sometimes called the "Rhind" papyrus. The Rhind papyrus is discussed in various places and its contents will not be elaborated on here. The interested student might consult Newman's *The World of Mathematics*,[1] Sanford's *History of Mathematics*,[2] Neugebauer's *The Exact Sciences in Antiquity*,[3] Aaboe's *Episodes from the Early History of Mathematics*,[4] and articles in various journals.[5,6]

The field of geometry was enlarged by many significant contributions from the Babylonians relative to mensuration and navigation. Evidence of these developments is found in the clay Babylonian problem and table texts which reflect much advancement over the work of the Egyptians. The table texts consisted of tables of squares, square roots, cubes, reciprocals of numbers, trigonometric ratios, and other useful numbers, whereas the problem texts contained solutions to sample problems, much like the illustrative problems worked in modern handbooks. Several hundred of these tables, or portions of them, have been discovered and are found in various museums in many countries. The Babylonians made far-reaching contributions which have had a large impact even on the mathematics of today. Their numeration system, for example, was based on 60, and, even to the present time, the use of Babylonian fractions has been kept alive and is used in the subdivision of degrees and hours. One can see that this system progressed to minuteness very quickly "on the right" and large numbers "on the left". The adoption of this same system by the Romans to express angle measure and time led to Latinized names. We have for the sixtieths the terminology *partes minutae primae* (the first small parts) and for the thirty-six hundredths *partes minutes secundae* (the second small parts) which are now known as minutes and seconds with reference to angles; also there are 60 *minutae* in an hour and 60 *secundae* in a *minuta*. Aaboe writes "Few people realize that when we say it is 2 hours 30 minutes and 10 seconds p.m. we are actually talking the language of the Babylonians of 400 years ago."[7] The Babylonians would have written, however, W ⁑ ≤ with no indication, except from the context, that they meant W* ⁑ ≤ , where we have used the mark * to denote a sexagesimal point.

The beginnings of geometry not only furnish many interesting historical anecdotes which can brighten class periods and provide topics for further

---

[1]James R. Newman, *The World of Mathematics*, Volume I (New York: Simon and Schuster, 1956), pp. 170 et seq.

[2]Vera Sanford, *A Short History of Mathematics* (Boston: Houghton Mifflin Company, 1930).

[3]O. Neugebauer, *The Exact Science in Antiquity* (Princeton, N J.: Princeton University Press, 1952).

[4]Asger Aaboe, *Episodes from the Early History of Mathematics* [New York: Random House, Inc. (and The L. W. Singer Company), 1964], p. 22.

[5]R. J. Gillings, "'Thing-of-a-Number' Problems 28 and 29 of the Rhind Mathematical Papyrus (B. M. 10057-8)," *The Mathematics Teacher*, LIV (February, 1961), 97-100.

[6]________ "Problems 1 to 6 of the Rhind Mathematical Papyrus," *The Mathematics Teacher*, LV (January, 1962), 61-9.

[7]Asger Aaboe, *op. cit.*

study, but also suggest ideas for teaching. One such idea comes from the fact that early geometry arose as the answer to a need of society; it came into being as man attempted to answer questions about locations, lines, angles, and areas. Geometry grew as a branch of mathematics to serve man, illustrating the outward-looking facet of mathematics. The experiences in our classroom become much more real if the student perceives that mathematics is invented to answer questions in a physical or intellectual realm, and especially if the student helps to invent some of it. There is a vast difference between the dogmatic, "Now I will show you how to bisect an angle," and the *invitation to create*, "If we are limited to the use of a straightedge and compass, can you invent a way to bisect this angle?" Through such procedures the teacher will learn new approaches and the student is helping mathematics grow—mathematics *in statu nascendi.*

A second helpful guideline to teaching suggested by geometry comes from the empirical nature of early mathematics. Early man experimented with quadrilaterals and circles and came to conclusions on methods to compute area which, in today's classroom, suggest the very natural use of laboratory approaches. Indeed, one is mistaken if he feels that the mathematician always just thinks as a cold logical engine and produces theorems. Most often he makes conjectures which have arisen from laboratory methods and checks numerically or otherwise to see if there is promise of possible theorems. The geometry classroom provides a multitude of opportunities for this kind of valuable experience for students. Laboratory approaches have been discussed in more detail in Chapters 2 and 3.

The teacher of geometry might take lessons from early beginnings in geometry for instruction at both the elementary and secondary level, indeed, even at the university level. Study guides should deliberately contain timely suggestions from which geometry may develop. The writers have found the use of guides such as these very effective; surely the reader and teacher may devise many more as new teaching situations arise.

1. In Fig. 6.1 are some examples of closed curves, one of which is a *circle*. How can you define a circle to distinguish it from the others?

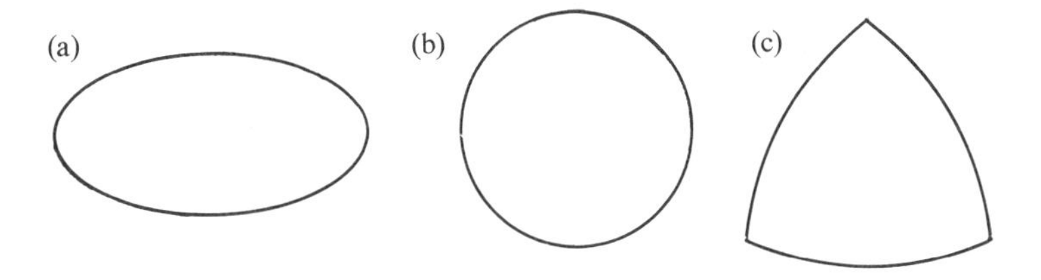

Figure 6.1

2. In each of the circles in Fig. 6.2 an angle has been drawn so the vertex lies on the circle and the rays of the angle go through points B and C

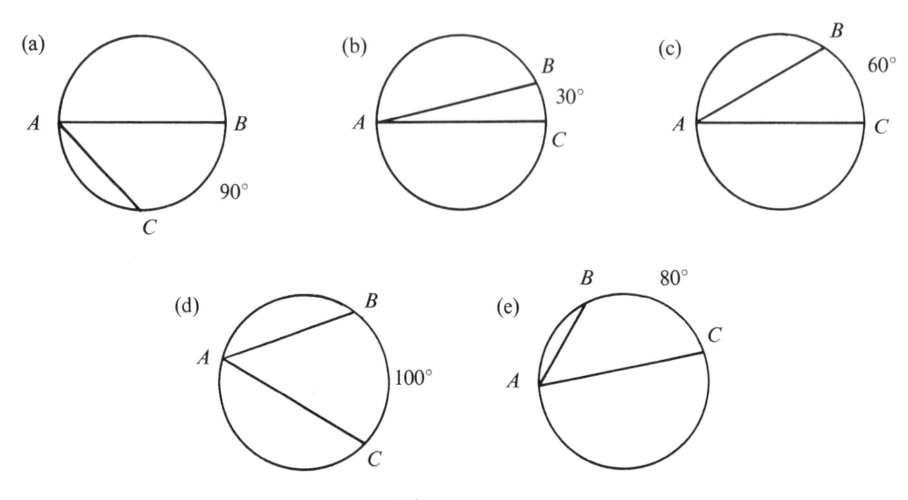

Figure 6.2

as shown. The magnitude of the arc BC is given. What is the measure of the angle BAC in each case? Do you note any promising relation between the measure of the arc and the measure of the angle BAC? (At some point it is well to ask, "What would be a good name for the kind of angle where the vertex is on the circle?" The student has already dealt with "central" angles.)

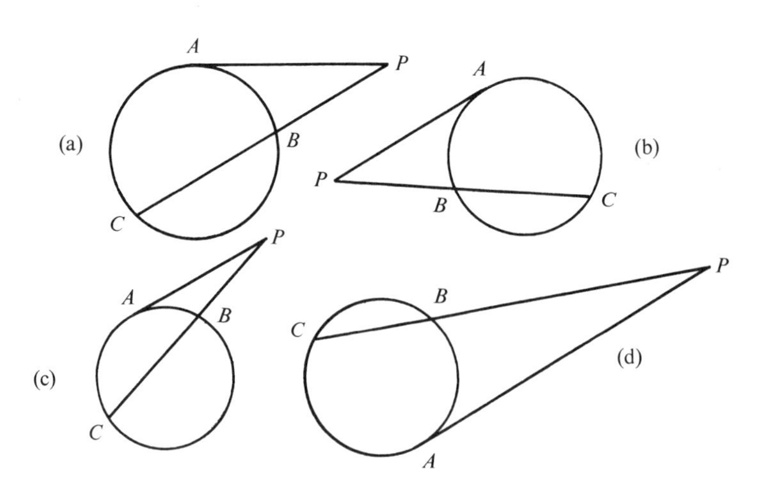

Figure 6.3

3. Note Fig. 6.3. With each circle there is a tangent and a secant from an external point. Are there any relations among the lengths of $\overline{AP}$, $\overline{PB}$, and $\overline{BC}$ or among $\overline{AP}$, $\overline{PB}$, and $\overline{PC}$ which are suggested as you tabulate their respective lengths? (This problem often gives rise to several conjectures.)

Even while the student is investigating examples and exploring mathematics new to him, *uses* of his findings, though perhaps still in the realm of conjecture, can be emphasized. The mathematics of parallel lines and related topics and the mathematics of angles are put to work immediately in exercises like the following. The student gains a respect for the idea of *angle* if he realizes that it is a concept of utmost importance. The teacher must bombard the student with uses of angles, but here are some samples in which the student uses them himself.

1. You have been hired to paint marks at a lot near a motel for slantwise parking (Fig. 6.4). What means can you use to make sure that the lines

Figure 6.4

are parallel? (Suggestions on corresponding angles and on elementary properties of parallelograms may come from this problem.)

2. A certain forest plot was guarded against fire by four lookout towers with positions as shown. One day smoke was seen and the observer at C reported that the smoke was located $30°$ from the north. A reported $120°$ from north. In which region was fire? On another day C reported $70°$ N and B reported $125°$ N. Where was the fire? (use Fig. 6.5).

Exercises like these provided on sheets duplicated for the student give immediate emphasis to the outward-looking aspect of mathematics and, at the same time, develop in a natural way the concept of the acute and the obtuse angle. Too, the student gains the idea of measuring angles from north as its done in navigation. Other interesting uses of angles will be discussed in later sections, but the reader should begin now to assemble ideas and plans for introducing a beginning concept.

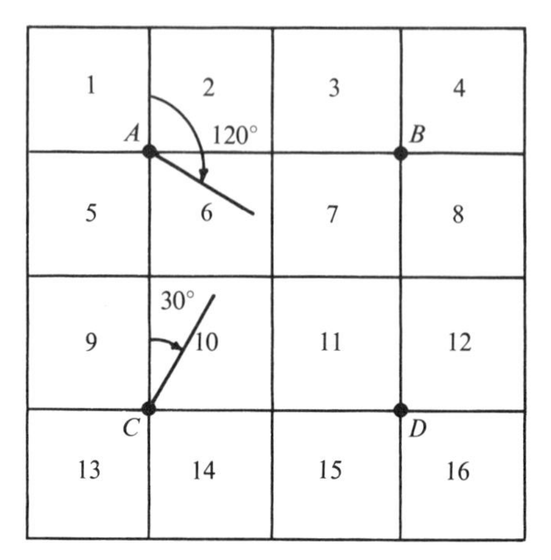

Figure 6.5

Exercises

1. There are several theorems in the geometry of right triangles which involve mean proportion. Select one of these and plan a series of laboratory exercises from which these relations will come as a conjecture.

2. The bisector of an interior angle of a triangle intersects the opposite side of the triangle and forms two segments. Are the lengths of these segments related in any way to the lengths of the sides which include the angle being bisected? Plan a laboratory approach.

3. If you have studied projective geometry, you have heard of the complete quadrilateral constructed from the points A, B, C, and D (see Fig. 6.6). The quotient of the ratios of the lengths

$$\frac{m(\overline{AH})}{m(\overline{HB})} = \frac{m(\overline{AG})}{m(\overline{GB})},$$

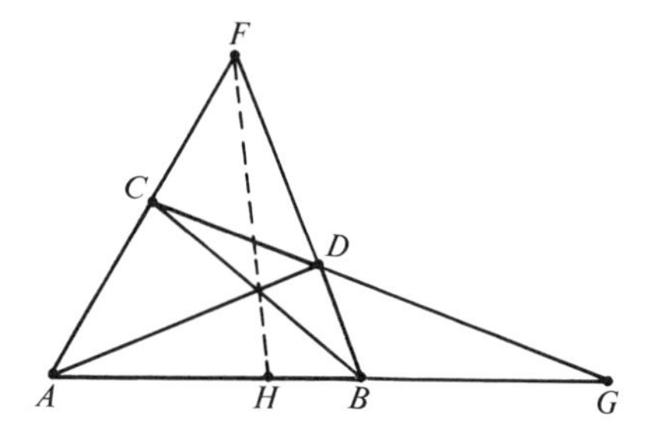

Figure 6.6

is interesting. Plan a series of figures for your students to study this "double ratio." Are there other lines on which the same relation is suggested?

4. Provide a road map for each student in your class. What is the angle from north which gives a straight-line path from Columbus to Cleveland? from Cincinnati to Akron? (choose other cities if you wish). Suggest straight-line paths which will help the student to become acquainted with obtuse angles.

It must be emphasized again that in the outward-looking facet of geometry as illustrated in Exercise 4 above, the student perceives that mathematics-at-work problems are not made to order in that the answers come out evenly; indeed, in such problems, there is development of concomitant skills such as locating positions, using a ruler, estimating fractional degrees, and emphasizing the measuring of angles from the north.

6.2. Logical Beginnings in Geometry

The earliest system of geometry was merely an accumulation of facts and, likewise, beginning weeks in secondary school geometry might well be a review of previously found facts assembled with a view to seeking logical connections between them. Laboratory experience with concrete objects, with drawings and constructions, and with results of computation reveal basic facts in geometry from which, with the teacher's help, a logical organization may emerge. For example, the students who explore may learn by experimental procedures interesting properties of geometric figures such as those suggested in the following:

(a) Properties of angles at the point of intersection of two straight lines
(b) Properties of angles when two parallel lines are cut by a transversal
(c) Properties of parallelograms
(d) Properties of nonparallel lines: two lines intersect in one point, three lines intersect in three points, four lines in six points, and so on
(e) Properties of right triangles
(f) The property of a triangle balancing at the point which is the intersection of the medians
(g) The constructing of inscribed and circumscribed circles about a triangle.

Also students find *nonmetric* theorems and constructions interesting. Two of these are:

(h) The Theorem of Desargues: if two triangles are so situated that the lines joining corresponding vertices are concurrent, then the corresponding sides of the triangles meet in points which are collinear.
(i) If the lines CP and EP as shown in Fig. 6.7 contain the additional

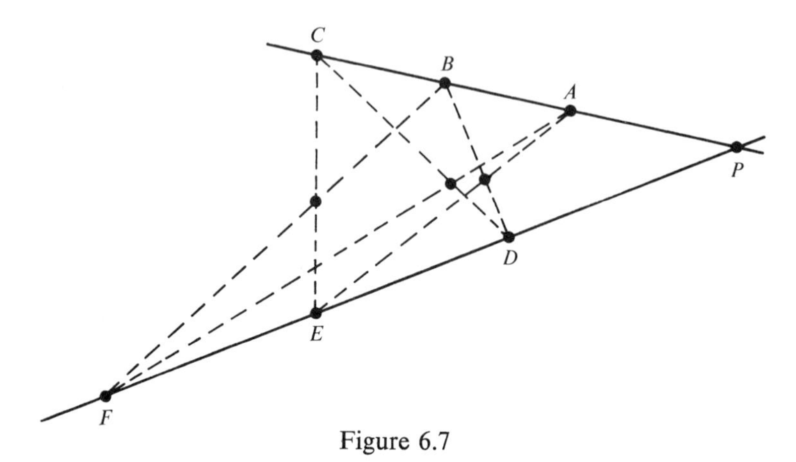

Figure 6.7

points indicated on the figure and if the pairs of lines $\overleftrightarrow{\mathbf{AE}}$ and $\overleftrightarrow{\mathbf{BD}}$ $\overleftrightarrow{\mathbf{AF}}$ and $\overleftrightarrow{\mathbf{CD}}$ $\overleftrightarrow{\mathbf{CE}}$ and $\overleftrightarrow{\mathbf{FB}}$ are drawn, then they intersect in points which are collinear.

Usually nonmetric properties as those mentioned in (h) and (i) remain unknown unless one studies projective or higher geometry, but elementary geometry students find them interesting, too. To introduce such properties, the teacher may simply point out, "Last night I was doing some drawing and it seems that every time I drew... [a figure like that in (i)], I obtained the same result." Then the teacher's question may be, "Can you find any cases in which this does not happen?" Junior high school students will try this over and over to find that they are becoming acquainted, perhaps for the first time, with the nature of an *invariant* property, or with the idea of "absoluteness" in geometry, or with the concept of something being a necessary consequence of given conditions.

The above examples suggested the gathering of facts physical means such as construction, measurement, and the use of models. One can learn new properties by the use of *computation* also and from facts empirically established one can reason out certain consequent facts. For example, one can introduce problems such as this:

Given: $\overline{AC} \cong \overline{BC}$

$\overleftrightarrow{\mathbf{AD}}$, $\overleftrightarrow{\mathbf{BD}}$ angle bisectors

$m\angle BAD = 25$

To compute: $m\angle C$, $m\angle D$.

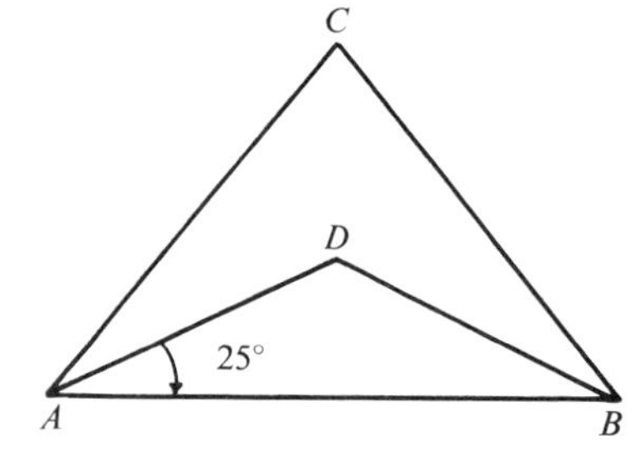

In the same figure, if $m\angle BAD = a$, the student can compute measures for $\angle A$ and $\angle B$ and begins to have the feel of making deductions, if this is pointed out by the teacher. Indeed, exercises such as these help to orient the student so that he seeks proof.

We see that the "geometric facts" which can be assembled can extend from the most simple observations on lines and angles to properties of much more intricate diagrams, and that it is possible for many of these to originate in the elementary and junior high school. The student in the late junior high school knows many facts in geometry which he may fit into a logical pattern and, of course, hints or suggestions to this end should be forthcoming. Indeed, as Professor E. H. Moore has suggested

> . . . the teacher should lead up to an important theorem gradually in such a way that the precise meaning of the statement in question, and further, the practical . . . truth of the theorem is fully appreciated . . . and indeed the desire for formal proof is awakened before the formal proof itself is developed. Indeed, much of the proof should be secured by the research of the students themselves.[8]

This quotation might well serve as a dependable guide for effective teaching.

More formal logical beginnings in geometry seem to have taken root with the Greeks. Their love of art, beauty, and logic and their philosophical tendency to organize knowledge as a pattern in which some steps were consequences of others were applied slowly to the geometry of the Egyptians and the Babylonians and to new findings of the Greeks. Thales (600 B.C.) is called the father of demonstrative geometry for

> It is believed that Thales presented his geometrical teaching in the form of isolated propositions not arranged in logical sequence but that the proofs were deductive, so that the theorems were not mere induction from a large number of special instances as probably had been the case with the Egyptians. The deductive character he thus gave to the science is his chief claim to distinction.[9]

W. W. Rouse Ball lists some of the theorems attributed to Thales as follows:

> The angles at the base of an isosceles triangle are equal. A triangle is determined if its base angles be given. The sides of equi-

---

[8]E. H. Moore, "On the Foundations of Mathematics," *Bulletin of the American Mathematical Society,* IX (May, 1903), 419. This is the presidential address given before the American Mathematical Society at its ninth annual meeting, December 29, 1902.

[9]W. W. Rouse Ball, *A Primer of the History of Mathematics* (London: Macmillan and Company, Limited, 1930), p. 3.

> angular triangles are proportionals. The angle subtended by a diameter of any circle at any point in the circumference is a right angle.[10]

Further development in Greek mathematics was made largely under the teaching of Pythagoras (ca 569-500 B.C.) and his followers. Indeed,

> Pythagoras made geometry the foundation of his teaching; moreover, he impressed on it that deductive character which it still possesses, and there is reason to believe that he arranged the leading propositions in a logical order. He himself probably knew and taught the substance of what is contained in the first two books of Euclid about parallels, triangles, and parallelograms, and was acquainted with a few other isolated theorems including some elementary propositions on irrational magnitudes. . . [11]

The reader may consult works on the history of mathematics for more details. The Pythagorean school continued to flourish and in the fifth century more geometry, including studies of curves, was developed. The geometry of the Pythagorean school was mixed with metaphysical ideas and with certain philosophical beliefs.

Hippocrates of Chios (ca 440 B. C.), thought to be a student of the Pythagoreans, founded the Athenian school in philosophy and mathematics with the help of Plato (ca 400 B. C.) and Eudoxus (ca 375 B. C.). It is believed that Hippocrates wrote the first textbook in geometry and proved some theorems on circles. Plato might have introduced some definitions, postulates, and axioms, and directed attention to the analytical method of proof from which a synthetic proof is constructed if the steps are reversible. Eudoxus is thought to have discovered much of what is found in Euclid's Book Five and established the "method of exhaustions."

In 323 B. C., Ptolemy of Egypt established a university at Alexandria which was the center of learning for one-thousand years. In the first hundred years, the University of Alexandria had on its staff in mathematics the three famous mathematicians of antiquity: Euclid, Archimedes, and Apollonius. Euclid (300 B. C.) made a great contribution in summarizing and assembling all the mathematical knowledge of his time, most of which appears in his *Elements*. Indeed, this monumental history-making, Euclidean structure contained new styles in writing proofs, each proof being presented as a logically correct succession of statements supported by axioms and postulates. The tenth book of Euclid seems to be the work of Euclid himself, but the other parts of the *Elements* contain the work of others systematized and organized in a logical manner. This had never before been

---

[10] *Op. cit.*, pp. 3, 4.
[11] *Op. cit.*, p. 6.

done. One cannot emphasize too much that Euclid's contribution was a peak of intellectual achievement. His was the most thorough presentation of mathematics as a deductive science up to this time and it was regarded so highly by the learned world that Euclid's *Elements* came immediately to be used as a text for the study of geometry. This method of organizing and developing geometry has become the model for other branches of mathematics and all mathematicians have tried to duplicate Euclid's logical style. Only in the last century was this accomplished in algebra and in analysis; only since about 1960 has one seen much formal deduction in high school algebra courses. Indeed, it was Pythagoras and Euclid along with their students who found the beauty and the intellectual challenge of this inward-looking facet of mathematics. It was they who originated mathematical method and with it raised the level of man's thought from accepting the results of crude experiments to the testing of conjectures by deduction. It was they who constructed a logical system in mathematics and caused scholars in other fields to try to arrange their subject matter in a like logical manner. Euclid introduced a *way of organization*—a pattern for others to follow, and as teachers of geometry, we have both the opportunity and the responsibility to develop this point of view in the minds and practices of their students. In mathematics we see logical system in a pure form, because in such systems emotion or prejudice should not influence reasoning. The mathematical consequences stand unchanged by any bias of the thinker. We must patiently, but persistently, train the mind of the student so that he not only respects this point of view concerning mathematics but will also desire to explore possibilities or logical relations among mathematical facts and principles which he has accumulated. The student should learn that most of the properties of a mathematical system are *consequences* of the primitive statements, and that once a system is formulated some conclusions arise *solely by deduction.* All findings do not come from laboratory exercises to be confirmed by deduction and the teaching of geometry provides an opportunity to introduce young men and women to those processes of thought by which man has achived his greatest intellectual triumph. There is indeed much freedom in mathematics at the beginning of a system but then *die iacta est* (Caesar's "The die is cast") can be said of the remainder. Cassius J. Keyser implies this in a brief way when he uses as a subtitle for his book *Mathematical Philosophy* the words "A Study in Fate and Freedom."[12] The respected principle that we reap what we sow is as valid in the mathematical classroom as it is elsewhere in life.

It should not be thought that once Thales, Pythagoras, Hippocrates, and Euclid showed the academic world this climax of intellectual achieve-

[12]Cassius Jackson Keyser, *Mathematical Philosphy: A Study of Fate and Freedom* (New York: F. P. Dutton and Company, Inc., 1922).

ment, the use of the empirical approach in mathematics disappeared. The Greeks continued to attack problems empirically and Archimedes, one of the greatest mathematicians of all time, used all devices at his command to investigate relationships such as counting, measuring, and weighing. Indeed, emphasis on logical procedures actually increases the value of the empirical method, for conjectures thus suggested can be tested by the deductive process. In the *Twenty-Fourth Yearbook* we read, "'We all start by being empiricists.' The genesis of the mature concept of proof lies in the deep empiricism, that is, in experimentation and observation in early childhood."[13]

One of the pressing problems of the teacher of geometry is to help students understand the idea of and develop the skill essential to the organization of a deductive proof. Some teachers find it effective to develop the idea of argument gradually to let the student give statements in his own way before formal styles of presentation are introduced. In fact, J. W. Young advises that, "In order to induce a pupil to think about geometry, it is necessary first to arouse his interest and then let him think about the subject in his own way."[14] Proof has been looked upon by some as "that which is convincing at the student's present level," with the intention that we should always attempt to raise that level. The copious use of "Why do you think so?" or "What is your reason for that statement?" can lead the student to consciousness of deductive reasoning. Consider these exercises:

(1) ℓ is a straight line. What is the measure of $\angle a$? How can you defend your statement?

(2) In Fig. 6.8 ℓ and ℓ' are straight lines. What relation exists between angles b and c? What reasons can be given for your conclusion?

(3) ℓ and ℓ' are straight lines. Examine the conjecture that $\angle 2 \cong \angle 3$ to see if it can be deduced from properties you already know. State each property being used.

(4) ℓ and ℓ' are parallel and they are cut by the line m. We shall take this to mean that $m\angle 1 = m\angle 2$. See if you can argue that $m\angle 2 + m\angle 3 = 180$. Support your steps in argument by reasons.

(5) ℓ and ℓ' are parallel and they are intersected by the line m. Support the conjecture that $\angle 3 \cong \angle 2$ Give reasons for each statement you make.

---

[13]The National Council of Teachers of Mathematics, *The Growth of Mathematical Ideas, Grades K-12* (Twenty-Fourth Yearbook) (Washington, D.C.: The National Council of Teachers of Mathematics, 1959), p. 112.

[14]J. W. A. Young, *Lectures on the Fundamental Concepts of Algebra and Geometry* (New York: The Macmillan Company, 1936), p. 5.

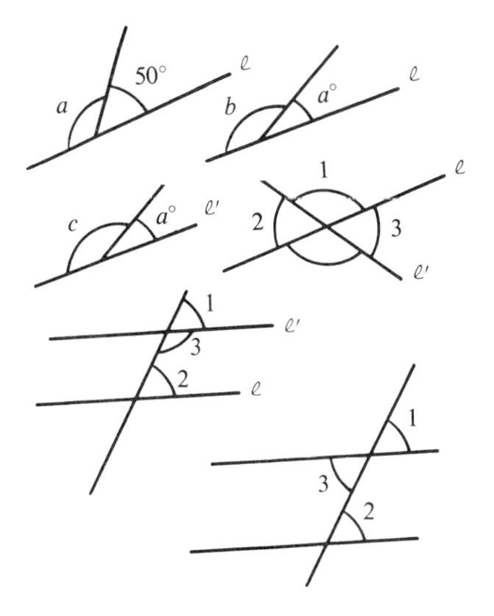

Figure 6.8

Exercises like the above included with other items early in geometry acquaint the student with defending his statements before he is asked to remember the formalities of proof and to duplicate the precise way in which the teacher wishes the proof to be written. A study guide used to introduce proof might look like this:

1. *Review*. Set up expressions for and compute the areas of each part of Fig. 6.9.

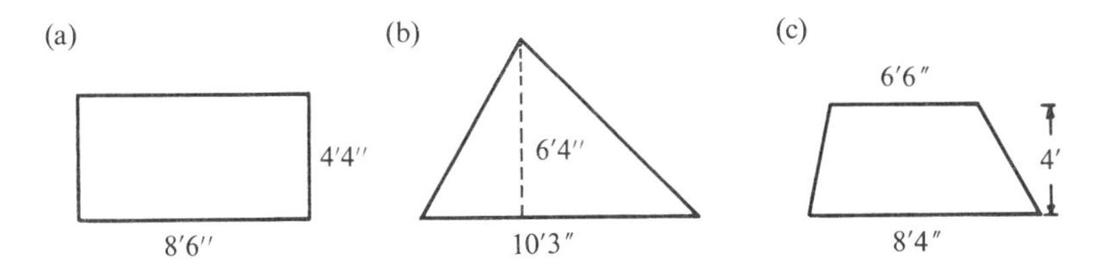

Figure 6.9

2. *Review*. Indicate the measure of each of the angles marked.
3. Exercises from the text.
4. If $\ell \parallel \ell'$ it is conjectured $m\angle 3 + m\angle 2 = 180$. See if you can confirm

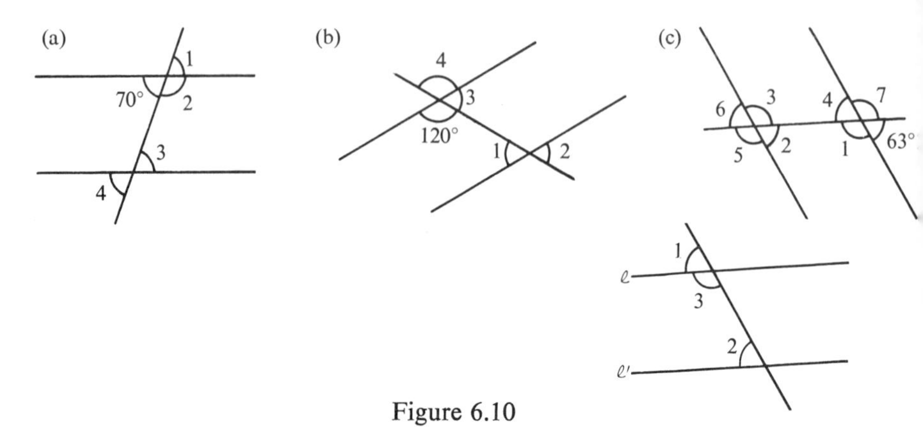

Figure 6.10

this conjecture by argument from definitions and agreements already accepted.

Just when the teacher will introduce the formal writing of proofs will depend much on the level and interest of students; they may ultimately invite a more formal approach.

A second way to help students see what is meant by proof is to have exercises on arranging steps in logical order. Sometimes proofs which are not in good order can be taken from quizzes and texts and used as exercises for students to straighten out. Erroneous proofs frequently provide materials for study in how to write good proofs.

Another way to help the student learn the technique of proof and not to be overwhelmed by argument (which may seem long to him) is to ask the student to work on one part of a theorem at a time. For example, the theorem "The line segment which joins the midpoints of two sides of a triangle is parallel to the third side and has a length equal to half that of the third side," has a proof which is rather long for some beginning students in geometry. Use a figure such as Fig. 6.11 as an example. On being invited to

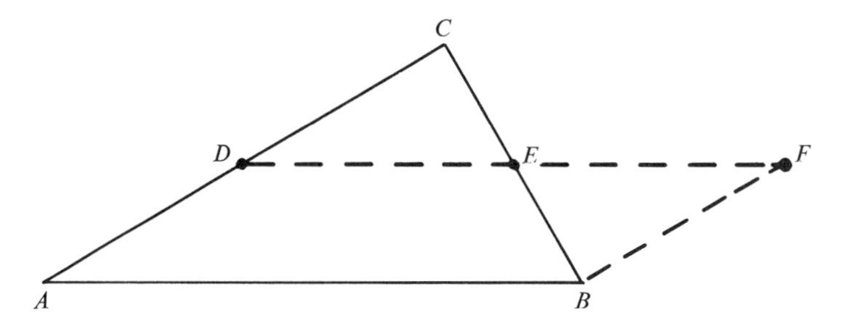

Figure 6.11

contribute an argument, no student in the writer's classes has ever suggested extending $\overline{DE}$ to F so that $\overline{DE} \cong \overline{EF}$. It seems necessary for the teacher to encourage this start. For beginning or slower classes a part of the longer theorem may well apper as a previous problem, simply to show that $\triangle DCE \cong \triangle FBE$. Then can arise such questions as, "Can we deduce anything from this?" and so on. Students have suggested that $\overleftrightarrow{\mathbf{DC}} \parallel \overleftrightarrow{\mathbf{BF}}$ which directs part of the next assignment—to write up a proof. "Let us see if we can deduce more" will invite students to go much further and often to do more than the teacher expects. An approach of this kind has far more value in the creativity it encourages and interest it develops than simply assigning the entire theorem as something to be studied from a book. Teachers may well use parts of the longer standard theorems as exercises before one comes to these theorems in the text. Another such geometry theorem is that which begins, "If a diameter of a circle is perpendicular to a chord. . . " Many students have difficulty constructing long proofs but they can be encouraged to devise proofs for *parts* of theorems. In this way they can begin to perceive the inward-looking aspect of mathematics, the beauty of which may be killed if proofs appear in packages too large or complicated.

In like manner the idea of indirect proof or the methods of using the contrapositive and *reductio ad absurdum* should be introduced with familiar theorems far in advance of theorems for the proofs of which indirect methods are necessary. Otherwise, the student has two difficulties: a new kind of proof and new theorems. More on indirect proof will appear later.

The understanding of a logical sequence of theorems can be sharpened when the time is opportune by asking students to prove a theorem, then prove each of the theorems used as reasons and each of the theorems used as supports in the proofs of these, and so on, until one finds the entire structure built on primitive terms and statements. The students can make charts to show these sequences and the logical basis for the given theorems. Deepened insight is experienced as a result of such activity.

In this section on "logical beginnings," it must be urged that teachers should not stop the examination of what they might recognize as wrong conjectures even though they may know they cannot be supported by deduction. Seldom, if ever, in the development of a topic should a teacher say, "Yes, that is right," or "No, you are in error." The student might learn much in pursuing a conjecture which, later, turns out to fit in the logical system. Indeed, Professor Howard Fehr says that "the student of mathematics rediscovers what his predecessors have created. He learns to ask meaningful questions, he makes hypotheses and subjects them to logical procedures for proof or disproof."[15]

---

[15]Quotation from an address by Howard Fehr at the Dallas (Texas) Meeting of the National Council of Teachers of Mathematics, 1959.

Although the logical (inward-looking) beginnings of geometry began to develop largely through the work of the Greeks, it is not to be expected that in our advanced civilization every student is drawn naturally to logic in geometry. This facet of the student's development must be encouraged and the student must often be shown that there is value for him in the study of the logical organization and methods of mathematics. Transfer of values from geometry to the problems of living is not automatic, and the teacher must "teach for transfer." In *General Education in a Free Society* we are reminded that, "One of the few clear facts about the unclear and much disputed question of transfer of powers from one subject to another is that it will tend not to take place unless it is deliberately planned for and worked for."[16] Elsewhere, with special reference to geometry, these writers urge that:

> The projection of the structure of geometry into areas of more immediate and often of more practical interest to the student should be taught explicitly. It is only in this way that there can be accomplished the "transfer" of mathematical values to other spheres of human interest, which is a primary concern of general education.[17]

What these "powers" and "mathematical values" are is also elaborated on in the *Harvard Report* in this manner:

> The ability to analyze a concrete situation into its elements, to synthesize related components into a related whole, to isolate and select relevant factors, defining them rigorously, meanwhile discarding the irrelevant; and the ability to combine these factors, often in novel ways so as to reach a solution, all are important features of the mathematical procedure.[18]

The practical guidelines proposed by 65 mathematicians suggest the values sought in mathematics to be as follows:

> The mental processes which suggest what to prove and how to prove it are as much a part of mathematical thinking as the proof that eventually results from them. Extracting the appropriate concept from a concrete illustration, generalizing from observed cases, inductive arguments, arguments by analogy and intuitive grounds for an emerging conjecture are mathematical modes of thinking.[19]

---

[16]Report of the Harvard Committee, *General Education in a Free Society* (Cambridge: Harvard University Press, 1950), p. 74.

[17]*Ibid.*, p. 165.

[18]*Ibid.*, p. 161.

[19]"On the Mathematics Curriculum of the High School," *The Mathematics Teacher,* LV (March, 1962), 192.

These are the qualities we must teach for transfer. Some suggestions on how this might be done will be made throughout the succeeding material.

Exercises

1. Using Fig. 6.12, prove the theorem that if a line bisects the exterior angle at the vertex of an isosceles triangle, then this line is parallel to the base. Now prove each of the supporting theorems and further supporting theorems until one arrives at primitive terms and primitive statements.

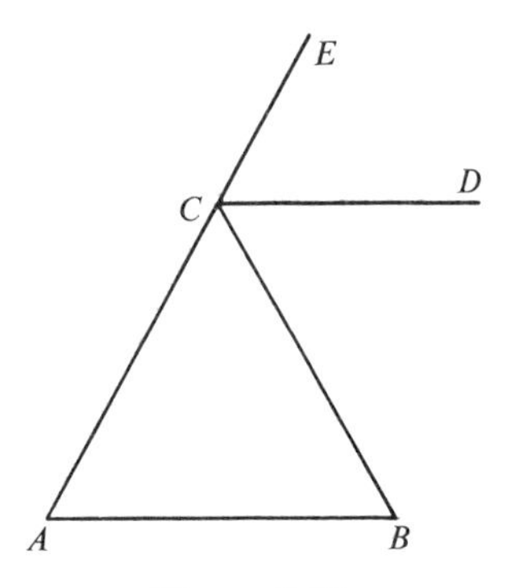

Figure 6.12

2. In like manner, prove that the sum of the interior angles in a triangle is 180°. Prove supporting theorems and reduce the entire structure finally to primitive terms and primitive statements. Different texts may have different sequences leading to this theorem. It will be interesting to learn if different sequences stem from different logical beginnings.

3. Consider the following sequence of definitions and theorems:

   T1. If two straight lines intersect, the vertical angles are congruent.
   D. Two straight lines are parallel if and only if, when cut by a transversal, the corresponding angles are congruent.
   T2. If two parallel lines are cut by a transversal, then the alternate interior angles are congruent.
   T3. If two parallel lines are cut by a transversal, then the interior angles on the transversal are supplementary.
   T4. The sum of the measures of the interior angles of a triangle is 180°.

   In the proof of T1 we make use of the postulate "if equal quantities are subtracted from equal quantities, then the remainders are equal," or, "if $a = b$ and if $c = d$ then $a - c = b - d$." Suppose now that the postulate reads that the "left-hand side is less than the right-hand side," or that "$a - c < b - d$," what effect, if any, does this new postulate have on the statements T1, T2, T3, T4? Here we experience the logical consequence of a postulate. One may even run into difficulty and see that this new postulate on subtraction of equals causes the entire set of postulates (which we have been assuming known from geometry) to be inconsistent.

6.3. The Nature of Deductive Thinking; The Syllogism

To teach for transfer and at the same time to develop in the student a deeper understanding of proof in mathematics, one must be thoroughly informed and skilled in the use of one of the main tools of deduction: the *syllogism.* The syllogism is one of the patterns of how people make deductions. For example, if we make the statement

(a) all birds have feathers

along with the assertion that

(b) all robins are birds

then our minds work in such fashion that we conclude that

(c) all robins have feathers.

The statement in (a) is called the major premise, that in (b) the minor premise, and that in (c) the conclusion. The terms "birds," "has feathers," and "robins" are called the middle, major, and minor terms, respectively. Some texts in geometry discuss the *syllogism* of classical logic as a pattern of deduction, while others introduce the more modern implication $p \rightarrow q$ and present in varying detail the deductive forms *modus ponens* (rule of detachment), the rule of the syllogism, the "rule for substitution of variables on a sentence," and combinations of these. All are important but the transfer value, or the real use of these forms, is evident only when the student sees these devices at work in argument. It is then a challenge to the teacher to prepare materials to demonstrate this use.

It has been found effective to introduce the syllogism first with nongeometric examples and then to ask the students to use syllogisms to construct arguments. For example, let us suppose that we wish to argue that a plan for serving milk in class should be adopted; we must then build a syllogism which will have for its conclusion

$\therefore$ A plan to serve milk in class should be adopted.

What premises will yield this statement as a conclusion? Premises which either can be supported or are accepted by general agreement. If a student says, "A plan of serving milk in class should be adopted because it is healthful," he is really suggesting this syllogism:

Any plan which promotes health should be adopted.
The serving of milk in class is a plan which promotes health.
$\therefore$ A plan for serving milk in class should be adopted.

Now one who is attempting to argue this conclusion already has in the above statements a valid syllogism but the conclusion will not be accepted unless the premises are also accepted. Hence, the arguer must make sure that his audience is in accord with the premises and, sometimes, to accomplish this, items of support must be given. Hence, the arguer might arrange his "case" according to this syllogistic outline:

> Any plan which promotes health should be adopted.
> The serving of milk in class is a plan which promotes health for
>
> > It will insure the consumption of more minerals and vitamins.
> > It will provide sustenance for students who awakened too late to eat breakfast.
> > It will encourage a certain amount of informality which often eases tension.
> > It will prevent a mid-morning "slump."
> > ∴ The serving of milk in class is a plan which should be adopted.

After the student has constructed a few such arguments, he soon realizes what argument really is. The major premise is often a well-accepted statement and usually the minor premise is the one which requires support. Sometimes both need support. In the example given above, different arguers might have suggested different premises, but the conclusion would be the same. Hence, a debater, a lecturer, an attorney, or a writer might have several syllogisms leading to the same conclusion. Indeed, a skillful arguer may have all of his syllogisms lead pointedly to the same conclusion if he is trying to drive a point home.

Also the writer or arguer will not simply state a syllogism and list his points of support. Rather he will clothe his syllogism with connecting threads of prose so it reads in a more smooth and accepted manner. The above syllogistic argument might be like this:

> Ladies and gentlemen, I have come to feel that we should adopt the plan of serving milk in class. Surely any program which promotes health should be arranged for, and the serving of milk in class does. Not only will this plan *insure* the consumption of more minerals and vitamins—and this would be enough argument —but it will also provide sustenance for students who did not awaken early enough to go to breakfast. And even if they have had breakfast, milk served in class would prevent that well-known "morning slump." Moreover, it might well help to provide a certain amount of informality which, in itself, eases tension and hence promotes health. I believe that the adoption of this plan is most urgent!

A consideration of the syllogism in geometry helps the student to look into the entire field of *argumentation* and can furnish glimpses of areas hitherto unexplored by the student and perhaps by the teacher, too. Students and teachers may well consult books on argumentation and debate on the uses of the syllogism, fallacies which may arise, and tests for the validity of syllogisms. Exercises for the students may include the construction of syllogisms to argue conclusions, the finding of syllogisms in prose arguments, the detecting of syllogisms in advertisements and the test of the validity of syllogisms.

Exercises

1. Write the major and minor premises and the conclusion of the valid syllogism which seem to be suggested by these statements.
   (a) He should be elected for he is a loyal member. (Answer: Anyone who is a loyal member should be elected. He is a loyal member. Therefore, he should be elected.)
   (b) Fruit should be served at every meal because it contains minerals.
   (c) $\triangle ABC$ and DEF are congruent for they have $SAS \cong SAS$.
   (d) $a < b$ for $a + c < b + c$
   (e) He should be commended for he saved the girl.
2. See if you can devise another pair of premises which will argue that "The serving of milk in class is a plan which should be adopted."
3. Write a syllogism with a supported premise to argue that the school sessions should last 200 days. You may make up supporting items, but have them realistic.
4. Write a syllogism with supported premise that no one should be excused from examinations. You may invent supporting items.
5. Write a syllogism with a few supporting items to argue that one should cooperate in the blood donor program. Now attempt to rewrite this so it is presented in a smooth prose style, but so that it still contains the syllogism. The result might appear like an editorial in a newspaper or journal.
6. Find a short editoral in the newspaper and see if you can detect syllogistic structure. Is there a major premise? A minor premise? A conclusion?
7. Select several advertisements in which you detect a syllogism. Example: What is the syllogism involved in "Use Purity soap; it floats"?

Experiences as those suggested above make the student aware of how much the syllogism is used in deductive argument. A brief discussion on the syllogism and a study of examples at frequent times will do much to help pave the way for the use of the syllogism in geometry and for student recognition of this form in his daily nongeometric experience.

Once the structure of the syllogism is known, the student should see that this is the very basis of geometry reasoning. Indeed, the student should be asked to write several proofs of geometry theorems to show this syllogistic structure. In the proof of the theorem that a diagonal divides a parallelogram (Fig. 6.13) into two congruent triangles, we make use of congruency by $ASA \cong ASA$, but the syllogistic form of the proof is

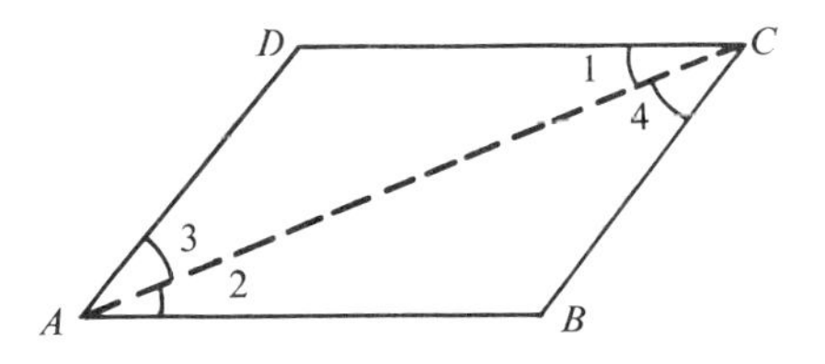

Figure 6.13

Any two triangles which have $\angle S \angle \cong \angle S \angle$ are congruent. Triangles *ABC* and *CDA* have $\angle S \angle \cong \angle S \angle$, for

(1) $\angle 2 \cong \angle 1$ (alternate interior angles)

(2) $\angle 4 \cong \angle 3$ (alternate interior angles)

(3) $\overline{AC} \cong \overline{CA}$ (identical)

$\therefore$ Triangles *ABC* and *CDA* are congruent.

Now each of the items (1), (2), and (3) could be expressed as conclusions of syllogisms since for (1), for example, we could have

> Any two angles which are alternate-interior angles made by a transversal cutting two parallel lines are congruent. $\angle 2$ and $\angle 1$ are angles which are alternate-interior and are made by a transversal cutting two parallel lines.
> $\therefore$ $\angle 2$ and $\angle 1$ are congruent.

Hence the geometric argument, in syllogistic style, can become about as complicated as one wishes. Some proofs in geometry are clearly composed of several syllogisms in which the conclusion of the first becomes the minor premise of the second.

The benefits of some attention to the syllogistic study of proofs are very great; it emphasizes that mathematics is the "science of necessary conclusions."

The study of the syllogism leads one to other forms of argumentation; geometry may be the only medium through which the student learns of some of these. It has been said that the teacher "is the student's only evidence outside the text that a great humanity exists,"[20] and the "humanity," in this case, may be methods of thought and argumentation which man has devised. Other forms to argue conclusions (called "constructive methods"

---

[20]William Arrowsmith, "The Future of Teaching," *Journal of Higher Education*," XXXVIII (March, 1967), 134.

by some) are those of analogy, *a fortiori*, and generalization or induction, The method of generalization is familiar; examples are used. The boy who urges that he should be permitted to go to the game "because all the other boys are going" is making use of *generalization*, One who argues, "James' dad is letting him go, so why can't I?" is using *analogy*. The term *a fortiori* describes the kind of analogy which is stronger even than that required for the particular desired conclusion in question. When Tommy, aged 16 suggests that "he should be allowed to drive the tractor because his brother, only 14, already does," he is using an *a fortiori* argument. There are several *a fortiori* arguments in the Bible.

To argue a conclusion by "destructive methods" one can use the syllogism, *reductio ad absurdum*, the method of residues, and the dilemma. Although it is not our purpose to elaborate on all these methods, experience teaches that students find much interest not only in devising examples of the various kinds of argument listed above, but also in originating dilemmas and in detecting the use of *reductio ad absurdum* in editorials, sermons, and, sometimes, in everyday speech.

A brief study of argumentation in the course of geometry pays dividends in helping the student see what argument and proof really are, one claim made by those who support the teaching of geometry. It is an effective way to transfer some of the characteristics of mathematical study to other fields of thought. With such pursuit the student becomes a more intelligent reader, listener, and thinker for he has learned more about argument than that connected merely with a geometric proof, although this is important too. It is urged that teachers not overlook the resources in texts on argumentation and debate to introduce students to this part of our practical and intellectual experience.

Geometry can provide an introduction to fallacies in argument as well. Indeed proofs handed in by students sometimes contain fallacies of "reasoning in a circle" or "it does not follow" which can be used to introduce *circulo probando* or *petitio principii* and *non sequitur*. Arguing from the converse and from the inverse are other kinds of fallacies, and, of course, the use of false analogy may be included in this category. Sometimes, in order to attack the argument made by an individual, the man himself is attacked. Such a fallacy which attempts to discredit the arguer, although the conclusion may be a valid one, is called the fallacy of *ad hominem*. A further study of fallacies is a valuable contribution to teaching for transfer in geometry.

The syllogism and other forms of argumentation are concepts and tools which have had simple beginnings and which can be easily understood by the student if permitted to develop as they did in classical logic. Through the invention of symbolic logic these forms take on new and refined stature along with other methods of argument. To introduce the refined and symbol-filled ideas first, as, let us say the form

$$[p \wedge (p \rightarrow q)] \rightarrow q$$

or

$$\begin{array}{ll} \text{premises} & \left\{\begin{array}{l} p \rightarrow q \\ p \end{array}\right. \\ \hline \text{conclusion} & \therefore q \end{array}$$

which we know as *modus ponens*, and prove that this is a tautology, with no previous experiences of the significance of it may tend to hide in the details of logical algebra the real and useful nature of these logical forms. Indeed, symbolic logic is a tool by which we can summarize, condense, and examine further the Greek-invented methods with an inward look. Symbolic logic should climax the study of argumentation and hence become the basis and tool for further analysis. Good expositions of such subjects are found in various sources, however, the real-life character of the syllogism and other forms, comes from their use in situations. The intellectual pursuit and invitation to devise methods which might replace the syllogism stem from symbolic or logical algebra. Lewis and Langford write

> As a result of improved methods of notation, we find ourselves in a lively period of new discoveries: an old subject, which has been comparatively stagnant for centuries, has taken on new life. We stand today with respect to logic, where the age of Leibnitz and Newton stood with respect to what can be accomplished in terms of number. . . [21]

In this subject which has taken on new life we now have the tool to show that the method of *reductio ad absurdum* is every bit as valid as that of the syllogism, a far cry from the suggestion made in a text of a generation ago that the indirect method was considered as a "kind of last resort" in proof.[22] Furthermore, proof (itself) of a problem in geometry, after the various parts have been assigned labels as used in symbolic logic, is now regarded by some writers as establishing that the symbolic logic statement of the implication to be tested is a *tautology.*

Exercises

1. Given: $\ell \parallel \ell'$ (i.e., $\angle 4 = \angle 3$)
To show: $m\angle 2 + m\angle 3 = 180$

---

[21] J. R. Newman, ed., *The World of Mathematics,* 4 vols. (New York: Simon and Schuster, Inc., 1956), p. 1860.

[22] George Wentworth and D. E. Smith, *Plane and Solid Geometry* (New York: Ginn and Company, 1913); p. 83.

First prove in the usual way and then write the argument with one syllogism with a supported minor premise.

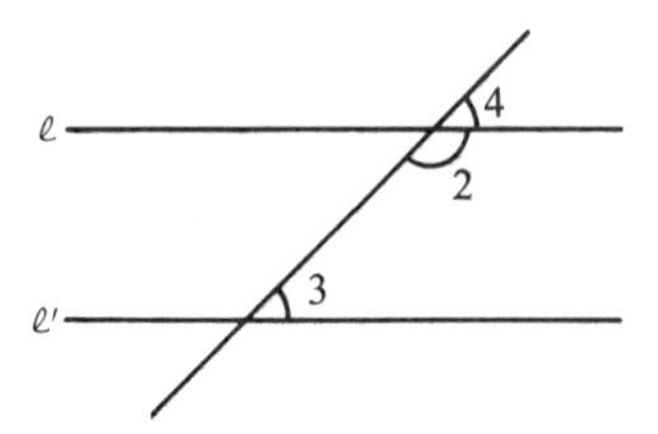

2. Given: *ABCD* a parallelogram
To show: $\triangle DEC = \triangle BEA$.
Write the proof in the usual form and write a one-syllogism argument with supported minor premise.

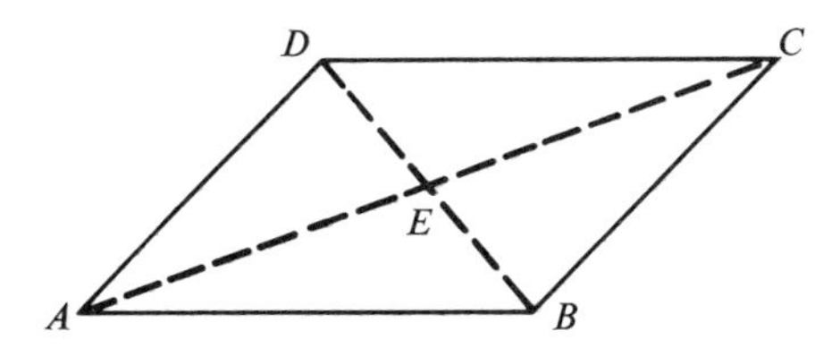

3. Here are some "constructive" arguments. How would you classify each?
(a) John's father said he could go—so may I go too?
(b) All good Republicans went to the meeting. George is a good Republican so I suppose he went too.
(c) The householder gave the maid a raise so I would expect that the matron received one also.
(d) All our class members are going to the party—can't I go too?
(e) If $a > b$ but $b > c$ then surely $a > c$.

4. Read the Scriptures, Luke 12: 28: "But if that is how God clothes the grass, which is growing in the field today, and tomorrow is thrown on the stove, how much more will he clothe you." What kind of argument is this? Do you know of other examples of this kind of argument in the Bible?

5. Here are some examples of "destructive" forms of argument or situations; classify them.
(a) If I wear the tie the students will be amused, but if I don't my uncle will be disappointed.
(b) "Shall we pay tribute to Caesar?"
(c) (Against representative government) Does a captain consult his crew everytime he wishes to change his course?
(d) "Anyone who is a good Democrat should be elected! Yes, I agree and you say that Harry Jones should be elected. But this does not follow because Harry Jones is not a good Democrat. He . . ."

6. Make up a nongeometric example of argument which is syllogistic; which is an analogy; which is an *a fortiori* form; which is generalization.

7. What kind of *fallacy* is found in each of these arguments?

(a) Mr. Jones is not a good contractor for the new building. He drives a Buick and, anyway, I do not like his laugh.

(b) Designing people are not to be trusted. Mrs. Henry is a designer. Hence Mrs. Henry is not to be trusted.

(c) Given: $\ell \parallel \ell'$, m a transversal.

To show: alternate interior angles are congruent

Proof: Let us prove that

$$\angle 1 \cong \angle 2$$

Now $m\angle 1 + m\angle 3 = 180$

and $m\angle 2 + m\angle 4 = 180$

$\therefore$ $m\angle 1 + m\angle 3 = m\angle 2 + m\angle 4$

But $m\angle 3 = m\angle 4$

$\therefore$ $m\angle 1 = m\angle 2$ (by subtraction)

$\therefore$ $\angle 1 \cong \angle 2$

Therefore, if two parallel lines are cut by a transversal, the alternate-interior angles are congruent (this came from a class exercise in geometry.) Q.E.D.

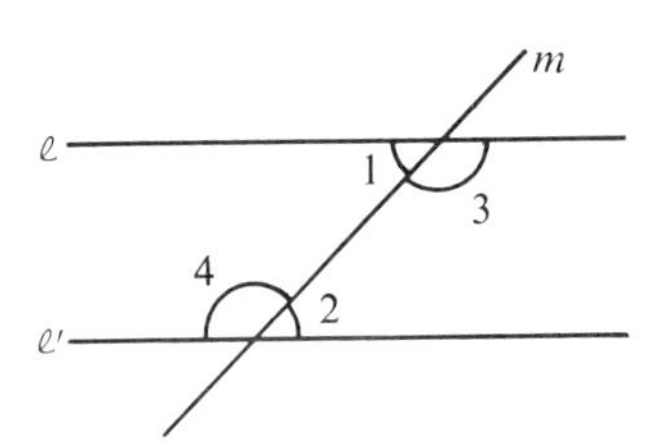

(d) In a picture by the famous French comedian, Sacha Guitry, some thieves are arguing over division of seven pearls . . . one says . . . "I will keep three." The man on his right says, "How come you keep three?" "Because I am the leader." "Oh. But how come you are the leader?" "Because I have more pearls."[23]

(e) "China is better than earthenware becase it is made of better clay; and the clay is better, because china is better than earthenware."[24]

---

[23] W. Ward Fearnside and William B. Holther, *Fallacy—The Counterfeit of Argument* (Englewood Cliffs: Prentice-Hall, Inc., 1959), p. 167.

[24] A. Wolf, *Textbook of Logic* (London: George Allen and Unwin, Ltd., 1961), p. 363.

(f) "Of course she doesn't dislike me. She told me that she doesn't. And I know she wouldn't lie to me about it, for she always tells the truth to people she likes."[25]

(g) Nothing is better than wisdom. Dry bread is better than nothing. Therefore, dry bread is better than wisdom.

6.4. Indirect Proofs

Indirect proof does not follow the syllogistic pattern; rather it is a destructive type method which proves that a certain result *is* valid by showing that all alternative results are *not*. Indeed, at one time indirect methods were considered less desirable than direct methods of proof and were used only when there was no other way. It can be proved in symbolic logic, however, that the indirect methods are logically equivalent to syllogistic methods and that, therefore, they are just as valid as the direct method. It is instructive to prove theorems in geometry both by direct and indirect methods if the theorem can be readily done both ways, for not only does this broaden the skill and concept of proof, but it also encourages the student to look with favor on both methods. There are several methods of indirect proof; two of these are the *reductio ad absurdum* and the *contrapositive* method.

The method of *reductio ad absurdum* appears in Euclid's *Elements* and it must have been a contribution of the Pythagoreans. The pattern of *reductio ad absurdum* is this:

(a) we wish to prove that $p \rightarrow q$;

(b) we state that either q is true or $\sim q$(not q) is true;

(c) we assume $\sim q$ is true;

(d) we then attempt to show that either reasoning from $\sim q$ along with p produces a contradiction or an "absurdity," or, we show that $\sim q \rightarrow \sim p$ and thus to have p (given) and $\sim p$ (deduced) simultaneously is a contradiction. We say in this case that "we have a contradiction of the hypothesis."

(e) if we are successful in (d) then we state that q remains as the *valid* alternate, and

(f) $\therefore \quad p \rightarrow q$ by *reductio ad absurdum.*

The method of *reductio ad absurdum* is based on the laws of Aristotle which may be stated as:

I. Whatever is, is (Law of Identity)

II. A thing either is or it isn't (Law of Excluded Middle)

---

[25]Stephen F. Barker, *The Elements of Logic* (New York: McGraw-Hill Book Company, 1965), p. 194.

III. A thing cannot both be and not be at the same time (Law of Contradiction)

The beginning of the *reductio ad absurdum* method uses Law II above and Law III is involved at the close. In high school teaching, however, it must either be postulated or proved by symbolic logic to accept the results of this method. In its simplest form this method of proof, to show that $p \rightarrow q$, ends by showing $\sim(p \wedge \sim q)$. Indeed

$$\sim(p \wedge \sim q) \leftrightarrow \sim p \vee q \leftrightarrow p \rightarrow q$$

and *reductio ad absurdum* is just as valid as the direct method, $p \rightarrow q$.

One should not wait until the indirect methods are needed to prove theorems for then the student, as we have indicated before, has two difficulties: the newness of the theorem and unfamiliarity with the method of proof. He may feel that the *reductio ad absurdum* method is something less than desirable. Indirect methods should be introduced as early as possible; this can be done easily in the simple parallel-line theorems:

Definition: Two lines are parallel if and only if, on being cut by a transversal, the corresponding angles are congruent.

Theorem: If two parallel lines are cut by a transversal the alternate-interior angles are congruent.

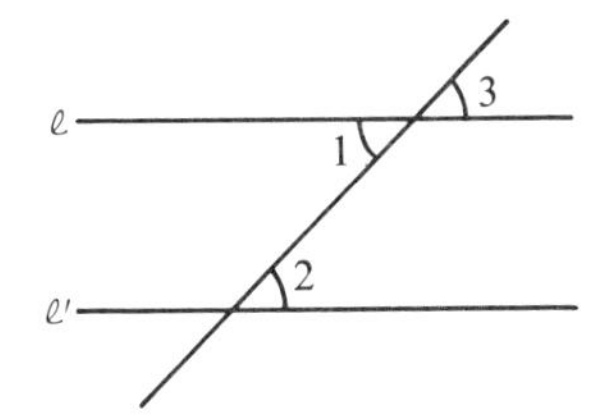

Given: $\ell \parallel \ell^1$

To show: $\angle 1 \cong \angle 2$

Proof: Either $\angle 1 \cong \angle 2$

or $\angle 1 \ncong \angle 2$ (Aristotle Law I)

Assume: $\angle 1 \ncong \angle 2$

Now: $\angle 1 \cong \angle 3$ (vertical angles)

$\therefore$ $\angle 2 \ncong \angle 3$ (logical identity)

$\therefore$ $\ell \nparallel \ell^1$ (corr $\angle s$ not $\cong$)

But this contradicts the hypothesis that $\ell \parallel \ell^1$ and we cannot have both $\ell \parallel \ell^1$ and $\ell \parallel \ell^1$ by Aristotle Law III.

$\therefore$ $\angle 1 \cong \angle 2$ (by *reductio ad absurdum*) Q.E.D

In the above proof we have contradicted the hypothesis. Another way to produce a contradiction would be as follows:

Assume:	$\angle 1 \not\cong \angle 2$	
But	$\angle 2 \cong \angle 3$	(since $\ell \parallel \ell^1$, given)
$\therefore$	$\angle 1 \not\cong \angle 3$	(logical identity)
But	$\angle 1 \cong \angle 3$	(vertical angles)

and we have a contradiction or an "absurd" situation. The other possibility is valid, or,

$\therefore \angle 1 \cong \angle 2$ (*reductio ad absurdum*) Q.E.D.

Sometimes students present as *reductio ad absurdum* in method a proof like the following:

Assume:	$\angle 1 \not\cong \angle 2$	
Btu	$\angle 2 \cong \angle 3$	(since $\ell \parallel \ell^1$, given)
and	$\angle 1 \cong \angle 3$	(vertical angles)
and	$\therefore \angle 1 \cong \angle 2$	(logical identity)

But this contradicts the assumption (! !).

$\therefore \angle 1 \not\cong \angle 2.$

This is not *reductio ad absurdum* at all, but it is really a direct proof contained in the three lines in the frame; indeed, the assumption that $\angle 1 \not\cong \angle 2$ is not at all necessary here. Such a proof should not be regarded as *reductio ad absurdum* in character.

Students and teachers alike should watch for nongeometric examples of *reductio ad absurdum* proofs. The lawyer who proves that his client was not on the scene of a robbery, but was several miles away, is really using reductio ad absurdum: either the accused was there or he wasn't; assume he was; is his arm long enough to reach where the crime was committed? absurd! Speakers will often seemingly agree with a proposal but carry if further to describe how absurd it could become. Here again is teaching for transfer; opportunities for students to explore theorems to see if they can be proved in different ways.

Exercises

1. With the same definition of parallelism as above, prove by *reductio ad absurdum* that if two lines are cut by a transversal then the interior angles in the same side of the transversal are supplementary.
2. Euclid's original Postulate 5 was stated as follows: If two lines are cut by a transversal and if the interior angles in the same side of the transversal

have a sum (of measures) *less* than two right angles, then the two lines, if produced, meet on that side on which the sum is less than two right angles. Now prove by *reductio ad absurdum* that if two parallel lines are cut by a transversal then the interior angles in the same side of the transversal are supplementary. Note that the interior angles must have the sum *less than* two right angles before Euclid's Postulate 5 can be used. You may use the usual equality and inequality axioms.

3. In the first example of proof by *reductio ad absurdum*, let these letters represent the following statements:

   p: $\ell \parallel \ell^1$

   q: $\angle 1 \cong \angle 2$ (alternate interior angles congruent)

   r: $\angle 1 \cong \angle 3$ (vertical angles congruent)

   s: $\angle 2 \cong \angle 3$ (corresponding angles congruent)

   Now the structure of the argument in the example is that

   $$[(\sim q \wedge r) \rightarrow \sim s] \wedge (\sim s \rightarrow \sim p)$$

   and this along with p is contradictory and therefore implies the statement q. Written more fully the argument is

   $$\{[(\sim q \wedge r) \rightarrow \sim s] \wedge (\sim s \rightarrow \sim p)\} \wedge p \rightarrow q.$$

   Use logical algebra to show that this logical expression is a tautology. First, one can use the transitive law in the conjunction enclosed in the braces.

4. In the second example of *reductio ad absurdum*, write the pattern in terms of logical symbols and see if you can show that it is equivalent to $p \rightarrow q$.

5. Consult editorials in newspapers and periodicals and listen to sermons or everyday conversation to find nongeometric examples of *reductio ad absurdum* methods of argument.

6. The exterior-angle theorem in geometry states that an exterior angle of a triangle is greater than either opposite interior angle. Use this and Euclid's Postulate 5 to prove by *reductio ad absurdum* that if the alternate interior angles of two lines cut by a transversal are congruent then the lines are parallel.

7. The theorem that if two straight lines intersect then the vertical angles are congruent can be written in this way:

   Given: a straight line $\wedge$ ℓ^1 a straight line intersecting at P

   To show: $\angle 1 \cong \angle 2$

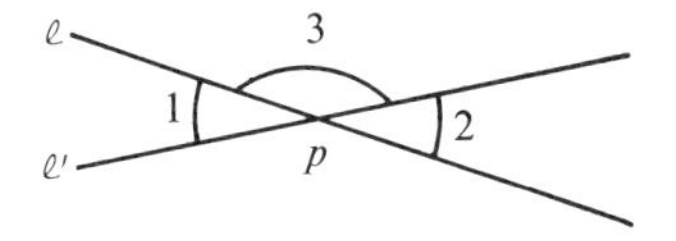

Recalling that a conjunction is true only when both parts are true, prove this theorem by the method of *reductio ad absurdum.*

8. Prove by *reductio ad absurdum* that if a^2 is an even natural number then a is even.

9. Prove by *reductio ad absurdum* that if the natural number a^2 is a multiple of three then a is a multiple of three.

10. Prove by reductio ad absurdum that

$$\phi \subset A, (A \neq \phi)$$

or that the null set is a proper subset of any set A, where A is not the null set.

The use of the *contrapositive* of an implication is of more recent origin. The first use of proving a theorem by the method of proving its contrapositive was found by Lazar to have been made by Hauber in 1829.[26] Since that time use of the contrapositive method has been growing. This pattern of proof is less involved than that of *reductio ad absurdum* although the productive steps in the proofs are similar and may be exactly the same. The contrapositive of $p \rightarrow q$ is $\sim q \rightarrow \sim p$ and these two implications are readily proved to be logically equivalent, hence the contrapositive method in proof is to show $p \rightarrow q$ by proving the equivalent $\sim q \rightarrow \sim p$. This explains why the contrapositive method is called an indirect method.

To illustrate the use of the contrapositive method, let us consider this problem

Given: $\ell \parallel \ell^1$ (p)

(that is, $\angle 2 \cong \angle 3$)

To show: $\angle 1 \cong \angle 2$ (q)

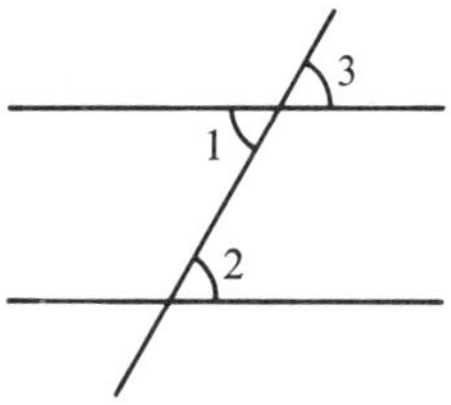

Proof: To use the contrapositve method we must prove that

$$\sim q \rightarrow \sim p$$

or, we must show that

[26] Nathan Lazar, *The Importance of Certain Concepts and Laws of Logic for the Study and Teaching of Geometry* (Menasha, Wisconsin: George Banta Publishing Company, 1938), p. 29.

$$\angle 1 \not\cong \angle 2 \rightarrow \ell \nparallel \ell^1.$$

Proof: $\angle 1 \not\cong \angle 2$ (assumption)

But $\angle 1 \cong \angle 3$ (vertical angles)

$\therefore$ $\angle 2 \not\cong \angle 3$ (logical identity)

$\therefore$ $\ell \nparallel \ell^1$ (corresponding angles are not congruent)

$\therefore$ $\ell \parallel \ell^1 \rightarrow \angle 1 \cong \angle 2$ (contrapositive law)

Q.E.D.

In the geometry class the students might well discover either by logical algebra or by Euler diagrams the logical equivalence of $p \rightarrow q$ and $\sim q \rightarrow \sim p$. Indeed, if $p \rightarrow q$ be denoted by the diagram at the right, meaning loosely that "every p is a q," we can consider many questions:

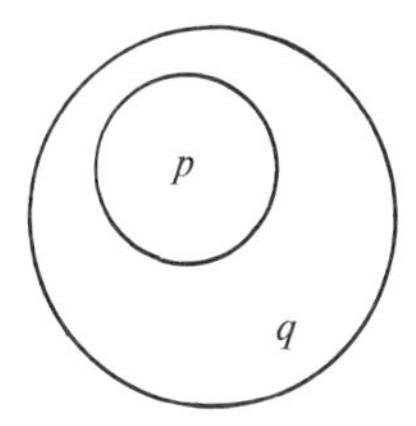

Does

(a) $q \rightarrow p$? (d) $\sim p \rightarrow q$?

(b) $\sim p \rightarrow \sim q$? (e) $\sim p \rightarrow \sim q$?

(c) $\sim q \rightarrow \sim p$? (f) $p \rightarrow \sim q$?

There is one which always follows: if $p \rightarrow q$ then $\sim q \rightarrow \sim p$ and from the diagram we see that

$$p \rightarrow q \leftrightarrow \sim q \rightarrow \sim p$$

This relation can be postulated and used in proof. Later, with the development of logical algebra, the above may be reclassified as a theorem.

The use of the contrapositive is a must in geometry for it not only deepens the students idea of proof, but the student can manufacture new theorems. For example, the contrapositive of the theorem

> If two lines cut by a transversal have the interior angles on the same side of the transversal supplementary then thc lines are parallel

is

> If two lines are not parallel then the interior angles . . . are not supplementary

which could well read,

> if two lines and a transversal form a triangle then the sum of the interior angles is not 180°

or

> the sum of any two interior angles of a triangle is not two right angles.

The early use of the contrapositive method provides resources for research and discovery; of course, one should not wait until a theorem arises for which there is no other method of proof.

Exercises

1. Prove by the use of the contrapositive method the theorem that if two parallel lines are cut by a transversal, then the interior angles on the same side of the transversal are supplementary.

2. Prove by the contrapositive method that if a^2 is a natural number and is even, then a is even.

3. How are contrapositive and *reductio ad absurdum* methods similar? How different?

4. Prove by logical algebra and also by truth (or existence) tables that

(a) $\sim(p \wedge \sim q) \leftrightarrow p \rightarrow q$
(b) $\sim q \rightarrow \sim p \leftrightarrow p \rightarrow q$

(a) represents simple *reductio ad absurdum* and (b) summarizes the contrapositive method.

5. Prove by logical algebra and also truth tables that

$$[(p \rightarrow q) \wedge \sim q)] \rightarrow \sim p$$

This form is known as *modus tollens* and is another way to look at indirect proof.

One should not leave a discussion of contrapositive methods without mentioning the value of student exploration of the contrapositive of theorems with several elements in the hypothesis and several in the conclusion. Also, care must be used lest logical errors be made. Consider first the implication

$$p \wedge q \rightarrow r$$

An example of a theorem which has this structure is the well-known vertical-angle theorem:

$$\ell \text{ straight} \wedge \ell^1 \text{ straight} \rightarrow \angle 1 \cong \angle 2$$

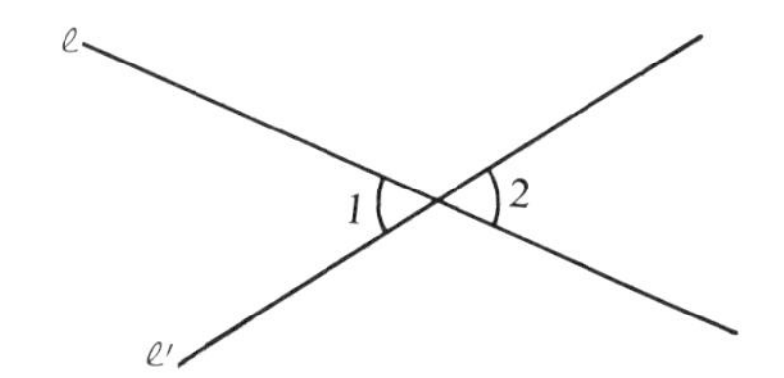

Without elaborating we see that there are three contrapositives, two of which are called *partial contrapositives*:

$$\left.\begin{array}{l} p \wedge {}^{\sim}r \rightarrow {}^{\sim}q \\ {}^{\sim}r \wedge q \rightarrow {}^{\sim}p \end{array}\right\} \quad \text{(partial contrapositives)}$$

$$ {}^{\sim}r \rightarrow {}^{\sim}(p \wedge q) \quad \text{(complete contrapositive)}$$

It can be shown that all of these contrapositive are equivalent to $p \wedge q \rightarrow r$ and, therefore, the implication $p \wedge q \rightarrow r$ can be proved by substantiating any one of the three contrapositives.

Exercises

1. Prove by logical algebra and by truth table that (a) $p \wedge {}^{\sim}r \rightarrow {}^{\sim}q$ and (b) ${}^{\sim}r \rightarrow {}^{\sim}(p \wedge q)$ are each equivalent to $(p \wedge q) \rightarrow r$.
2. Prove that if ℓ *is* a straight line and if $\angle 1 = \angle 2$, then ℓ^1, then ℓ^1 is *not* a straight line. This is proving that one of the partial contrapositives is valid.
3. Prove that $p \wedge q \wedge r \rightarrow s$ is equivalent to $p \wedge q \wedge {}^{\sim}s \rightarrow {}^{\sim}r$.

When studying contrapositives it is interesting to investigate implications, as

$$p \wedge q \rightarrow r \wedge s$$

Are partial contrapositives equivalent to the original implications? What partial contrapositives can be formed which are equivalent to the original implication? This problem can become the center of some research on the part of interested students. In geometry, the use of the contrapositive method is more effective and practical if the theorems have only one element in the conclusion; if there are several one can simply state several theorems and examine each by contrapositive methods. For example, the theorem

> If a line segment joins the (p) midpoint of one side of a triangle and also the (q) midpoint of another side of the triangle, then (r) is parallel to the third side and (s) has a length equal to half that of the third side,

which takes on the symbolic form

$$p \wedge q \rightarrow r \wedge s,$$

can be broken into the two theorems

$$p \wedge p \rightarrow r$$

$$p \wedge q \rightarrow s$$

and then the *two* partial contrapositives

$$p \wedge {}^{\sim}r \rightarrow {}^{\sim}q \text{ and } p \wedge {}^{\sim}s \rightarrow {}^{\sim}q$$

proved will be equivalent to proving the original theorem. The reader may wish to explore further the contrapositives of more complicated theorems. In the secondary school, simple theorems are most often used. However, there exist here many opportunities for developing and studying new theorems. It is surprising, too, that some theorems, hitherto considered as independent, are related by being the contrapositive of each other.

Exercises

1. Consider the theorem:

   Given: $\overline{CA} \cong \overline{CB}$

   $\angle 1 \cong \angle 2$

   To prove: $\overline{AD} \cong \overline{BD}$

   $\angle 3 = \angle 4$

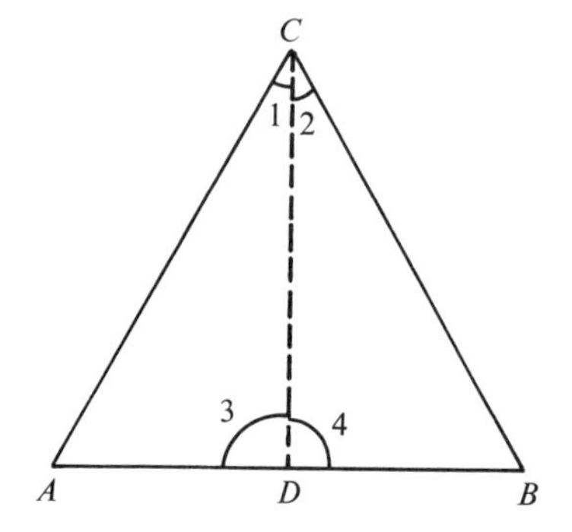

   State this as two theorems and then state a partial contrapositive for each.

2. Had Euclid known of the contrapositive, what new theorem could have come directly from his (a) Postulate 5, (b) his theorem that if alternate-interior angles are congruent then the lines are parallel?

3. In a previous exercise it was proved that $p \wedge q \rightarrow r$ and its partial contrapositive $p \wedge {}^{\sim}r \rightarrow {}^{\sim}q$ are equivalent, Study the partial contrapositives of $p \wedge q \rightarrow r \wedge s$ for equivalence with $p \wedge q \rightarrow r \wedge s$.

   (Ans. $p \wedge {}^{\sim}r \rightarrow {}^{\sim}q \wedge s$ is not equivalent. Is $p \wedge {}^{\sim}(r \wedge s) \rightarrow {}^{\sim}q$?)

6.5. Converses and Inverses

The preceding section dealt with the rich sources of new theorems and investigations provided by the study of contrapositives of theorems. Perhaps more interesting are examinations of *converses* and *inverses* of theorems, for one may not be as confident of the validity of these as he is of certain partial contrapositives of theorems. The partial contrapositive $p \wedge {\sim}r \rightarrow {\sim}q$ is a valid statement if the original implication is valid, but the partial converse, $p \wedge r \rightarrow q$ may not be.

One need not dwell long on the definitions of converses. The implication

$$p \wedge q \rightarrow r \wedge s$$

has the *complete* converse

$$r \wedge s \rightarrow p \wedge q$$

and four *partial* converses

$$p \wedge r \rightarrow q \wedge s$$
$$p \wedge s \rightarrow r \wedge q$$
$$r \wedge q \rightarrow p \wedge s$$
$$s \wedge q \rightarrow r \wedge p.$$

There is no assurance that if a theorem is true (or valid) then any of its converses will be true; indeed, this constitutes the challenge of research on converses.

Even in the subject of converses the development can grow from the study and suggestions of students. A guide sheet used during the study of parallels might include exercises like this:

1. We have just proved that $\ell \parallel \ell^1 \rightarrow \angle 1 \cong \angle 2$. Is it possible to prove that $\angle 1 \cong \angle 2 \rightarrow \ell \parallel \ell^1$? Try it.

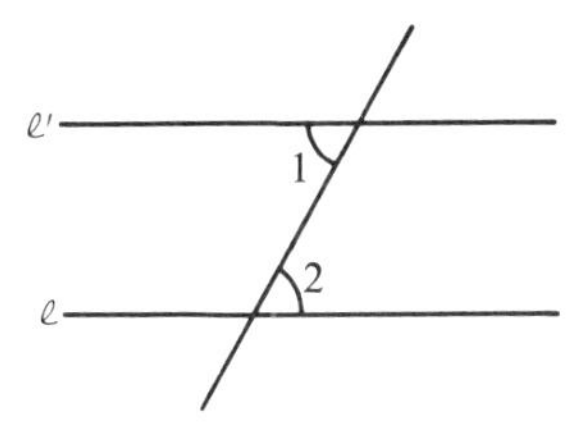

2. If you are successful in Exercise 1, then we have *two* theorems:

$$\ell \parallel \ell^1 \rightarrow \angle 1 \cong \angle 2 \text{ and } \angle 1 \cong \angle 2 \rightarrow \ell \parallel \ell^1$$

or

$$p \rightarrow q \text{ and } q \rightarrow p$$

What term might be used to describe the relation between these two theorems?

It is highly improbable that anyone will suggest "converse." Some may say "the backwards relation," the "reverse," the "invert," but here teacher can mention what others have come to call this relation: the converse.

Later the teacher can ask that if one has the statement $p \wedge q \rightarrow r$ then would the expression $p \wedge r \rightarrow q$ be considered a converse? how? why? If so, is there another converse?

Exercises

1. Write in symbolic form the converses of (a) $p \rightarrow q$; (b) $p \wedge q \rightarrow r$; (c) $p \wedge q \wedge r \rightarrow s \wedge t$

2. Consider the theorem $p \wedge q \rightarrow r \wedge s$ where

(p) D is the midpoint of $\overline{AC}$

(q) E is the midpoint of $\overline{BC}$

(r) $\overleftrightarrow{\mathbf{DE}}$ $\quad$ $\overleftrightarrow{\mathbf{AB}}$

(s) $\ell(\overline{DE}) = 1/2\ \ell(\overline{AB})$.

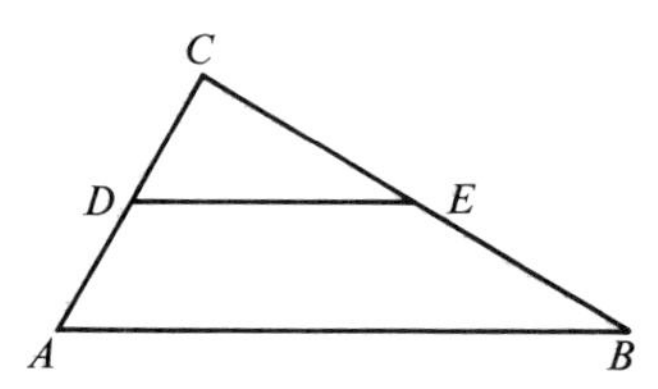

Although it may seem tedious to do this (mathematics is not a "spectator" activity) write the five converses of this theorem. See how many you can prove to be valid.

3. The theorem $p \wedge q \rightarrow r \wedge s$ where

p: $\overleftrightarrow{\mathbf{OP}}$ goes through the center of the circle

q: $\overleftrightarrow{\mathbf{OP}} \perp \overleftrightarrow{\mathbf{AB}}$

r: $\overline{AP} \cong \overline{BP}$

s: $\widehat{AQ} \cong \widehat{BQ}$

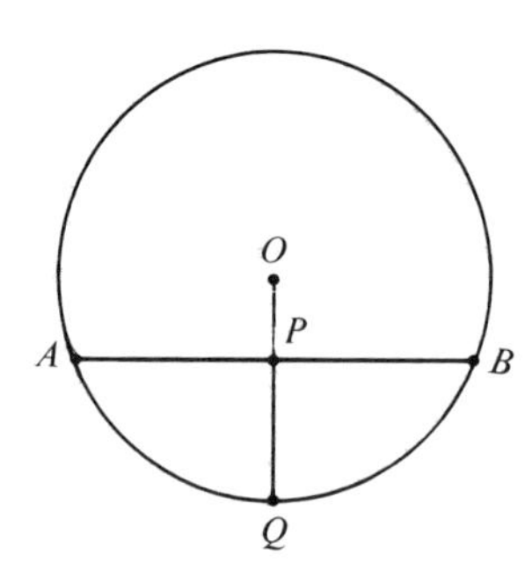

provides interesting converses also. Examine several of these to see if they can be deduced as theorems.

4. Consult a geometry text to find more theorems with two elements in the hypothesis and two elements in the conclusion from which converses can be formed. See if the converses can be proved as theorems.

5. Consider the theorem

$$p \wedge q \rightarrow r$$

where

p: $\angle C$ is a right *angle* in $\triangle ABC$

q: $\overleftrightarrow{\mathbf{CD}} \perp \overleftrightarrow{\mathbf{AB}}$

r: AD: DC:; DC: DB where AD means $\ell(\overline{AD})$, etc.

See if the partial converses are also valid implications.

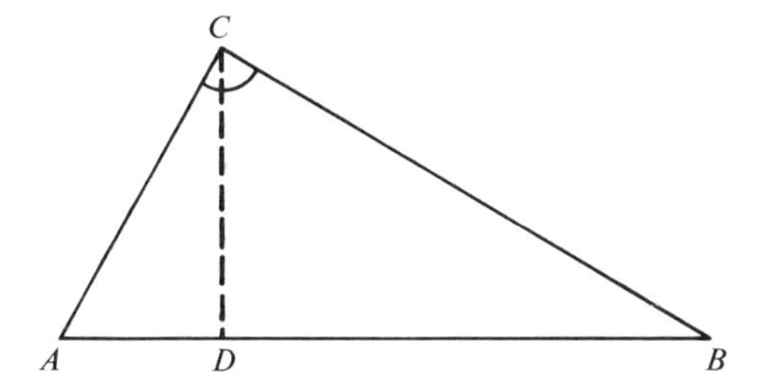

6. Consult a geometry text to find more theorems which contain two elements in hypothesis and one element in the conclusion.

7. Consult a geometry text to find several theorems with one element in the hypothesis and one in the conclusion. These are good to provide a beginning in the study of converses.

8. The Theorem of Desargues states that if two triangles are so situated that

   p: lines joining pairs of corresponding vertices are concurrent

   then

   q: the pairs of corresponding sides produced meet in points which are collinear.

   Write the converse of this theorem. Try to rewrite the theorem above so there are more elements in the hypothesis and in the conclusion.

9. The Theorem of Pascal describes this property:

If

p: a hexagon

q: is inscribed in a conic

and if

r: the pairs of opposite sides are produced to meet in points

then

s: the three points are collinear.

Written symbolically we have

$$p \wedge q \wedge r \rightarrow s$$

Suggest several converses of this theorem. One of these converses is used to construct a conic through five given points.

The study of converses in geometry helps the student to become interested in formulating and investigating converses in other branches of mathematics. Theorems in the calculus and analysis often have several parts in the hypothesis and are challenges in the study of converses.

The *inverse* of the theorem $p \rightarrow q$ is defined to be $\sim p \rightarrow \sim q$. This is recognized as the *contrapositive* of the *converse* of $p \rightarrow q$. Indeed, if the converse $p \rightarrow q$ is true, then so is the inverse of $p \rightarrow q$.

We shall not elaborate further on the subject of inverses. Each theorem, as

$$p \wedge q \rightarrow r \wedge s$$

has the *complete* inverse $\sim(p \wedge q) \rightarrow \sim(r \wedge s)$ or $\sim p \vee \sim q \rightarrow \sim r \vee \sim s$ and these *partial* inverses:

$$p \wedge \sim r \rightarrow \sim q \wedge s$$

$$p \wedge \sim s \rightarrow r \wedge \sim q$$

$$\sim r \wedge q \rightarrow \sim p \wedge s$$

$$\sim s \wedge q \rightarrow r \wedge \sim q$$

The inverses of any given theorem can become the subject of investigation by students.

Simple inverses are interesting because the public is often asked to believe that the inverse of an implication is always true. In the statement "If you use NEW toothpaste, your teeth will sparkle," the advertiser may be hoping that one will think that "If you do *not* use NEW toothpaste, your teeth will *not* sparkle." Students can bring advertisements to class and study their contrapositives, converses, and inverses.

Students and teachers interested in further reading on contrapositives, inverses, and converses might consult Lazar,[27] Butler and Wren,[28] and Wiseman.[29]

---

[27]Nathan Lazar, *op. cit.*

[28]Charles H. Butler and F. Lynwood Wren, *The Teaching of Secondary Mathematics* (New York: McGraw-Hill Book Company, Inc., 1960), pp. 481-5.

[29]John D. Wiseman, Jr., "Complex Contrapositives," *The Mathematics Teacher,* LVIII (April, 1965), 323-6.

6.6. The Use of Applications in Geometry

The study of applications of geometry and situations and uses from which more geometry can be developed help to give "sense experiences" to students. Edgar Dale, in the November 1966 Ohio State University School of Education *The News Letter*, urges that

> "We must never forget that children have eyes, ears, noses, and muscles, and that they like to use them" . . . (and) . . . in our present legitimate and necessary concern for building concepts we may neglect the sense experiences out of which abstractions and principles grow."[30]

Including applications in our courses in geometry helps to motivate the study of geometry for those students who derive a sense of purpose in mathematics by seeing mathematics at work. Too, the use of applications emphasizes in a very convincing way that in most instances it is the "outward call" which causes mathematics to develop. Logical organization comes later.

By applications of geometry we shall mean the use of theorems and facts of geometry to answer questions about physical things which may not always be objects which occupy space but may be drawings. In this sense, the measurements of distances, the determining of latitude position, geometric constructions, constructions of maps, the solution of problems in navigation, and the like, would all be applications of geometry.

In a brief section of this kind one cannot mention many applications, but a few will be discussed in addition to the usual circle square, pantograph, proportional dividers, and others, and these will not be presented in extreme detail. It is not intended that every one of these should be used in a year's course in geometry, but many of them can enliven the usual work in geometry.

The Egyptian Level

This instrument contrument contains an isosceles triangle, as shown in Fig. 6.14, such that the plumb bob points to the midpoint of the base if and only if the base is horizontal. The Egyptian level can be constructed by students and it should be exhibited along with the bubble-type level. The pyramids of Egypt were built with this kind of level.

In examining the geometry of the Egyptian level the student may be asked through the use of study-guide sheets or by other means (a) what facts should be considered as *assumed data* and (b) what *conjectures* should be

---

[30]Edgar Dale, "Growth through Rich Experience," *The News Letter* (School of Education, The Ohio State University, Columbus, Ohio), XXXII (November, 1966), p. 1

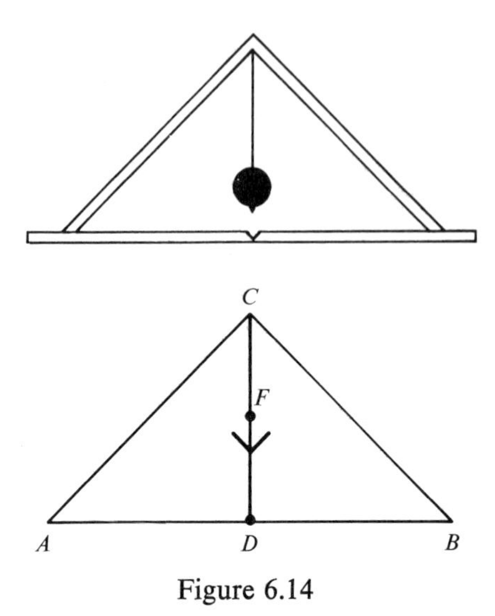

Figure 6.14

tested? We see here a severalfold opportunity for the student to formulate a mathematical problem from a physical situation. First of all, in an actual model of an Egyptian level, it is highly improbable that $\overline{AC} \cong \overline{BC}$ hence the method of the mathematician is to idealize the situation by assuming that these two sides are congruent; moreover, the point D is assumed to be the midpoint of $\overline{AB}$ and $\overleftrightarrow{\mathbf{CF}}$ is assumed to be vertical. We formulate as a *conjecture to be tested* the statement that if $\overleftrightarrow{\mathbf{CF}}$ intersects $\overleftrightarrow{\mathbf{AB}}$ at D then at D then $\overleftrightarrow{\mathbf{AB}}$ is horizontal. All physical problems must be treated by idealizing certain facets of the situation if they are to be studied mathematically; this method of attack should be emphasized in our teaching. Indeed, even in problems of falling bodies it is assumed (the situation is idealized) that $s = 16t^2$ describes the distance-time relation correctly. In examining physical situations mathematically one gains practice in selecting what can be regarded as *assumed data* and in detecting and formulating what *conjectures* are to be *tested*; the student gains insight into deciding what part of the problem he can idealize and what part he must compute or test. That the student usually cannot do this easily is supported by recalling the difficulty which many have as they attack word problems in algebra or in calculus, differential equations, and applied mathematics. There is real need here for a methodology which encourages these skills.

In classroom practice, the suggestions readily arise that $\overline{AC} \cong \overline{BC}$ and D a midpoint may be regarded as assumed data. Finally, someone will extract the mathematical problem from the situation by saying that if $\overleftrightarrow{\mathbf{CF}}$ is vertical (which we assume it always is) and it intersects $\overleftrightarrow{\mathbf{AB}}$ at D, then we must see if we can prove that $\overleftrightarrow{\mathbf{AB}}$ is horizontal or that $\triangle ADC$ is a right triangle. How can this be formulated mathematically? The suggestion is

$$\begin{array}{lll} \text{Assumed Data:} & \overline{AC} \cong \overline{BC} & (p) \\ & D \text{ midpoint of } \overline{AB} & (q) \\ \text{Conjecture to be Tested:} & \overleftrightarrow{\mathbf{CD}} \perp \overleftrightarrow{\mathbf{AB}} & (r) \end{array}$$

The student can readily show that this conjecture can be deduced from the assumed data by the use of the SSS-congruence postulate. Note that this theorem, $p \wedge q \rightarrow r$, has two partial converses which might also be examined. Inverses and contrapositives might also be studied and stated. From the geometry of the Egyptian level we learn that geometry is more than an intellectual creation, thus the outward look is motivating to some students because they see a real reason for studying mathematics. As Edgar Dale says, "Note that the word *motivation* has the same root as *emotion*—it means "to move."[31]

Exercises

1. Examine the partial converses of the theorem related to the Egyptian level. Can they be proved? Hence are the partial inverses valid? Why?
2. The above application made use of congruent triangles and can be used early in geometry. Some instruments and methods of doing various practical things make use of similar triangles. Consult references, encyclopedias, histories of mathematics, and learn the use and principle of the (a) alidade and plane table, (b) the pole-shadow method of determining the height of an object, (c) the proportional divider, (d) a Boy Scout method for estimating distances. Consider at what point you could introduce these to a class in geometry.
3. In like manner consult references to the instrument known as the pantograph; explain its use and its mathematical principle. How can it be proved that it does what it is designed to do? Students can make these.

The Cross-Staff

This interesting instrument which was used several hundred years ago is also known by several other names: Jacob's staff, the baculum, and the arbalete, The term cross-staff is most descriptive because this device consisted of a rod which was about four feet long which carried a sliding cross-bar

[31] *Ibid.*, p. 2.

(Fig. 6.15). The bar AB was always perpendicular to the rod CD and $\overline{AE} \cong \overline{BE}$.

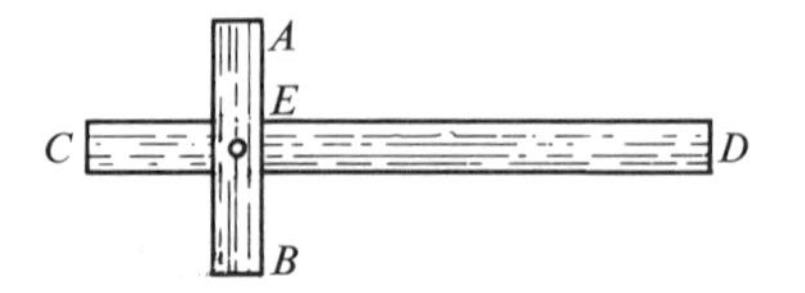

Figure 6.15

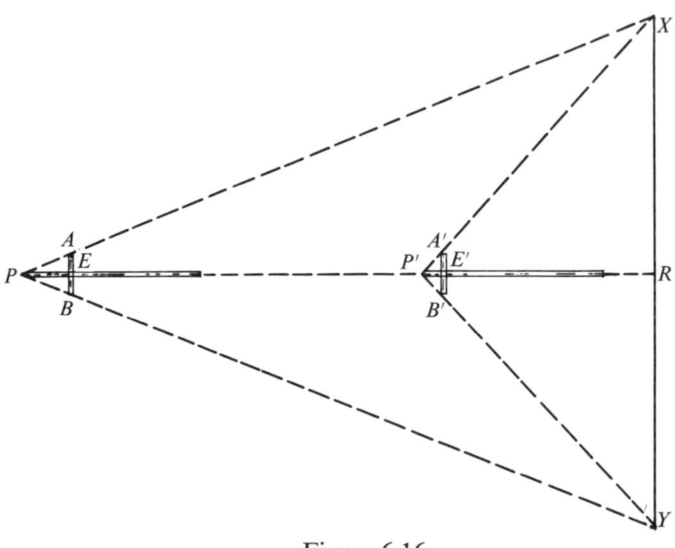

Figure 6.16

To determine the distance between X and Y the users of the cross-staff would begin by finding a certain position, say P, at which the cross-bar could be adjusted so that both X and Y were just in the line of sight of A and B, respectively. Then the operatiors would walk forward in a straight line which they thought was perpendicular to XY to some point P' and slide the cross-bar so that the man at the end could just see again the positions labeled X and Y while sighting along the ends of the bar and the end of the rod. How far the cross-staff was carried from P to P' and the lengths of $\overline{PE}$ and $\overline{P'E'}$ yield the length of $\overline{XY}$ which is to be determined. Indeed, it can be shown that $\ell(\overline{XY}) = \ell(\overline{PP'}) \times \dfrac{\ell(\overline{AB})}{\ell(\overline{PE}) - \ell(\overline{P'E'})}$.

The teacher who wishes to have the student assist in devising how this apparatus can be used can prepare a guide something like this (Fig. 6.17):

1. An ancient instrument used in determining "inaccessible" distance was made by a rod and a cross-bar which was perpendicular to the rod at

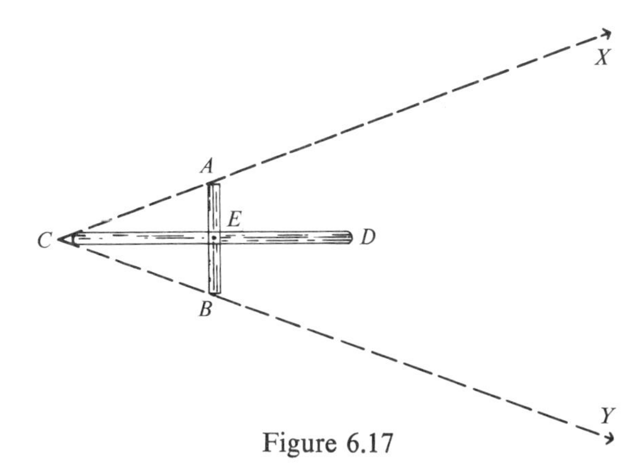

Figure 6.17

the center of the bar. The bar AB slides back and forth along rod CD. It is said that by sighting along $\overleftrightarrow{\mathbf{CA}}$ and $\overleftrightarrow{\mathbf{CB}}$ at two trees (X and Y), let us say, and then by walking toward the line joining the trees and taking another position to sight in a similar manner at the trees, one can determine the distance between them. Can you outline in more detail a method which will accomplish this?

2. Should any condition be placed on the line determined by the trees X and Y in Number 1? Is it necessary that $\overleftrightarrow{\mathbf{XY}}$ be perpendicular to $\overleftrightarrow{\mathbf{CD}}$?
3. As one moves closer to the "inaccessible line" $\overleftrightarrow{\mathbf{XY}}$ does $\overline{CE}$ become longer or shorter?
4. The lengths of what segments do you think will play a part in the length of $\overline{XY}$? Do you think the length of the *cross-bar* will affect the length of $\overline{XY}$?
5. If $\overline{PP'}$ has the length 20 feet, $\ell(\overline{AB}) = 2$ feet, $\ell(\overline{PE}) = 3$ feet, and $\ell(P'E') = 1$ foot can you compute $\ell(\overline{XY})$? Drawing lines from A and B parallel to $A'P'$ and $B'P'$ so they intersect on PE may help.
6. If the men walked 40 feet with the same cross-staff and $\ell(\overline{PE})$ was 3 feet whereas $\ell(P'E')$ was 2 feet then what is $\ell(\overline{XY})$?
7. See if you can develop an expression for $(\overline{XY})$ in terms of the lengths of segments $\overline{PP'}$ $\overline{AB}$, $\overline{PE}$, and $\overline{P'E'}$.

In some classes one may wish to suggest the cross-staff as an instrument which a scout might use: this often creates immediate interest. It can be made by students and its use can be demonstrated in the classroom. Note what mathematics can be studied with two sticks! How can one determine the distance between two unaccessible objects by the use of two sticks and a tape measure? One might receive several suggestions, one of these may be precisely that of the cross-staff. All the suggestions will probably be applications of similar triangles. Also through the determination of inaccessible distances and subsequent measurement directly one may introduce the concept of "percent error." Experiences in geometry may encompass many items of other parts of mathematics as well as parts of history, language, and social sciences. For example, the mental picture of a small town in the Middle Ages with the alchemist's shop, men in the street carrying a Jacob's staff, merchants in shops determining bills on "casting tables" or on their medieval counting boards, others in offices using Napier's rods can provide the student with an enriched background in *history*, too. Applications in mathematics can play a significant role in helping the student to create a picture of the world in which he lives.

Perspective Drawing

In the discussion of parallel lines, it is interesting to think for a moment about the "parallelism" of rails on a track or of the sides of roads, and even the top and bottom edges of buildings. Of course, we know they do not meet, but they *seem* to meet at a point on the horizon. Students find introduction of perspective drawing both interesting and useful. Indeed, latent talent often appears during such work as this. Also, such work can be used to demonstrate ideas of affine geometry in which one point on each line is differentiated from the others and is called the *ideal* point—that point at which parallel lines intersect. If one takes a little artist's license and introduces a mixture of orthographic and perspective projection simultaneously, he can show the perspective of a brick as that indicated in Fig. 6.18. Ver-

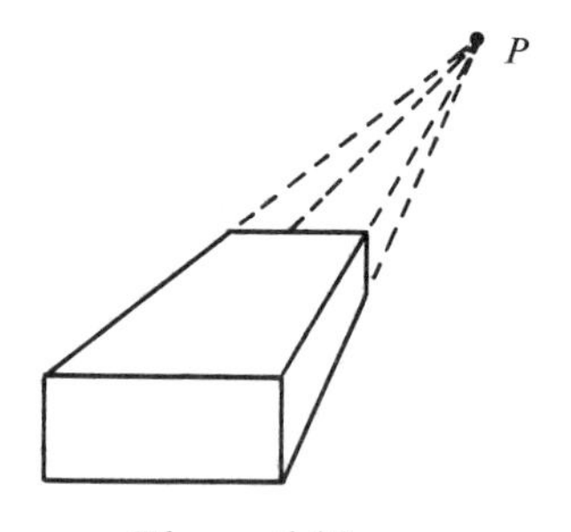

Figure 6.18

tical lines on the object to be represented remain vertical in the drawing and lines horizontal remain horizontal; parallel lines on the object which recede from the observer meeet in the drawing at the center of perspection P which is also called the vanishing point. Beginning activities for students might include study guide exercises like those below. Also there is opportunity for student experimentation and contribution. A short time spent on this topic each day for a brief period will open an entirely new field for the student.

Exercises

1. Construct the perspective of two bricks side by side; three bricks side by side.
2. Construct the perspective of three bricks one on top of the other.
3. Construct the perspective of six bricks arranged in three rows on top of each other.
4. As you look along a row of equally spaced telephone poles receding from you to the vanishing point do all the poles appear to have the same height? What is the appearance of the equal distances between them?
5. Hence see if you can devise a method to construct the *perspective* of two equal-size bricks showing one behind the other (Fig. 6.19)? The essential

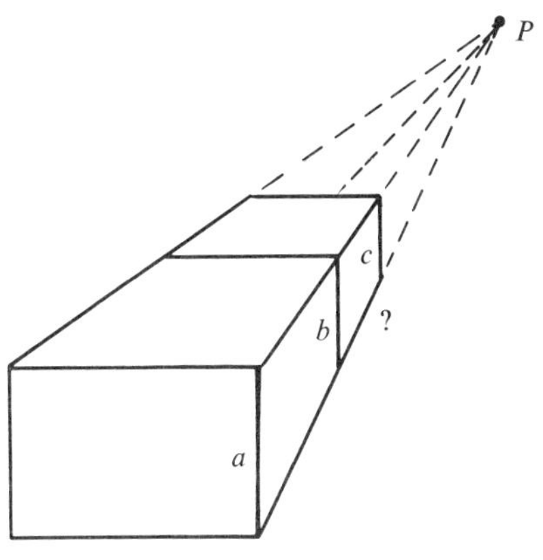

Figure 6.19

question is: given lines a and b, where and how should the vertical line c be constructed so it represents the perspective of the far end of the second brick which is assumed equal in length to the first brick?

One does not know what mechanical method will emerge from the class discussion, but here is one which is valid (Fig. 6.20): Let verticals a and b be given. Now from the top of a draw a line through the midpoint of b. Where this line intersects the bottom line going to the vanishing point, erect the line c; c is the required line. Hence, a succession of equal-size bricks appears as shown in Fig. 6.21. Now the student can draw streets with equal-length slabs, along with telephone poles, buildings with equal-spaced windows, and so on. One result might appear as we note in Fig. 6.22.

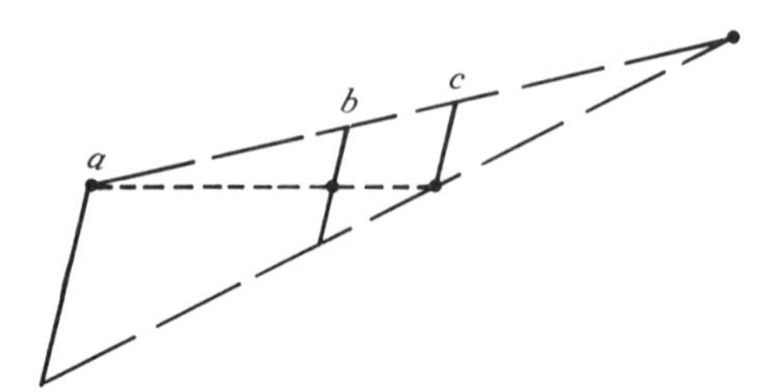

Figure 6.20

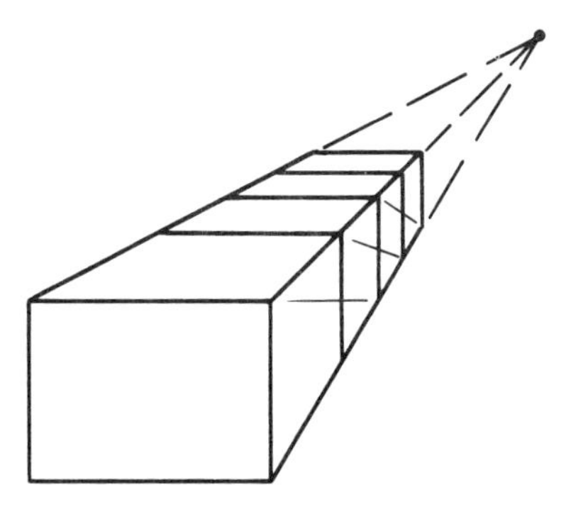

Figure 6.21

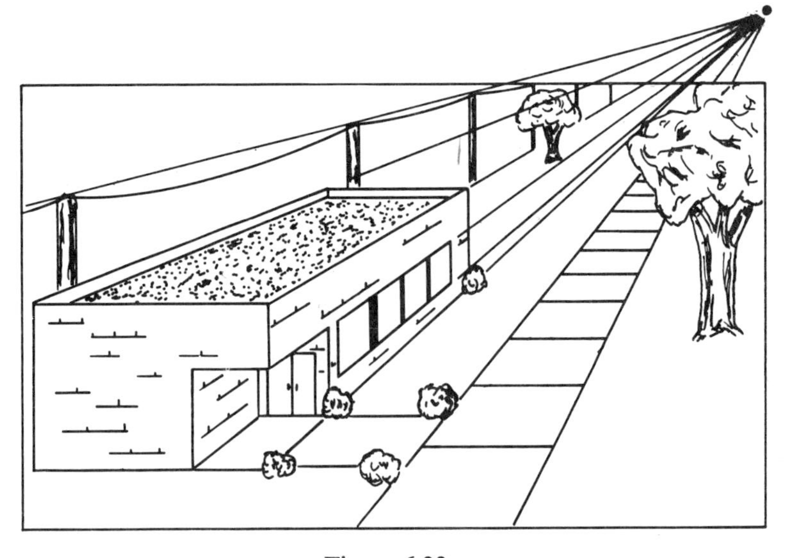

Figure 6.22

Guidelines are erased and the perspective is "framed." Of course, there are variations which the student may introduce; the vanishing point may be contained in the picture on the horizon.

What mathematics have we encountered so far? The mechanics of constructing the perspective of equal-spaced poles uses a converse of the well known theorem that the line segment which joins the midpoints of two sides of a triangle is parallel to the third side and has a length equal to half that of the third side. What is the mathematical formulation of the problem? Class discussion finally centers on this:

Assumed Data: $\overleftrightarrow{\mathbf{AB}} \parallel \overleftrightarrow{\mathbf{CD}}$ (p)

$\ell(ED) = 1/2\ \ell(\overline{DC})$ (q)

Conjecture to be Tested: $\overline{BD} \cong \overline{FD}$.

or that D is the midpoint of BF (r)

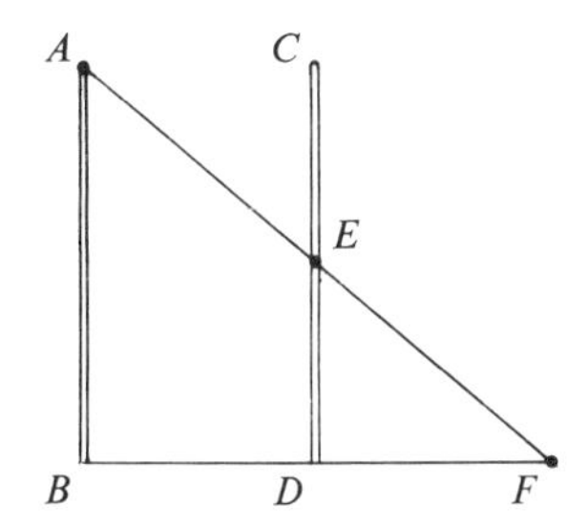

This conjecture can be deduced easily and we can prove also that E is the midpoint of $\overline{AF}$. Here we make use of the *complete converse* of $r \wedge s \rightarrow p \wedge q$. The complete converse of this important theorem is related to perspective projection and, therefore, the work of the artist. This is another example of the outward face of mathematics; the student sees practical application of *converses*.

Another interesting problem for the student to contemplate is the construction of the perspective of the peak of a gabled roof (Fig. 6.23). The gable at the end nearest the observer and the vertical b are made at will, but how is the line segment c constructed so it is perspectively correct? This may be the first time the student has been asked to determine the position of a point which he cannot do by other means (such as by measurement, or by other physical assistance) but must locate it by the intersection of lines constructed in a certain manner.

The next stage in progress in drawing consists of making *two-point* perspectives. Again verticals remain vertical, but now two sets of parallels receding from the observer are shown on the perspective. Fig. 6.24 shows the two-point perspective of several bricks and there seem to be good starting points for students. Bricks can become buildings, streets and power

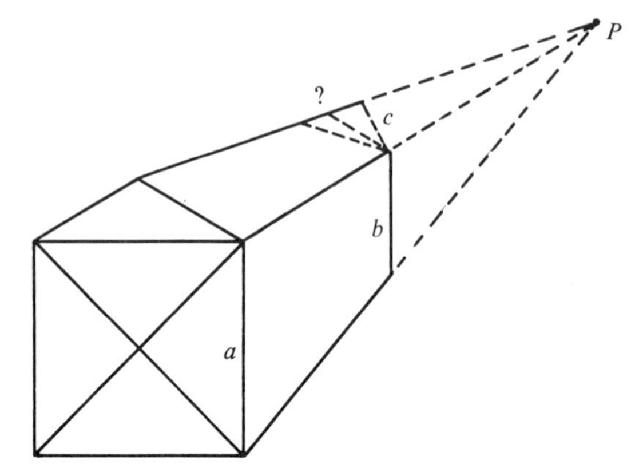

Figure 6.23

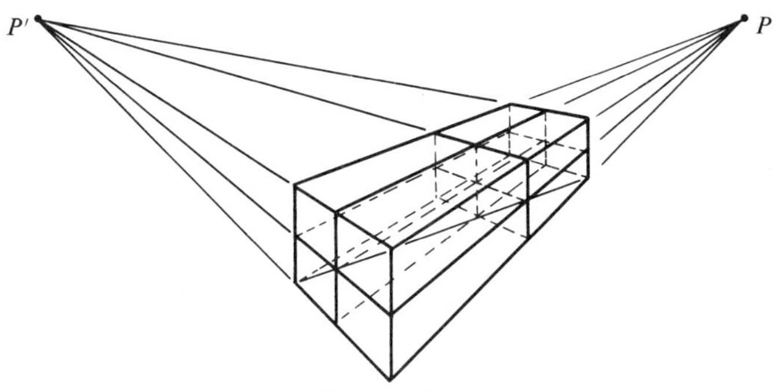

Figure 6.24

lines and poles can be added, and one can produce a pleasing drawing in a short time. One can the consider gabled buildings and the work becomes increasingly interesting.

Students can experience work in the field of drawing by adding color through the use of colored pencils, water colors, and other means. Some students draw pictures of their buildings at home and some have done perspectives of streets. It is not long until drawings which arise from this phase of geometry appear on classroom bulletin boards. Indeed, the artist creates many of his pictures first with a mechanics of perspectives which stems from the appearance that parallels seem to meet on the horizon and from the use of some interesting converses of well known theorems.

Exercises

1. Construct the two-point perspective of three equal-size bricks, one behind the other; now let there be six bricks in two rows of three each.

2. Construct the two-point perspective of four bricks with two on top of the other two.

3. Construct a two-point perspective of a bulding with a gabled roof. Add a street, trees, poles, et cetera, and then frame your picture with a rectangle. Keep the horizon in the picture.

4. Some students construct the perspective of a building, assume the position of the sun, and then construct the shadows of these buildings and other parts of the drawing.

5. Examine a newspaper for furniture advertisements and check the pictures which the commercial artist has produced to see if the drawings are in perspective. These may be mounted with perspective lines shown.

Determining Latitude

It is assumed that students can recall that at some moment on June 21 (or thereabouts) the sun shines directly on the Tropic of Cancer, on March 21 it shines directly on the Equator, and so on; if not, this topic will furnish a good review for him and he will see how useful it is to know on what parallel the sun is shining directly. Here, also, a new area of knowledge is opened to the student. One might begin simply in this way:

1. It is June 21 and Tommy finds that when the sun is at its highest point, he must look *straight up* to see it, that is, the altitude of the sun is 90°. What is his approximate latitude position? (Refer to Fig. 6.25).

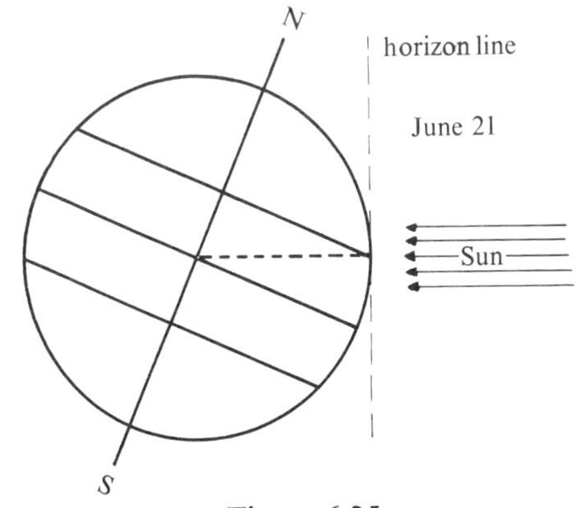

Figure 6.25

2. If it is December 21 and he must look straight up to see the sun, what is his approximate latitude position?

3. If it is March 21 and the angle of elevation, or the altitude, of the sun is $90°$, what is the approximate latitude position?
Now the student can consider how he would work this problem:

4. It is March 21 and Tommy finds that he must look south to see the sun and that the angle of elevation (or altitude) of the sun is $70°$. What is his latitude position? Students should be able to support their answers by reasoning. One approach to this problem is explained with the help of Fig. 6.26. The observer is stationed at C and he must raise his line

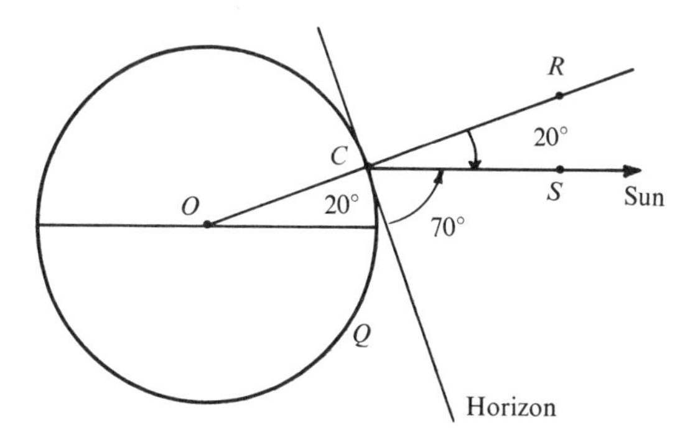

Figure 6.26

of sight from the horizon through $70°$ to see the sun. Now the horizon line is perpendicular to the *radius* $\overline{DC}$ and hence, the $m\angle SCR = 20°$ and, therefore, the angle at the center of the earth, $\angle QOC$ has the measure $20°$ also. Therefore, the arc CQ has the measure $20°$ and the position of the observer is $20°$ north latitude.

What mathematics is behind this solution? Note that $\overleftrightarrow{OQ} \parallel \overleftrightarrow{CS}$ and $\angle QOC \cong \angle SCR$ because they are *corresponding angles*. Hence this problem could be raised early in the course with alterations and refinements made at intervals.

Just how to determine when the sun is at the highest point in its arc and how to measure the angle of elevation of the sun can lead to homemade instruments to determine these data. A pole placed in the ground or a pencil upright on a table will make shadows whose lengths will vary as the sun rises; the shadow is shortest when the sun is at its zenith. Observations taken at intervals will fix the time of noon. Simultaneously, angles of elevation of the sun can be determined either by sighting with crude angle-measuring devices (using smoked or dark glasses), or with an astrolabe (which can be homemade), or with a sextant. To determine the parallel on which the sun is shining directly on a given day at noon, one needs a copy of the *Nautical Almanac*. The fact that a book like this exists demonstrates to the student that for *some* mathematics is a business and that navigators and others are very dependent on the outward-looking face of mathematics.

One more illustration of this follows:

> The captain of ship finds that he must look north to see the sun at an angle of 40° from the horizon. This is on a day on which, according to the *Nautical Almanac*, the sun is shining directly on the 20th north parallel. What is the latitude of the ship?

Figure 6.27 for this problem shows that since the line of sight to the sun, $\overleftrightarrow{CS}$,

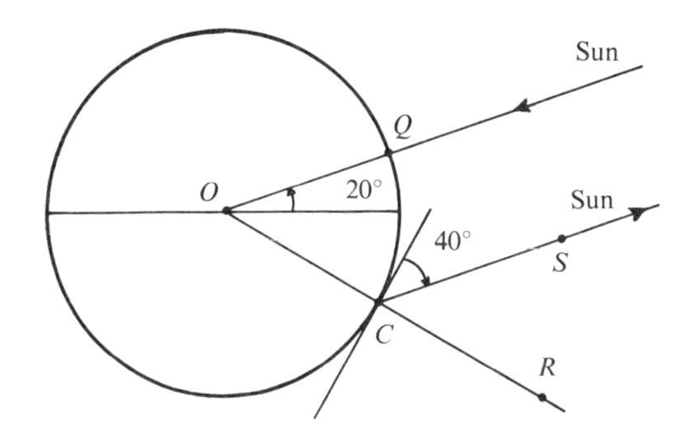

Figure 6.27

makes an angle of 40° with the horizon line then $m\angle RCS = 50°$ and, therefore, $m\angle COQ = 50°$ since they are corresponding angles of the two parallel lines $\overleftrightarrow{\mathbf{OQ}}$ and $\overleftrightarrow{\mathbf{CS}}$ both of which point to the sun. Since Q is on the 20th parallel, then the latitude of C is a 30° south. The captain, therefore, is justified in recording that his latitude position is 30° south.

Corresponding angles (formed by two parallel lines when cut by a transversal) being congruent helps the captain of a ship to determine his latitude. The outward facet of mathematics has a vast outreach.

Exercises

1. The sun is known to be shining directly on the 22nd south latitude parallel. An observer must look south with an angle of elevation having measure 35°. What latitude should the observer report? (Ans. 33° north latitude.)

2. The sun is shining directly on the 18th north latitude parallel. An observer must look toward the south with an angle of elevation of 85°. What is the observer's position? (Ans. 23° north latitude.)

3. Consult a text on the history of mathematics to learn the construction and use of the *astrolabe*. Students can make one of these.

4. The *sextant* is an improvement on the astrolabe. Note its construction, mathematical principles, and use.

5. Students often ask, "Now that we understand how one can determine the latitude of an observer's position, what methods help to tell the *longitude*?" See what you can learn about this.

Using the Parallelogram Law: Navigation

After the parallelogram has been studied, it can be shown that it is of interest to the physicist and engineer because of its relation to the *parallelogram law* which is used in the study of vectors. Laboratory studies can be made which substantiate the physicist's confidence in the parallelogram law.

Study guides might introduce this subject in this manner:

1. Robert is pulling on a wagon with a force of 40 lbs. and Johnny is pulling with a force of 30 lbs. Will the wagon move or will it stand still? If it moves, which way will it go? It will be acted on by what force?

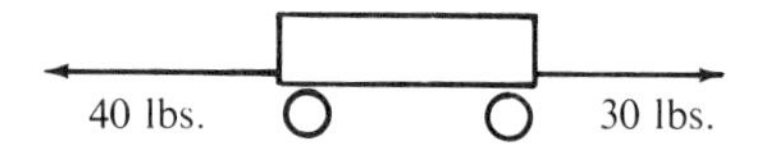

2. A force of 50 lbs. west and 30 lbs. east will produce what net effect?
3. A force of 40 lbs. north along with another force of 30 lbs. north will produce what net effect?
4. The wind is blowing toward the west at 20 miles per hour on a plane whose propellor is revolving such that the velocity in still air would be 100 miles per hour toward the west. What is the resulting speed? If the wind were reversed, what would be the resulting speed?
5. A motorboat's propellor is exerting a force so that the boat should speed at 70 miles per hour. It is moving against a current of 5 miles per hour. What is the resulting speed?

These exercises are very simple as beginning problems must be for the student to acquire complete understanding in elementary matters and a feeling of confidence. From the above cases there develops the idea of *net effect*, or *resulting force*, and finally the term *resultant* which may need to be teacher-introduced at an appropriate time.

The students may now consider the resultant of forces at right angles to each other. What is the resultant or "resulting effect" of a force of 30 lbs. to the north along with one of 40 lbs. to the west? How shall we answer this question? Archimedes would have used physical things such as weights, spring balances, or both, and this we can do, too. After trials with these and other forces at right angles, the class can become confronted with the question "is there an *arithmetic treatment* of the magnitudes of these vectors which seems to yield the same magnitude of resultant as that obtained by experimental means?" With teacher help, both the Pythagorean relation and the solution by scale drawing should come to light. The latter reviews the scale-drawing of the junior high school with a new and significant purpose. Examples of problems for students, after the usual cxploratory, student-centered, student-suggested investigations are made, are these:

Exercises

1. Find both by scale drawing and the use of the Pythagorean Theorem the resultants of the following pairs of vectors whose lines of action are perpendicular to each other. To describe completely the resultant one must tell its direction relative to one of the original vectors or relative to some standard direction.

   (a) 40 lbs. north and 20 lbs. west
   (b) 30 lbs. east and 10 lbs. south
   (c) current exerting a 10 mile per hour velocity to the east on a boat with adjustments made so it would go 60 miles per hour to the north in still water
   (d) wind exerting a velocity in a plane of 30 miles per hour to the northeast (45° from north) and the plane adjusted so it would go 150 miles per hour southeast.

2. How might one find the resultant of a force of 30 lbs. exerted toward the east and a force of 20 lbs. exerted simultaneously to the northeast?

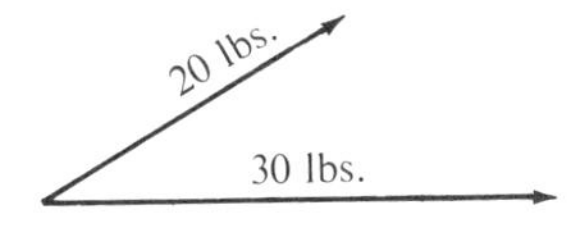

3. Suggest a way to see if your solution in No. 2 has promise of being correct. You may wish to use spring balances to measure forces and protractors to measure angles, and check your work experimentally.

Experimental work and scale drawings suggested by Exercises 2 and 3 will lead to the *parallelogram* law for finding the resultant of two vectors (or maybe, for your class, "George's" Law, or "Anne's" Law, which should afterwards be associated with a law which has been known for sometime). Students now see why the physicist has been interested in the parallelogram for many years. Moreover, exercises on guide sheets can lead him to see the navigator also uses parallelograms. Guides might contain a sequence of problems like these:

Exercises

1. A ship is sailing in a direction 20°N (which always means 20° east of north —angles given will be measured clockwise from north) with a speed of 80 miles per hour. The wind is blowing at 20 miles per hour from 150°N, i.e., from the southeast. What is the speed and the direction of the ship? Reproduce by a scale drawing with angles carefully indicated. (Ans. approx. 37 mph., 20°N)

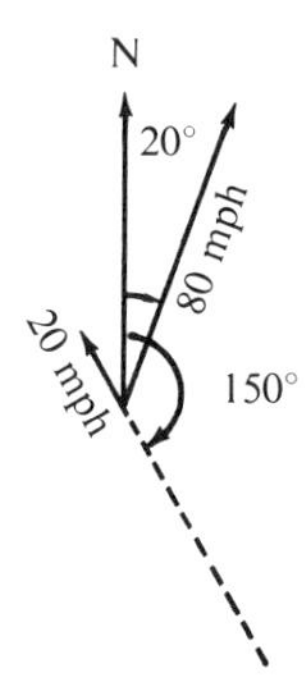

2. A plane is headed 80°N at an air speed of 150 miles per hour; the wind is blowing at 30 miles per hour from 340°N. What is the resulting ground speed and direction (or course) as determined by the given data?

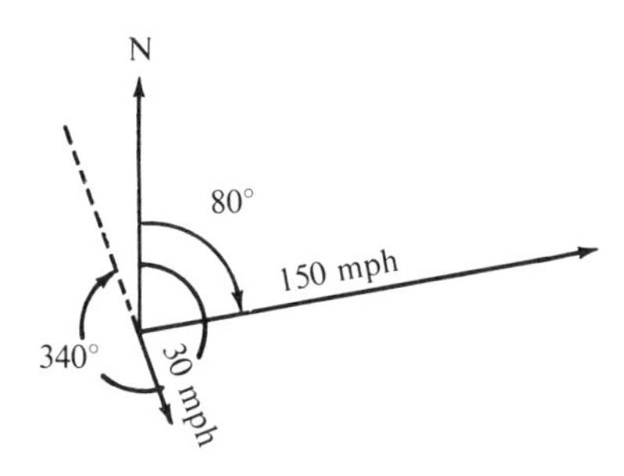

3. Navigators who use methods like those employed in Nos. 1 and 2 have sheets of paper with an interesting convenience printed on the page: a *compass rose* as in Fig. 6.28, which takes the place of the protractor. Use this compass rose to help determine the resultant of the vectors 20 mph at 30°N and 150 mph at 170°N. Simply place the arrows in the direction indicated with the initial points at the center of the compass rose, but draw the arrows to scale; centimeter rules are very handy for they are graduated in tenths.

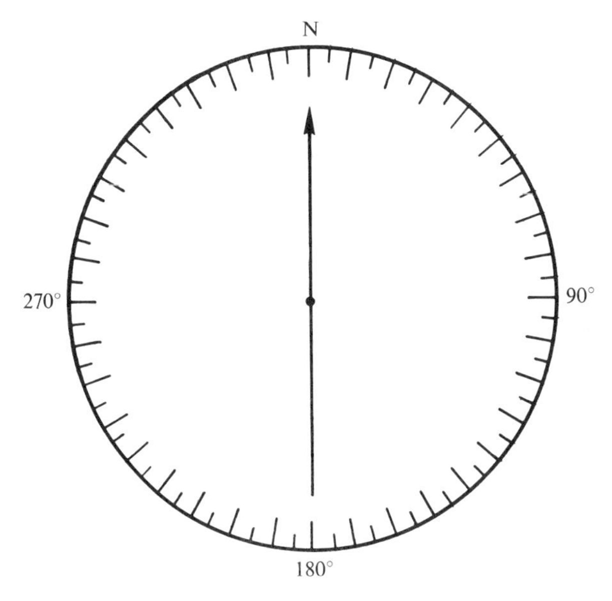

Figure 6.28

4. Use the compass rose to assist in determining the ground speed and course of a plane with true air speed of 200 mph and heading 80°N acted on by a 50-mph wind from 10°N.

The next step in introducing the use of parallelograms in navigation is to show how courses are plotted and how the navigator attempts to keep a record of where he is. Here the student not only reviews arithmetic, scale drawing, and the language of latitude and longitude, but gains an insight into other uses of mathematics and an appreciation that some men make their livelihood in it. He can now observe a plane in the sky or a ship at sea and feel a little kinship with the man whose responsibility it is to keep a record of position and checking to see if he is correct. To bring this picture of navigation into the classroom prepare duplicated copies of "pretended" plotting paper. Standard $8\ 1/2 \times 11''$ sheets are adequate, although navigators' plotting paper sheets are much larger. The plotting sheet appears as shown in Fig. 6.29 and is supposed to represent approximately a portion of a Mercator projection of the earth's surface. The vertical lines represent meridians and the horizontal lines represent parallels. There is an assigned distance of one degree arc between two adjacent lines on the chart and the subdivisions indicate 10-minute (arc measure) distances. Along the meridians each 10 minute interval represents ten nautical miles, since a nautical mile is defined to be a minute of arc along a meridian.

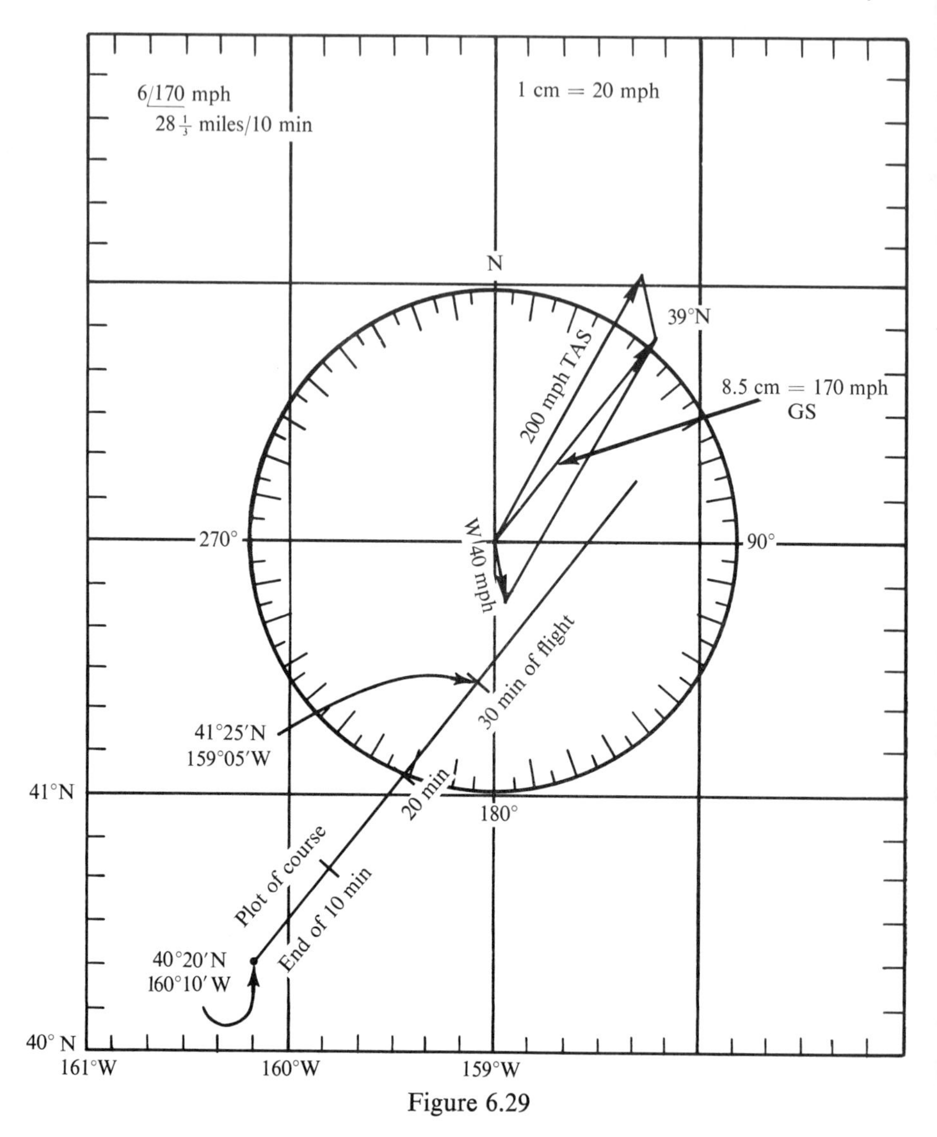

Figure 6.29

Now work a problem to illustrate the kind of exercises which can be done in a course in geometry. Of course, the navigator has instrument and other corrections to make which we shall not consider:

> A plane leaves the point 40° 20′ north latitude, 160° 10′ west longitude at a true air speed of 200 nautical miles per hour with a heading 30°N. The wind is blowing at 40 nautical miles per hour from 350°N. What are the ground speed and the course?

Plot the path of the plane in 10 minute intervals. Determine the position at the end of 30 minutes.

The first part of the problem is done as before, using the parallelogram law on the compass rose. One obtains an approximate ground speed of 170 nautical miles per hour with a course of 39°N. Now to plot the course as well as the navigator can determine it, he uses the plotting-paper grid and places a point on it labeled with latitude 40° 20′N and longitude 160° 10′ W, conveniently chosen so the plane's path does not move off the sheet as he indicates it. He now draws a line from the initial point parallel to the course line of the vector parallelogram. Theoretically, the plane is traveling on this line. The ground speed divided by six yields the distance in nautical miles covered each 10 (clock) minutes. The scale on the meridian is then used to determine how this distance is represented on the map. A compass helps to transfer this map-distance for each 10 minutes of flying time to the line of flight on the map. The end point of three such markings will indicate the position of the plane at the end of 30 minutes of flight. The position can be read from the plotting sheet. The approximate position, as the chart shows, is 41° 25′N and 159° 65′W (see Fig. 6.29).

To do the above requires practice and precision involving skills. The method of approximating one's position by means of the vector parallelogram law as described above is known as *deduced reckoning*, which was at one time abbreviated *ded. reckoning*, later written as *dead reckoning*. In modern times, however, the navigator uses computing devices to determine his course and ground speed. Also, radio navigation is used to determine and check positions, and, of course, there are celestial navigation methods, and piloting. Students interested in further study of navigation may consult texts and government manuals on the subject.

Exercises

1. Position of takeoff: 43° 10′N, 63° 20′W. True air speed: 180 (nautical) mph, heading 85°N; wind 45 (nautical) mph from 10°N. What are the ground speed and the course (or track)?
2. Plot the course of the plane in No. 1 at 10 minute intervals.
3. If the time of the takeoff in No. 1 is 10: 10 a.m. and the plane is in difficulty at 10: 25 a.m., what latitude-longitude position should the navigator report?
4. What is the length of a nautical mile? How is it defined?
5. Note that if you are the navigator on a ship you can check what deduced reckoning indicates as your latitude and longitude position. How?

Variations in problems can result from wind change; then a new vector parallelogram is needed and the path of travel of the plane changes direction, also the distance traversed every 10 minutes may change.

Another type of problem which makes use of the vector parallelogram is this:

> The wind is blowing at 30 mph from 270°N and a plane wishes to go from Cincinnati, Ohio (39°08′ N, 84° 30′ W) to Cleveland, Ohio (41° 30′ N 81° 42′ W). The course to be followed is determined by the angle which the straight line from Cincinnati to Cleveland makes with north. If the true air speed is 180 nautical miles per hour what must be the heading of the plane? What will be the ground speed? (Answers: heading about 40°N; ground speed about 198 nautical miles per hour.)

Books on navigation will suggest other problems such as how to find the direction of the wind while in flight, how to find the point of no return when one knows his fuel supply, and others.

The few applications of geometry which have been discussed in this section should serve to suggest others.

6.7. Vector Geometry

The parallelogram law for vector addition provides another avenue for an inward look in mathematics. Properties of the parallelogram show that we need not always construct a *parallelogram* to obtain the resultant of two vectors but that the "sum" of the two vectors a and b may be found by placing the initial point of b at the terminal point of a and obtaining the vector sum $a + b$ by drawing the arrow from the initial point of a to the terminal point of b. This suggests, in order to have more formal postulational beginnings, a *definition* for the sum of the vectors a and b.

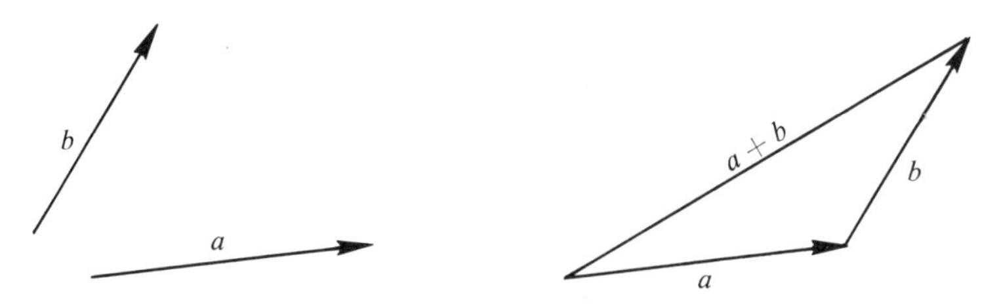

Indeed to make use of vectors in geometry and to maintain the logical spirit in geometry one must define the term *vector*.

1. A vector is an abstraction; it is a set of arrows each member of which has equal length and the same direction. The figure shows *two* sets of arrows: the vectors a and b.

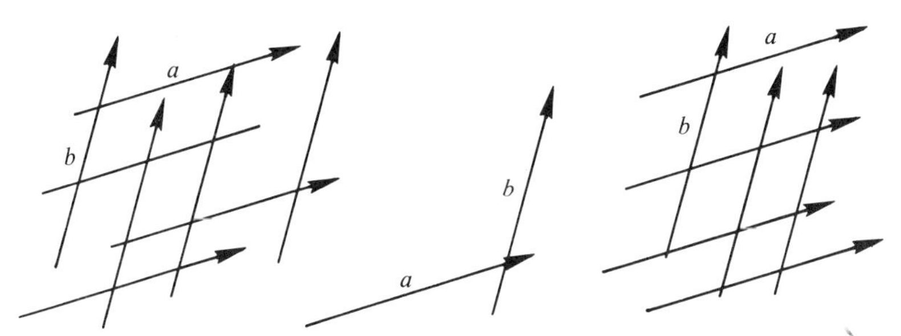

2. In actual practice we use a single arrow from each class as a *representative* of that class. For example, to add the vector b to the vector a we select a representative arrow from the (vector) set a and a representative arrow from the (vector) set b, and draw a diagonal arrow. This resulting arrow is defined to be the representative arrow of the vector $a + b$. This idea should be emphasized in teaching.

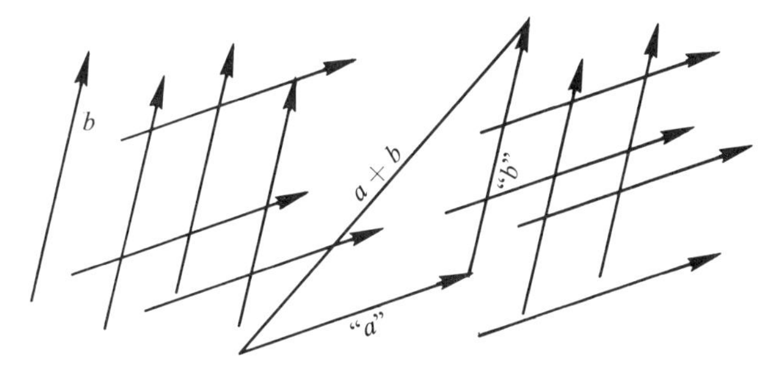

3. The definition of ka, where k is a real number, may be made for $k > 0$, $k = 0$, $k < 0$. One might study the set of vectors with the usual multiplication by scalars under the operation of addition.
4. Two vector sums as $ka + \ell b$ and $ma + nb$ are equal if and only if $k = m$ and $\ell = n$.

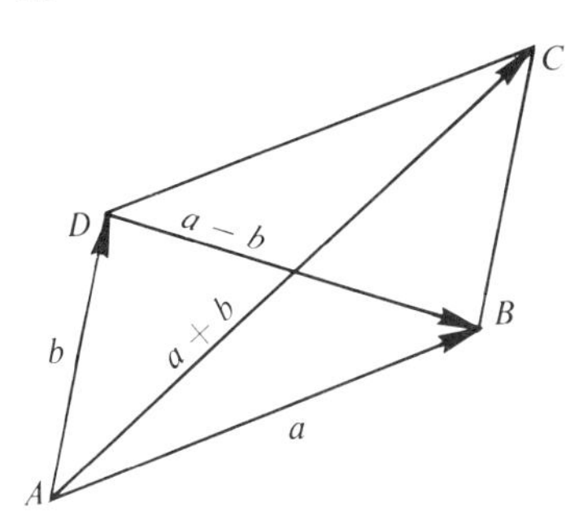

Several conjectures which ordinarily arise in geometry can be investigated quite easily by vector methods while others require more special techniques which will not be discussed here. One example is the proof that that diagonals of a parallelogram bisect each other:

Assume: $ABCD$ a parallelogram

Conjecture: The diagonals bisect each other.

Deductive test: Let the sides of the parallelogram be indicated by vectors a and b as shown.
Now $\overline{AC}$ will be represented by the vector $a + b$ and the segment from A to the midpoint of $\overline{AC}$ by the vector $1/2(a + b)$. Similarly the segment $\overline{DB}$ is represented by the vector $a - b$ and the segment from D to its midpoint is represented by the vector $1/2(a - b)$. From A the vector path to the midpoint of $\overline{DB}$ is $b + 1/2(a - b)$; also from A the vector to the midpoint $\overline{AC}$ is $1/2 (a + b)$. But $b + 1/2(a - b) = b + 1/2\,a - 1/2b = 1/2(a + b)$ and vectors from A to the midpoints of the two diagonals are the same; hence, the midpoints are identical and the diagonals intersect at their midpoints for they have this point in common.

Another vector method to test deductively the same conjecture is as follows:

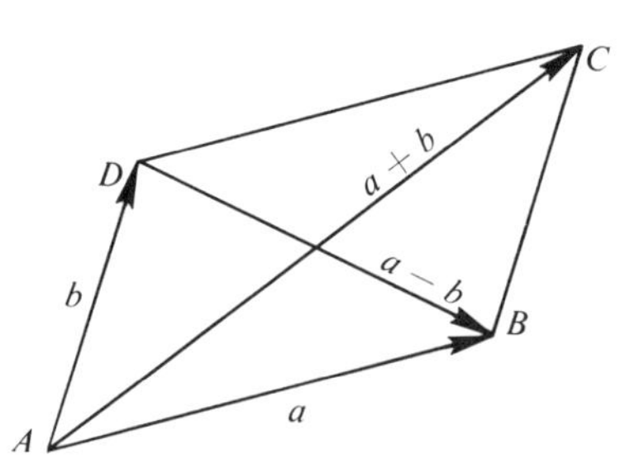

Deductive test: Let $\overline{AB}$ be represented by the vector a and AD by the vector b as shown. Then $\overline{AC}$ is represented by $a + b$ and $\overline{DB}$ by $a - b$. Now the point of intersection of $\overline{AC}$ and $\overline{BD}$ is somewhere on $\overline{AC}$, hence, we shall let the vector from A to this point be represented by $k(a + b)$, where k is a scalar to be determined. Likewise, the intersection point is somewhere on $\overline{DB}$ and its vector representation beginning at A can be indicated as $b +$

$m(a - b)$ where m is a scalar to be determined. But since there is only *one* point of intersection of the diagonals their vector representations from A must be equal, that is

$$k(a + b) \text{ must equal } b + m(a - b).$$

From this we have the imposition that

$$k(a + b) = b + ma - mb$$

and that

$$ka + kb = ma + (1 - m)b.$$

For these vectors to be equal, we must have the coefficients of vector a equal to each other and those of vector b equal to each other; hence

$$k = m \text{ and } k = 1 - m.$$

For these relations to be true simultaneously, we must have

$$k = 1 - k$$

or

$$2k = 1$$

or

$$k = m = 1/2.$$

Hence, the point of intersection is at the midpoints of the diagonals. Q. E. D.

The test illustrated above uses an excellent method of deduction by the use of vectors whose use is to be encouraged. Teachers and students interested in the use of vector methods in geometry may look further in such books as Schuster's *Vector Geometry*,[32] Coffin's *Vector Analysis*,[33] Bieberbach's *Analytische Geometrie* (in German),[34] and others. Most texts in vector methods have a section on uses in geometry. For proofs on perpendiculars, areas, angles, and other features of geometry, however, it is desirable, and sometimes necessary, to enter a more involved study of vectors than discussed in this section.

---

[32]Seymour Schuster, *Elementary Vector Geometry* (New York : John Wiley and Sons, 1962).

[33]Joseph G. Coffin, *Vector Analysis* (New York: John Wilely and Sons, 1911).

[34]Ludwig Bieberbach, *Analytische Geometrie* (Berlin: B. G. Teubner, 1932).

Exercises

1. Prove by vector methods the theorem in geometry that the line which joins the midpoints of two sides of a triangle is parallel to the third side and that the length of the join is half that of the third side.

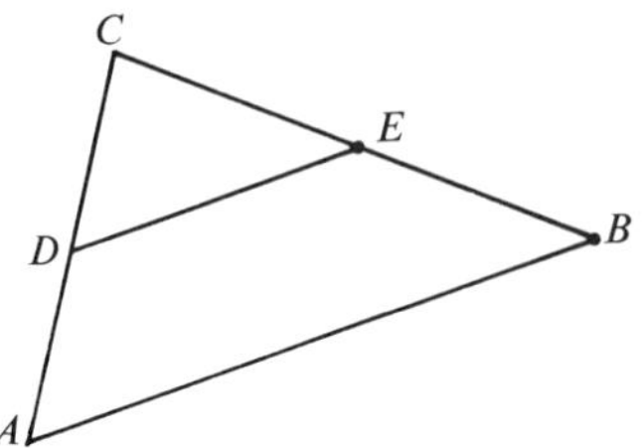

2. Use vector methods to show that if $ABCD$ is a parallelogram and E is the midpoint of $\overline{DC}$ then F trisects $\overline{AC}$ and also $\overline{BE}$.

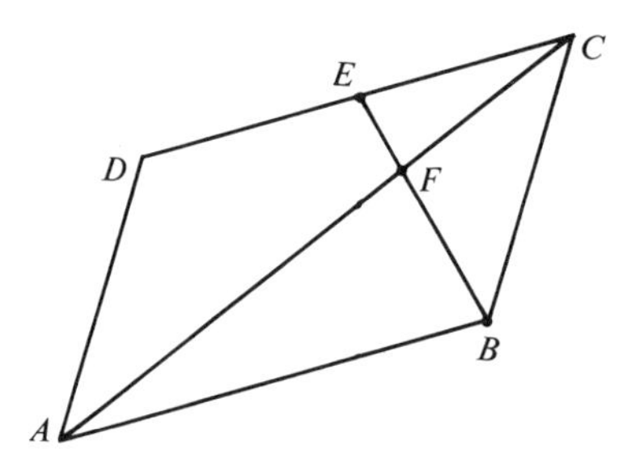

3. Make use of vector methods to show that in the quadrilateral $ABCD$ the joins of the consecutive midpoints form a parallelogram.

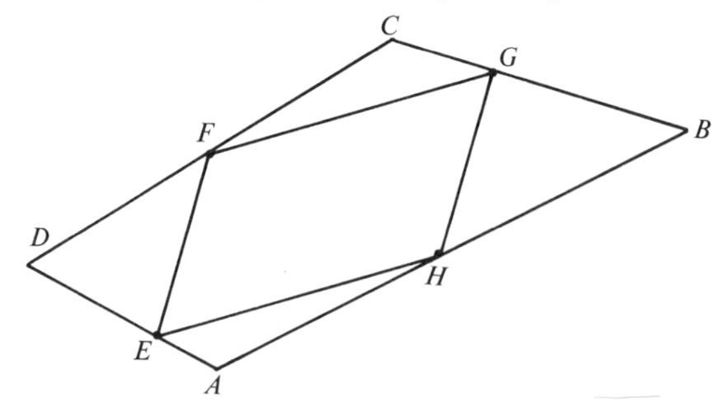

4. Investigate by vector methods whether or not the points of intersection of the medians of a triangle trisect each other.

Although the above problems can be done by the synthetic methods of high school geometry, it deepens insight to do them by other ways. Indeed, we find in the *Goals for School Mathematics* the statement that

On a large scale, mathematics is a unified subject in which each part

> may benefit from systematic investigation with different starting points. . . . they lead to the same results through different paths often of different lengths. They illuminate different aspects of the results.[35]

This statement refers to geometry and algebra as "unified by analytic geometry" but applies equally well to all different approaches through different sets of postulates.

6.8. Geometry in Space

Formal study in synthetic geometry of three dimensions in the form of courses in solid geometry for high schools seems to be passing from the curriculum. Instead, there is a tendency to introduce geometric concepts of both the plane and space in the elementary and junior high school and to study the two simultaneously in secondary school courses.

Surely enough "geometry of three dimensions" should be included in the secondary school program to have the student become aware of how proofs of theorems concerning plane figures change as another dimension is added (Fig. 6.30). Also it is good to see how our language and conceptions

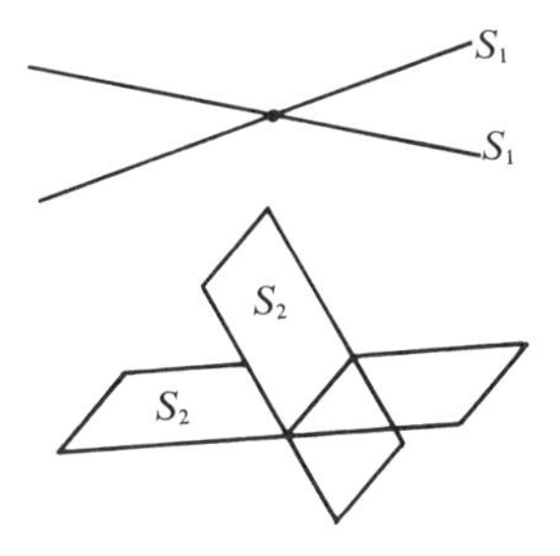

Figure 6.30

may continue although imagery stops. If S_i means a space of i dimensions then we note, for example, that the mental pictures of

2 nonparallel *lines* ($S_1's$) in a *plane* (S_2) meet in a *point* (S_0)
2 nonparallel planes ($S_2's$) in *space* (S_3) meet in a *line* (S_1)
2 nonparallel spaces ($S_3's$) in a hyperspace (S_4) meet in an ordinary (S_3) space

perhaps stop with the second statement but the language goes on. One should also include in the secondary scondary school program some geometry of the sphere so that properties of the spherical surface can be gained

[35]Cambridge Conference on School Mathematics, *Goals for School Mathematics* [Boston: Houghton Mifflin Company (Published for Educational Services, Incorporated), 1963], p. 15.

by Euclidean methods. There should be an examination of what occurs to a plane geometry theorem if a third dimension is added. For example, "If two straight lines intersect then the vertical angles are congruent" suggests that in three-dimensional space "If two planes intersect the vertical dihedral angles are congruent."

There are interesting applications in space geometry also; one of these is map making. Maps are flat, but the earth is approximately spherical. Is there error introduced when the earth is "flattened" to appear on a map? This may be the first time that students have thought of this problem or have realized that it *is* a problem. Again, the entire field of cartography (map-making) lies before them and the outward facet of mathematics has looked into another direction.

One of the simpler maps is that made when one thinks of a lamp placed at the center of the earth casting shadows of the earth onto a flat piece of paper, as in Fig. 6.31, tangent to the earth at the South Pole Q. Only the most elementary ideas of trigonometry are needed here. The angle θ is the

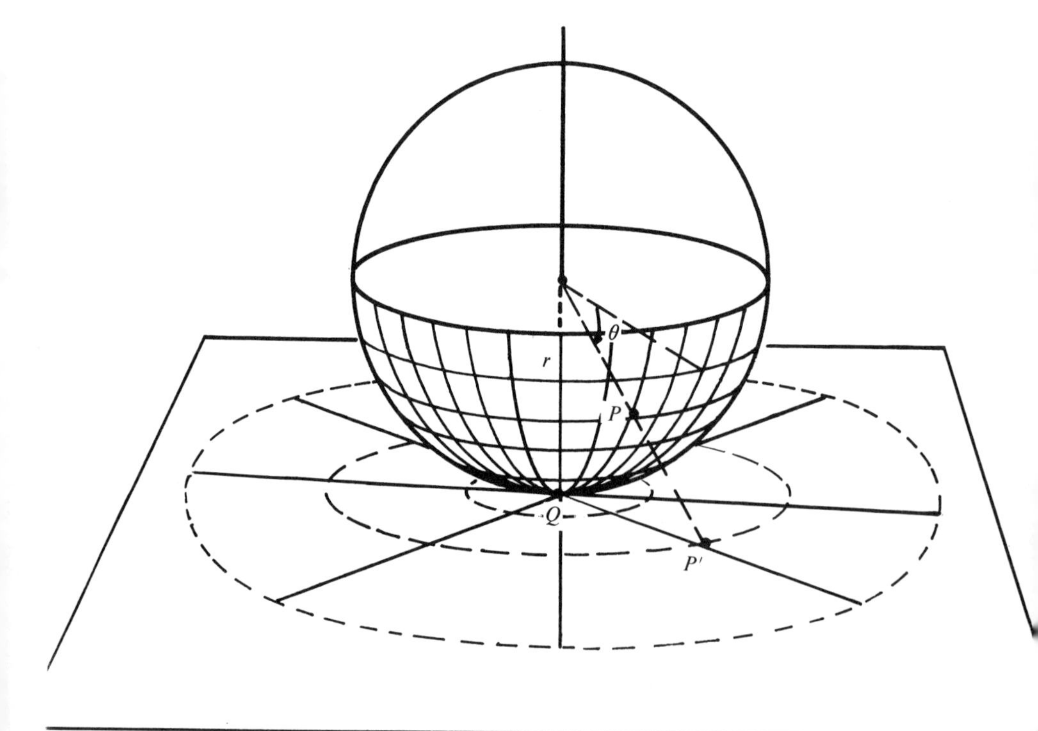

Figure 6.31

south latitude position of P and if the radius of the globe be r then the length of $\overline{QP'}$ is $r \cot \theta$ or $(r/\tan \theta)$. The meridians of the earth map into a pencil of straight lines emanating from the point Q and the parallels of earth map into concentric circles with centers at Q. On such maps straight-line segments are maps of portions of great circles of the sphere and, hence, straight-line distances on the map represent paths of shortest distances. A map constructed in the manner shown is called a gnomonic projection and the one shown in Fig. 6.32 is a north polar gnomonic projection.

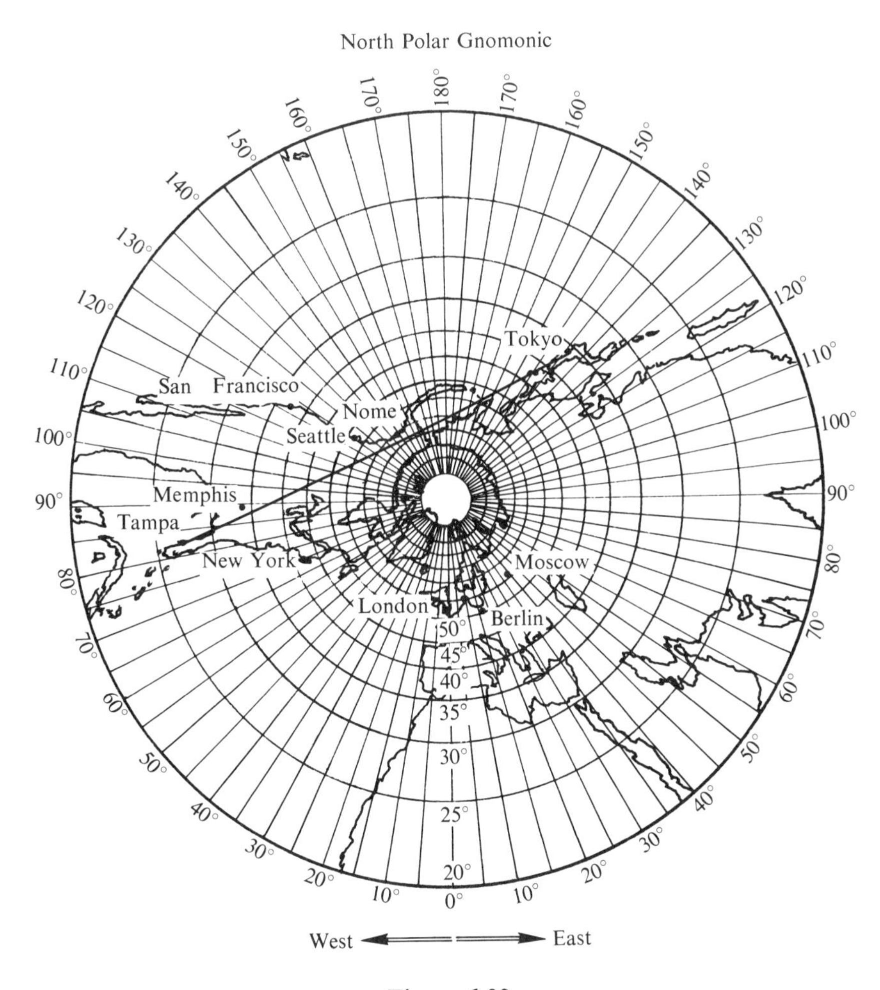

Figure 6.32

Exercises

1. In the gnomonic projection above, what is the map of the south geographic pole? The equator? Where is distortion the least? The greatest?

2. The concentric circles which are maps of the parallels vary in radius. Is the length of the radius in direct proportion to the south latitude? If not, how does the radius vary with the south latitude?

3. To determine the path of shortest distance between, say, Tampa, Florida, and Tokyo, Japan, one draws a straight line connecting these two points on the gnomonic map. Determine from the map the latitude at which the path crosses each meridian.

4. Consider the radius of the earth to be 5 centimeters and construct the framework for a south polar gnomonic projection on an $8\frac{1}{2} \times 11''$ sheet of paper. Recall that the radius of the 70th parallel (north latitude), for example, will be $2 \cot 70°$. Place some of the principal cities on your map.

5. Students might be interested in other kinds of projections; these are usually described in texts on navigation and in geography. The cylindrical projection is readily understood and made by high school students. In Fig. 6.33

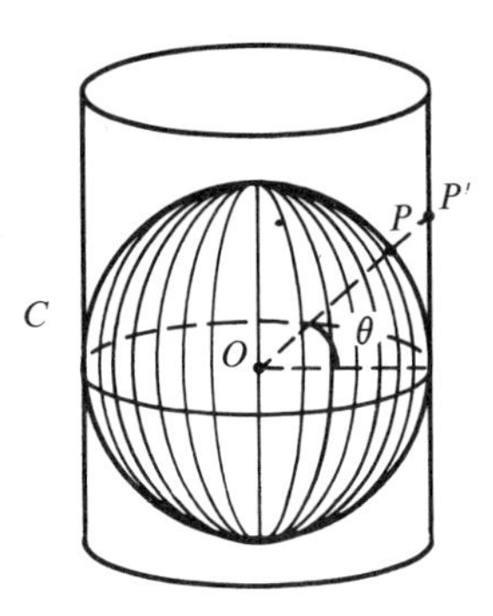

Figure 6.33

C is a cylinder tangent to the earth at the equator. θ is the north latitude measure in degrees. $\overline{OP}$ produced intersects the cylinder, in P' and the cylinder is cut along a meridian with the result that parallels become further and further apart and meridians are equally spaced (Fig. 6.34). If the radius of the earth be taken as r how far is the 10th parallel from the equator on the map? The 20th parallel? The 30th parallel? Where is distortion the least? The most?

6. Consult texts on mathematics to learn the nature of a stereographic projection. If a point P has north latitude θ and longitude ϕ determine the polar coordinates of the stereographic map of P (Fig. 6.35). (Ans. $2r \tan (\theta + \pi/2)/2, \theta$.

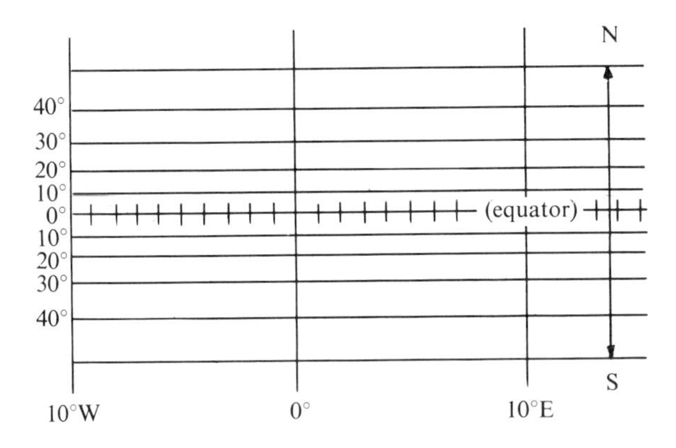

Figure 6.34

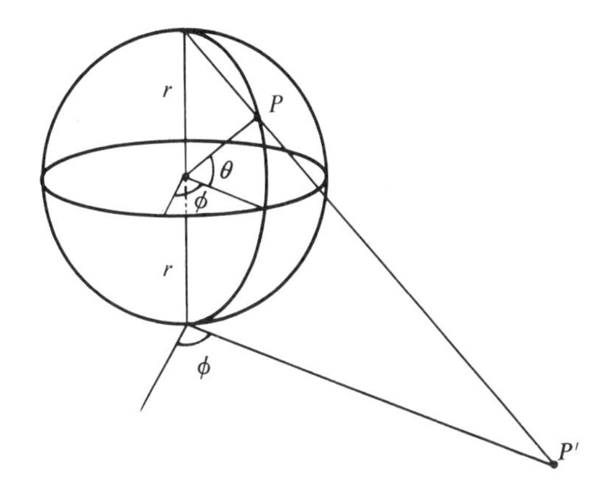

Figure 6.35

7. The stereographic map is said to be *conformal*. What does this mean?

8. List several other kinds of projections and their uses in geography and commerce.

The cylindrical type of projection altered in certain ways makes the Mercator projection devised by Gerhard Kremer in 1569 and is used by navigators. The name Mercator is the Latinized form for Kremer. Figure 6.36 is an example of a Mercator projection. In the Mercator projection, north is always "up" hence is not always changing direction as it does on the gnomonic maps. In actual use the navigator can transcribe the points to be crossed for a shortest distance route from the gnomonic projection to the Mercator projection and then measure the angle which must be maintained

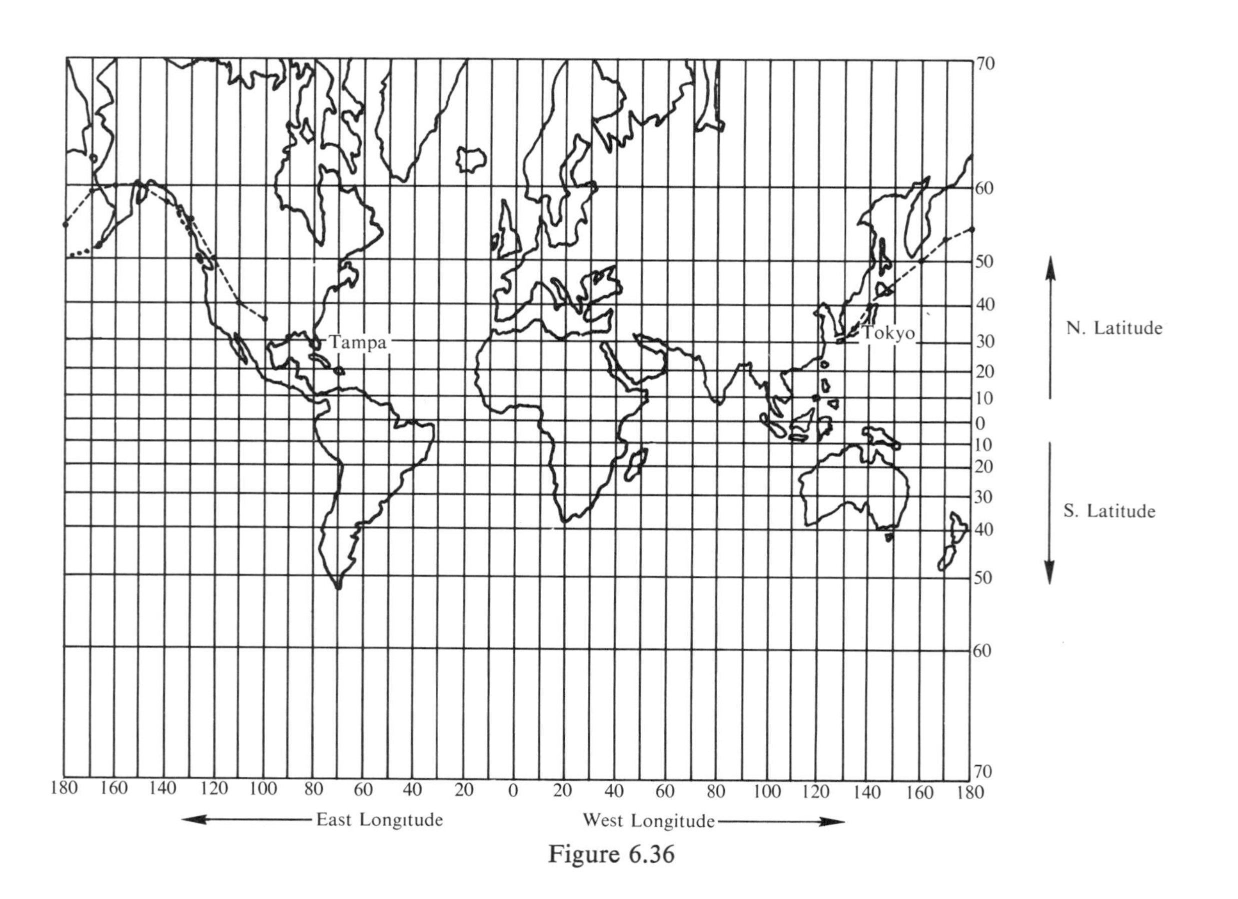
N. Latitude
S. Latitude
70
60
50
40
30
20
10
0
10
20
30
40
50
60
70
Tokyo
Tampa
180
160
140
120
100
80
60
40
20
0
20
40
60
80
100
120
140
160
180
East Longitude
West Longitude

Figure 6.36

with north to follow this approximate shortest path route. The gnomonic projection furnishes the points for the path of shortest distance, but the Mercator projection is used to "sail by." The points determined preceding Exercise 3 have been transcribed to the Mercator map.

Exercises

1. From the Mercator map what course must be maintained to carry one from the 120th meridian west to the 130th meridian west on the way from Tampa to Tokyo? (Ans: about 288°N)

   How will one know when he crosses a meridian? Note that this problem is related to one of those in navigation: finding what course must be pursued under given conditions of wind or current.
2. Use the north polar gnomonic to find points on each meridian which must be crossed to pursue the shortest distance path between New York and Seattle. Now transcribe these points to the Mercator projection and indicate what heading must be maintained between the point on the 90th meridian west and that on the 100th meridian west.
3. If you are in a plane and the true air speed is 100 mph and the wind is 30 mph at 40°N, what heading must be maintained to go from the first point to the second in Exercise 2?
4. A triangle determines a circumscribed circle, an inscribed circle, and three described circles. Describe a similar situation for a tetrahedron. Are there other notions in plane geometry which can be carried into a space of three dimensions? Suggest several.

6.9. Some Different Kinds of Geometry

A discussion of this topic is presented in many volumes on geometry. The purpose of this section is not to discuss different kinds of geometry in detail, but, rather, to suggest that there *are* different kinds and to point out how they came to be. Teachers should be on speaking terms with various geometries and should permit no student to believe that there is just one geometry, Euclid's. Indeed, the invention of non-Euclidean geometry caused an intellectual revolution which completely and dramatically undermined the idea that *Euclid was truth*, a veritable crisis in the history of thought. Morris Kline suggests the far-reaching value of the study of different geometries by advising that

> Apparently we should constantly re-examine our firmest convictions (Euclid), for these are most likely to be suspect. They mark our limitations rather than our positive accomplishments. On the other hand, non-Euclidean geometry also shows the heights to which the human mind can rise. In pursuing the concept of a new geometry, it defied intuition, common sense, experience, and the

most firmly entrenched philosophical doctrines just to see what reasoning would produce.[36]

To gain a glimpse of different kinds of geometries, it will be helpful to present a brief historical review, hoping that the student may explore much further all these facts in the excellent books on geometry which are now available.

Euclid's geometry presented in his *Elements* was based on five postulates, five axioms, and twenty-three definitions. These are found in various sources, one of which is Eves and Newsom's *Introduction to the Foundations and Fundamental Concepts of Mathematics*.[37] It is noticed immediately that Euclid's Postulate 5, or the original parallel postulate, is different from the parallel postulate found in modern texts. Actually Euclid's Postulate 5 became the first object of inquiry and study not long after the appearance of the *Elements* in 300 B. C. The length and type of statement of Postulate 5, which says essentially

> If two lines cut by a transversal have the sum of the two interior angles on the same side of the transversal less than two right angles then the two lines, if extended, will meet on that side of the transversal on which the sum of the interior angles is less than two right angles.

made it seem to some that it should be a theorem deducible as a consequence of the first four postulates. Men wrestled with this problem for years. Cassius Jackson Keyser writes that this postulate "had the fortune to be an epoch-making statement—perhaps the most famous single utterance in the history of science."[38] From the various attacks on the problem of Postulate 5 and from investigations carried on by mathematical scholars came new kinds of geometry which were consistent under strange postulates, searching studing studies into the foundations of mathematics and postulate systems in general, as well as improvements in the postulate system of Euclid.

Important contributions were made by Saccheri (1733), Lambert (1766), Bolyai (1832), Lobachevsky (1830), Legendre (1833), Riemann (1854), and many others. Saccheri denied the equivalent of Euclid's Postulate 5 and thought he reached contradictions which he said, "vindicated Euclid." From our knowledge of logic, however, we see that this would have meant that Euclid's Postulate 5 was a logical consequence of the first four postulates. Saccheri was found to be in error, for one of his

---

[36]Morris Kline, *Mathematics for Liberal Arts* (Reading, Massachusetts: Addison Wesley Publishing Company, 1967), p. 476

[37]Howard Eves and Carroll V. Newsom, *An Introduction to the Foundations and Fundamental Concepts of Mathematics*(New York: Holt, Rinchart and Winston, 1958) pp. 32-4.

[38]Cassius Jackson Keyser, *op. cit.*, p. 113.

"contradictions" did not exist and the other came as a result of a tacit assumption equivalent to Postulate 5 but which he failed to recognize. In addition to his work on parallels Saccheri produced a set of theorems which use only the first four postulates of Euclid, that is, theorems which retain properties studied in (Euclidean) geometry, but which are not based on the Fifth Postulate or on Playfair's postulate of the uniqueness of a line through a point parallel to another line. Such theorems make up what is called *absolute geometry*, or, *pangeometry*. Saccheri's theorems on the Saccheri quadrilateral make up a part of this geometry as well as the first 28 propositions of Euclid's Book 1; Eves describes this body of geometry by saying

> The deductive consequences of the Euclidean postulational basis with the parallel postulate extracted constitute what is called *absolute geometry*; it contains those propositions which are common to both Euclidean and Lobachevskian geometry.[39]

Another discussion of absolute geometry, or pangeometry, is found in Manning's *Non-Euclidean Geometry*.[40] Lobachevsky wrote a small book on pangeometry[41] and Bolyai did much work in this field. Coxeter includes an entire chapter on absolute geometry in his *Introduction to Geometry*.[42]

Lobachevsky and Riemann invented two new geometries by using negations of the Playfair form of the parallel postulate which says that "through a point P can be drawn one and only one line parallel to a given line ℓ." Lobachevsky's assumption (negating Postulate 5) that "through a point P there can be drawn *more than one line* parallel to a line ℓ" and Riemann's hypothesis (negating Postulate 5 also) that "through a point P *no* line can be drawn parallel to a line ℓ" both yielded consistent geometries. So now we have, thus far, *four* geometries: Euclidean, Absolute, Lobachevskian, and Riemannian.

It may be of special interest to teachers that although Legendre is well-known in his investigation of the "Fifth Postulate question" and made valuable contributions, he also wrote a text in geometry called *Elements de Geometrie* (1794) which was received favorably both in Europe and America. In this text he rearranged theorems and topics to attempt to improve Euclid pedagogically. Eves and Newsom report that the English translation of the *Elemente de Geometrie* made by Thomas Carlyle (the Scottish writer) in 1822

---

[39]Howard Eves, *A Survey of Geometry, Volume I* (Boston: Allyn and Bacon, 1963), p. 336.

[40]H. P. Manning, *Non-Euclidean Geometry* (Boston; Ginn and Company, 1901).

[41]Howard Eves and Carroll V. Newsom, *op. cit.*, p. 64.

[42]H. M. S. Coxeter, *Introduction to Geometry* (New York: John Wiley and Sons, Inc., 1961), Chapter 15.

enjoyed 33 American editions![43] There had been a previous English translation in 1819 by John Farrar of Harvard University.

Exercises

1. Examine several texts in high school geometry to see what parallel postulate is used. Is it that of Euclid? Of Playfair? Still another form? (Some texts *prove* that there can be one line through a point P parallel to a given line ℓ and then *postulate* that there can be no more than one.)
2. Study a text in high school geometry to see how far the writer goes without the assumption of the "uniqueness of parallels"; these theorems comprise absolute geometry.

With a background of different kinds of geometry the teacher has a well of rich resources from which to create exercises for his students which will emphasize goals in the study of mathematics. He can surely show that mathematics is the science of necessary consequences and as postulates are altered, or even subtracted, both the teacher and students can feel as Bolyai did when he wrote, on using postulates different from those of Euclid, "out of nothing I have created a strange new universe." Students should be encouraged to make investigations along with the current work in geometry. Teachers will find benefit in pursuing references given in such works as Coxeter's *Introduction to Geometry* and Eves and Newsom's *An Introduction to Foundations and Fundamental Concepts of Mathematics* (listed in the bibliography). The preparation of guide sheets to encourage some of these investigations is a gratifying experience. These might contain items as found in the problems below.

Exercises

1. Already in our course we have proved that the diagonals of a rectangle are congruent; this was done in several ways. In 1733 a mathematician whose name was Saccheri made some studies on a figure with these initial properties:

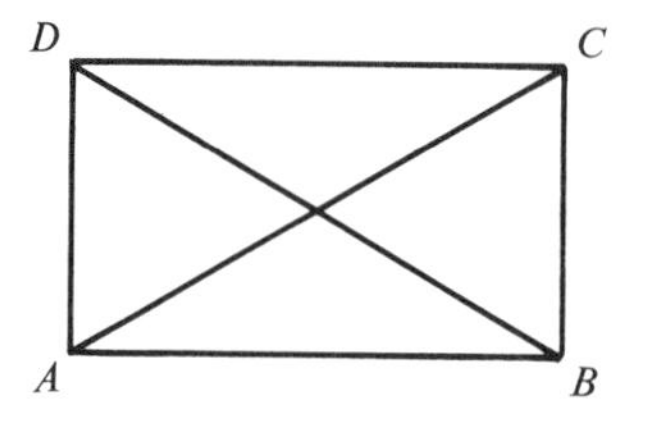

[43]Eves and Newsom, *op. cit.*, p. 61.

(a) $\overline{AD} = \overline{BC}$

(b) $m\angle A = m\angle B = 90$

Can you prove under this hypothesis that the diagonals are congruent? This figure has come to be called the *Saccheri quadrilateral*.

2. Saccheri planned to prove all he could about the Saccheri quadrilateral without using the assumption that we have made about parallels: that through a point P there exists no more than one line parallel to ℓ. Under this limitation can you prove that the diagonals bisect each other?

3. Can you prove the existence of any relation between $\angle ACD$ and $\angle BDC$? If so, write the proof.

4. In Problem 2, if the "uniqueness of parallels" is denied, can one show that $\angle ACD$ and $\angle BDC$ are right angles? Why or why not?

5. Let us suppose that in the Saccheri quadrilateral we have E and F are *midpoints* of segments $\overline{CD}$ and $\overline{AB}$ respectively. Explore this figure for some possible relations and see if these can be deduced.

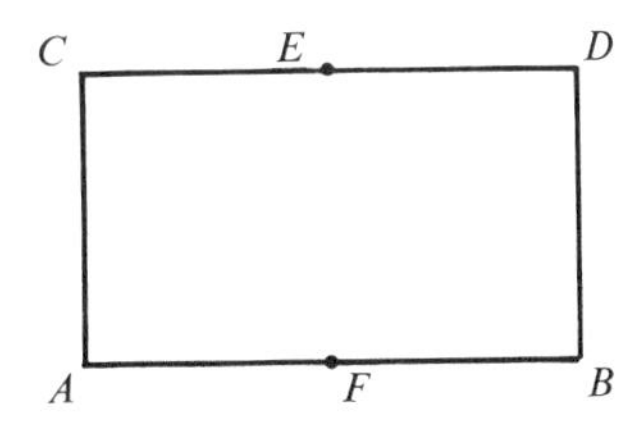

6. Given the triangle ABC and D and E midpoints, let perpendiculars be drawn from A and B to $\overline{DE}$. See if you can prove that the quadrilateral formed is a Saccheri quadrilateral.

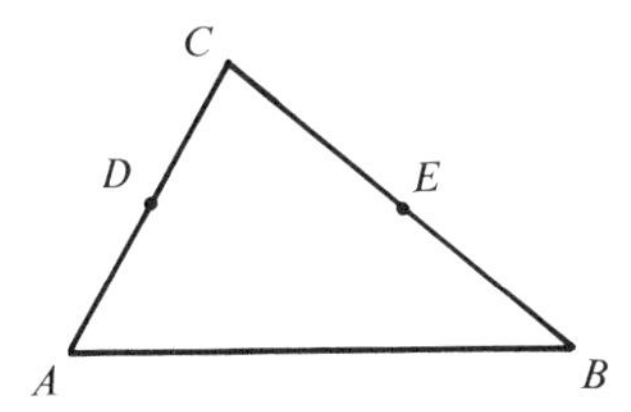

Saccheri used the quadrilateral above as a stepping stone from which he made three hypotheses which led eventually to the three geometries which we now call Euclidean, Riemannian, and Lobachevskian. The hypotheses of Saccheri were that either $\angle C$ and $\angle D$ (which he proved congruent) were both equal to, greater than, or less than a right angle. These hypotheses yielded, respectively, the three geometries listed above. Had Saccheri not been so eager to "vindicate Euclid" he might not have thought so readily that he had reached contradictions and could well have become the first non-Euclidean geometer.

The examination of Euclid's Postulate 5 clearly revealed that there were theorems in geometry logically equivalent to that postulate. Of course, Euclid's Postulate 5 and the Playfair form are equivalent, that is, the four postulates of Euclid plus Playfair's Postulate (along with ability to use any of the first 28 theorems of Euclid) can be used to prove Euclid's Fifth Postulate as a theorem and vice versa. Here is a means of emphasizing the inward look of mathematics also. One may introduce exercises such as those below:

Exercises

1. Prove that "if a line cuts one of two parallels it also cuts the other" is equivalent to the Playfair hypothesis.
2. Already we have proved that the Playfair Postulate yields the theorem that the "sum of the interior angles of a triangle is that of two right angles." See if you can prove that this angle-sum property taken as an assumption will produce the Playfair statement as a conclusion.
3. Show that Playfair's Postulate and Euclid's Fifth Postulate are equivalent.

The history of the development of geometry provides another fruiftful source of investigation and projects in geometry, especially with respect to attempts that have been made to improve Euclid pedagogically. The work of Legendre has been mentioned. It is always very profitable for purposes of review and to strengthen one's insight into structure and dependence to trace the logical support of theorems back to their postulational origin. Such a study of the interior-angle-sum theorem for triangles can yield rewarding results. Charts made by the student to make explicit the geometric principles on which this theorem depends are helpful. If they include the proposition that "an exterior angle of a triangle is greater than either opposite interior angle" then the sequence (through this section of theorems) is probably patterned after that of Euclid; if, however, the genealogy of this theorem includes the proposition that "if a line is perpendicular to one of two parallels then it is perpendicular to the other," the sequence is similar to that of Legendre. A study of sequences in geometry was made by Whitmore in his doctoral dissertation.[44]

Exercises

1. Consult several texts in geometry and see if the sequence of theorems leading to the angle-sum theorem for triangles is that of Euclid or of Legendre. Is some other sequence evident?

---

[44] Edward Whitmore, "A Study of Sequences in Elementary Geometry," doctoral dissertation, The Ohio State University, 1956.

2. There is another sequence which begins by *defining* two lines to be parallel if the corresponding angles are congruent. The interior angle-sum theorem follows. Do any of your texts use this sequence?

There are still other kinds of geometries, several of which bear mentioning. If roughly speaking, one deletes from the set of Hilbert's postulates those on congruence of nonparallel line segments and angles, but retains those on order and parallels and incidence, he obtains a geometry called *affine* geometry. This is an interesting geometry and some of the theorems studied in the secondary school (Euclidean) program are really affine theorems in the sense that they can be supported in a system of postulates not necessarily Euclidean. A method of differentiating between Euclidean and affine properties will be introduced later when geometries are considered as classified by means of certain kinds of transformations.

If the postulates on order and on parallels are deleted so that there are no considerations of order of points on a line and there are no parallels, that is, all lines meet, then the geometry is *projective geometry.* Projective geometry is an incidence geometry; its figures are made with straight lines and every pair of lines intersect in a point; all points and lines are alike, there are no "ideal" points or "ideal" lines. We became slightly acquainted with projective geometry when we discussed perspective drawing, in exercises involving Desargues' and Pascal's Theorems, and in exercises on the complete quadrilateral. Projective geometry is a demanding discipline in its own right. The teacher should develop background which includes some of its important theorems and ideas such as the theorems of Desargues, Brianchon, Pascal, cross-ratio, harmonic division, the complete quadrilateral, duality, and geometric construction. A still more general geometry in which straight lines lose their distinctiveness (that is, a "straight" line and a simple curve are equivalent) but in which there are still considerations of inside, outside, separation, and so on, is that discipline called *topology.*

Some of the above topics can be of immediate use in the classroom. One who understands the relation between and classification of geometries (as diagrammed)[45] will recognize, for example, that a given geometry

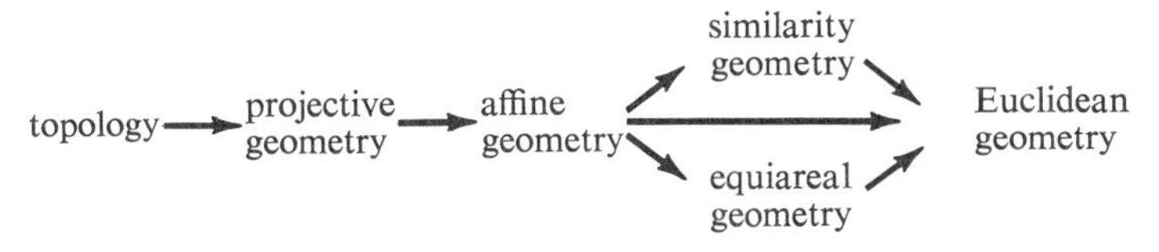

becomes "the next geometry" by selective changes in the postulational structure and can point out the evolution of certain theorems from other

[45]Bruce E. Meserve, *Fundamental Concepts of Geometry* (Cambridge: Addison-Wesley Publishing Company, Inc. 1955) p. 216.

theorems which originated in simpler settings. Theorems on the complete quadrilateral in projective geometry can evolve into theorems on trapezoids and parallelograms in affine geometry and then remain valid in Euclidean geometry. Certain theorems in conics which involve only incidence of lines and which arise in projective geometry can become theorems on parabolas, ellipses, hyperbolas, or pairs of interesting straight lines in affine geometry and on the circle in similarity geometry. Indeed, the theorems valid in projective geometry are valid in each successive geometry for it does not change the validity of the first theorems to add more postulates. The earlier theorems, however, produce other special theorems due to the effect of the additional postulates. We must help our students perceive this broad expanse of geometries created by different sets of postulates and this suggests the desirability of extending our insights into the field of geometry, guided by topics which arise in the classroom and by the work of other scholars and teachers pioneering on the frontiers of knowledge.

Exercises

1. In Section 6.1, we perhaps made the conjecture that in the complete quadrilateral shown in Fig. 6.37 that the double- (or cross-) ratio,

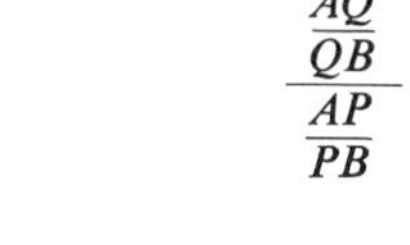

$$\frac{\frac{AQ}{QB}}{\frac{AP}{PB}}$$

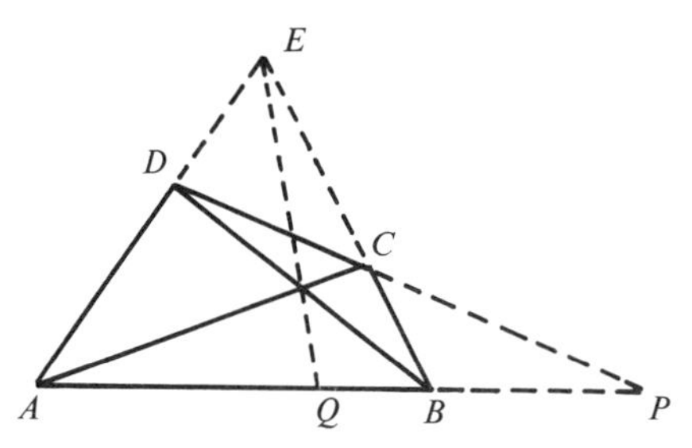

Figure 6.37

has the value -1, or, if *absolute values* or *length measures* are used, the cross-ratio is 1. This figure involves only lines meeting each other to form points and points "joining" each other to form lines; it is a theorem in projective geometry. This theorem is valid also in affine, similarity, equiareal, and Euclidean geometry. In affine geometry *parallels* are permitted, hence the quadrilateral *ABCD* might become the *trapezoid ABCD* (Fig. 6.38).

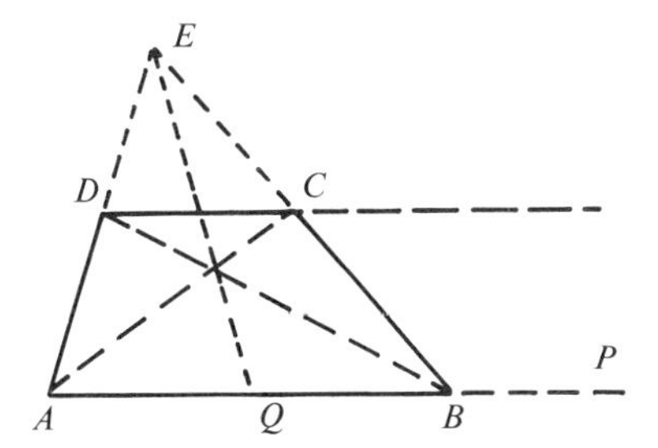

Figure 6.38

The point P will be at infinity. State the conjecture for this trapezoid.

2. In Exercise 1 see if you can determine the position of Q if the cross-ratio is 1, but P is at infinity. At the moment the expression is *indeterminate*, but perhaps you have learned how to attempt to study such indeterminate forms.

3. What theorem(s) does the Pascal Theorem become in affine geometry in which there is a differentiation among conics? In similarity geometry in which there is differentiation between ellipses and circles? In Euclidean geometry? Is this at all related to the result in (i) of Section 6.2?

4. Draw three circles with radii of different lengths (Fig. 6.39) and then construct pairs of common external tangents and note their points of intersection. Do you sense any interesting property about these points? This conjecture on *circles* suggests what other conjecture?

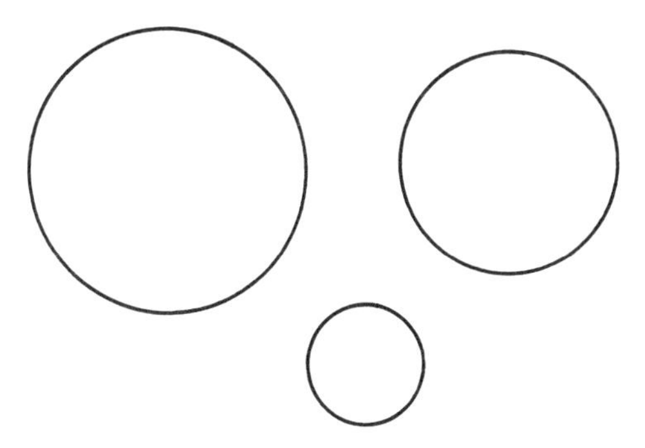

Figure 6.39

5. "Midpointness" enters the evolutionary development of geometry as we make assumptions sufficient to give affine geometry. In projective geometry the most we can say about the diagonals of a quadrilateral is that they

have what relation? In *affine* geometry, however, the quadrilateral may be what kind of figure and what can then be said about the diagonals?

If the teacher and class are interested in theorems from other geometries it is helpful to make a large wall chart with properties which belong to the various geometries and encourage students to formulate hypotheses which may be examined.

6.10. Different Geometries Classified by Transformations

This is, indeed, a fascinating topic which uses linear transformation to classify and probe into various geometries to learn what properties are valid in the given geometry. By using transformations one can cut through the discipline of geometry and see that there are many subgeometries, each with its own features. In fact, Felix Klein, in his inaugural lecture on accepting the chair of mathematics at the University of Erlangen, Germany, in 1872, defined a geometry as that set of properties which remains invariant under a group of given transformations. Euclidean geometry is the set of properties which remain invariant under the group of transformations which are (then) called Euclidean; likewise, affine geometry is the set of properties left invariant under transformations which are called affine, and so on. The various kinds of transformations are named in such a manner that the properties which they leave invariant are generated by systems of postulates which have the same name. There are no discrepancies in the two points of view of geometries, the creating of the geometry by generating theorems from postulates and the assembly of a set of properties left invariant under transformations, for it is possible to construct the same geometry from either standpoint.

This study is well treated in Meserve's *Fundamentary Concepts of Geometry*;[46] one might note also his article, "Euclidean and Other Geometries."[47] These different geometries might well be pursued further for teacher insight, but it may also interest advanced students in high school for there is opportunity for review of fundamental ideas and skills. The exercises are designed to illustrate the preceding discussion. The two points of view of transformations, the active and the passive, will not be discussed here. Previous to the use of transformations to help study geometry, however, students should learn what they are, what they do, and some of their properties. Much of this can be done by the use of study guides.

---

[46]Bruce E. Meserve, *op. cit.*

[47]Bruce E. Meserve, "Euclidean and Other Geometries," *The Mathematics Teacher*, LX (January, 1967), 2-11.

Exercises

1. Under the transformation

$$T: \begin{cases} \bar{x} = 3x + 5y + 4 \\ \bar{y} = x + 2y - 1 \end{cases}$$

the points whose names are (0, 0), (10, 0), (2, 8) are mapped into what points, respectively?

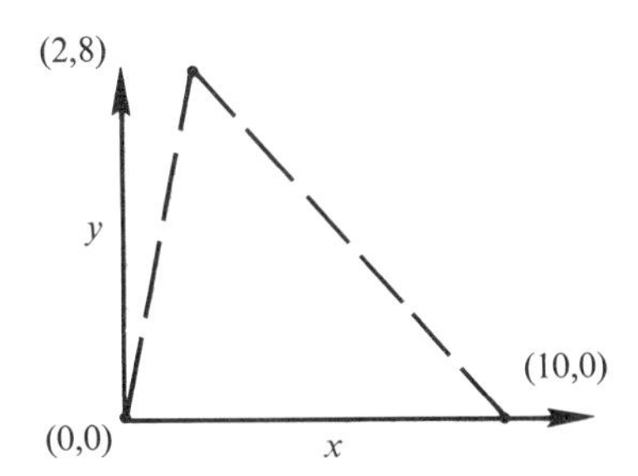

2. What is the area measure of the given triangle in Exercise 1? You may recall that one way to compute the area is by the use of this expression which involves a third-order determinant:

$$A = \frac{1}{2}\begin{vmatrix} 0 & 0 & 1 \\ 10 & 0 & 1 \\ 2 & 8 & 1 \end{vmatrix}$$

Is the *area* of the new triangle the same or different?

3. Assume two points on the coordinate plane and determine the *slope* of the line which goes through them. Now apply the above transformation to the points and see if the slope has changed.

4. Is the *parallelism* of two lines invariant under the given transformation? Plan and carry out a study to investigate.

5. Is the *perpendicularity* between two lines invariant under the given transformation? Plan and carry out a study.

6. Is the *length* of a line segment invariant under the given transformation? Investigate.

7. See if the *collinearity* of three points is invariant under the given transformation.

8. Investigate to see if the property of *circularity* is invariant under the given transformation. To do this easily, one can assume $x^2 + y^2 = r^2$ to be the equation of a circle. Solve the transformation equations for x and y and place their respective values in $x^2 + y^2 = r^2$. Is the resulting relation between $\bar{x}$ and $\bar{y}$ still the equation of a circle?

9. The investigations made above were perhaps worked out with *number* as coordinates. At best our claims that certain properties are invariants have only what status? What can be said about properties we suggest are *not* invariant? What must be done too see if the conjectures are theorems?

The transformation given in Exercise 1 is an example of an *equiareal* transformation. The general form is

$$T = \begin{cases} \bar{x} = ax + by + c \\ \bar{y} = dx + ey + f \end{cases}$$

where $|ae - bd| = 1$. The properties which we studied in the previous exercises and which we found were invariant are properties which belong to equiareal geometry. In this geometry triangles are congruent if they have the same area. Congruence, referring to equality of lengths and equality of angle measures, as we ordinarily understand it, does not exist. There are no perpendiculars or circles in this geometry and hence there are no theorems about any properties of these kinds. One can readily see the opportunities here in using this source of investigation. In advanced classes, students have imposed the condition on the general transformation

$$\begin{cases} \bar{x} = ax + by + c \\ \bar{y} = dx + ey + f \end{cases}$$

that area be invariant and have obtained $|ae - bd| = 1$ as a necessary condition, along with some review of fundamental skills and ideas with determinants. Also it is, in general, easier by the use of transformations to tell if a given property belongs to a certain geometry than by reasoning from the postulates.

Some transformations are placed here for reference.

1. Euclidean transformation:

$$\begin{cases} \bar{x} = ax + by + c \\ \bar{y} = dx + ey + f \end{cases}$$

where

$$ae - bd = 1, \quad a = e, \quad b = -d$$

(or where $ae - bd = 1$ and $a^2 + d^2 = 1$ and $b^2 + e^2 = 1$)

2. Equiareal transformation:

$$\begin{cases} \bar{x} = ax + by + c \\ \bar{y} = dx + ey + f \end{cases}$$

where

$$|ae - bd| = 1$$

3. Similarity transformation:

$$\begin{cases} \bar{x} = ax + by + c \\ \bar{y} = akx + bky + f \end{cases}$$

where

$$a^2 + b^2 \neq 0, \quad k^2 = 1$$

4. Affine transformation:

$$\begin{cases} \bar{x} = ax + by + c \\ \bar{y} = dx + ey + f \end{cases}$$

where

5. Projective transformation:

$$\bar{x} = \frac{ax + by + c}{hx + jk + 1}$$

$$\bar{y} = \frac{dx + ey + f}{hx + jk + 1}$$

where

$$\begin{vmatrix} a & b & c \\ d & e & f \\ h & j & 1 \end{vmatrix} \neq 0$$

A Euclidean transformation may represent a rotation of geometric figure about an origin, it may represent a translation, or one of these transformations followed by the other.

The study of various kinds of geometries as created by different kinds of transformations not only enriches the outward-looking aspect of mathematics (mapmaking, photography, the physics of deformation are but a few examples) but it also helps the student witness the panorama of geometry. One cannot refrain from illustrating one facet of this wide view as pointed out by Modenov and Parkhomenko as they see "length" generalizing into "cross-ratio." They write

> We thus see a steady process of generalization. Under orthogonal mappings, lengths are preserved; under similarity mappings, lengths in general, are not preserved, but ratios of lengths are; under affine mappings, ratios are not, in general, preserved, but ratios of segments on the same line are; under projective mappings, ratios are not, in general, preserved . . . even for segments of the same line, but *ratios of ratios* (italics ours) of adjacent segments on a line are.[48]

---

[48]P. S. Modenov and A. S. Parkhomenko, *Geometric Transformations, Volume* 2, *Projective Transformations* (New York: Academic Press, 1965), p. 40.

Somehow geometry must be taught so that our students perceive and attempt to organize and state generalizations as the one quoted above.

The geometric interpretations of what transformations do to the coordinate axes in general is interesting and can be developed by the use of carefully prepared guide sheets. For example, after the student is aware, that the transformation

$$\begin{cases} \bar{x} = x + 3 \\ \bar{y} = y - 8 \end{cases}$$

moves the origin (and, therefore, the axis system) *three* units to the *left* and *eight* units up, one can raise questions as these on this "passive" point of view of transformations:

1. Let us study the transformation

$$\bar{x} = 3x$$
$$\bar{y} = y$$

Under this transformation the point (4, 3) takes on what new name?

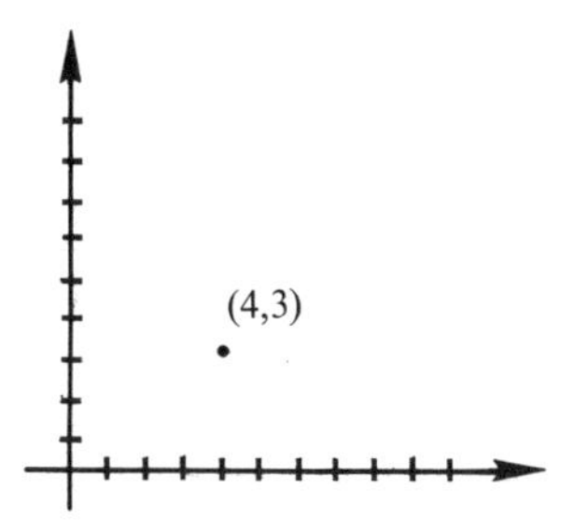

2. If the point remains in the same position, what change must take place in the x-axis to give the point its new name? Draw the new axis system over the original one.

3. Hence what change takes place on the x-axis under each of these transformations? In each case draw the new axis system over the old one.

(a) $\begin{cases} \bar{x} = 4x \\ \bar{y} = y \end{cases}$ (b) $\begin{cases} \bar{x} = 6x \\ \bar{y} = y \end{cases}$ (c) $\begin{cases} \bar{x} = \frac{1}{2}x \\ \bar{y} = y \end{cases}$

4. Given the transformation

$$\begin{cases} \bar{x} = 3x - 4 \\ \bar{y} = y \end{cases}$$

the point whose name is (3, 2) takes on what new name? If it remains unmoved, what change must take place in the axis system? Superimpose the new axis system on the old one.

5. The transformation

$$\begin{cases} \bar{x} = 4x - 5 \\ \bar{y} = y \end{cases}$$

produces what change in the axis system? Draw the new axis system and test your answer for the point $(-1, 3)$.

6. Draw on the original system the axis changes produced by the transformation

$$\begin{cases} \bar{x} = 3x - 4 \\ \bar{y} = 2y + 7 \end{cases}$$

Test your answer by the use of the point whose name is $(3, 4)$.

7. Suggest what the transformation

$$\begin{cases} \bar{x} = -3x - 2 \\ \bar{y} = -2y + 7 \end{cases}$$

does to the axis system. Draw. Check your answer by using the point $(2, 1)$.

8. Summarize the effect on the axis system of the transformation

$$\begin{cases} \bar{x} = ax + b \\ \bar{y} = cy + d \end{cases}$$

for $a > 1$, $c > 1$; $0 < a < 1$, $0 < c < 1$; and $a < 0$, $c < 0$, where b and d are positive in all cases. If b and d are negative in all cases, what effect is produced?

We have yet to study the effect on the axis system of a transformation of the kind

$$\begin{cases} \bar{x} = 3x + 2y + 7 \\ \bar{y} = 2x + 5y - 2 \end{cases}$$

Scale changes on the respective axes remain the same but the translation is different at different places on the coordinate plane. For example, at the point whose name is $(3, 1)$ the transformation becomes

$$\bar{x} = 3x + 2(1) + 7 = 3x + 9$$
$$\bar{y} = 3(3) + 5y - 2 = 5y + 7,$$

or, each unit in the x-axis becomes replaced by three shorter units and each unit on the y-axis is replaced by five shorter units, then the origin of the original system is moved *nine* new units to the left and *seven* new units down. The reader should discuss the changes in the axis system in the neighborhood of other points. It is remarkable that if the coefficients in the transformation

$$\begin{cases} \bar{x} = ax + by \\ \bar{y} = dx + ey \end{cases}$$

are "just right," namely $a = e$, $b = -d$, and $a^2 + b^2 = 1$, the scale changes and translations at different points caused by the transformation force the axis system to behave as though the transformation had been rotation about the origin (0, 0).

This discussion of transformations shows us how we can determine if a certain property is valid in a certain geometry and illustrates the Erlangen idea of geometries. Teachers and programs, however, sometimes make use of transformations in another way which is described in a later section.

6.11. Improving the Postulational Basis of School Geometry

The critical study of Euclid's geometry produced significant changes and creations other than various kinds of geometries and different sequences. Indeed, not only did the Postulate 5 controversy shake the very foundations of geometry and ultimately give rise to the study of axiomatics, but also new kinds of rumblings arose against the postulate system of Euclid. These suggested that from the five postulates *as stated by him* Euclid could simply not have proved his 465 theorems. Euclid had made some assumptions which he had not stated—he was guilty of omissions in his logical structure. One of these assumptions had to do with betweenness or order of points on a line and the other with continuity of a line (that there were no gaps in the line). Euclid assumed (tacitly) that if AB were produced to C then B would be between A and C and never C between B and A. He also assumed, but did not state, that if two lines intersected there would always be a point. It is felt by some teachers that these are some questions to be raised after a student has had a little experience in geometry, to see if the text being used defines lines to have these properties. To look at such matters too early in a course, perhaps, may tend to kill geometry because the student is not yet ready for such niceties. We must never forget that *proof* is that which is convincing to the student at his level and logical intricacies should be introduced as they can be appreciated.

Although others sought to improve Euclid's logical structure, one of the most influential contributions was that by David Hilbert in his *Grundlagen der Geometrie* (1899) on present-day mathematics. He erected a postulational structure to overcome the shortcomings of Euclid by the use of five primitive terms (point, line, on, between, congruent) and fifteen postulates stated in five groups under the headings:

Postulates of Connection (Incidence)
Postulates of Order
Postulates of Congruence
Postulate of Parallels
Postulate of Completeness

These are listed and discussed in the various treatments of geometry already mentioned.

Hilbert stated postulates for Euclidean geometry explicitly rather than reading into the old postulates of Euclid meanings which were not stated. This given rise to concepts in modern geomtry texts: postulates on betweenness, more fundamental theorems on the incidence of lines and points, and postulates on congruence rather than the congruence by superposition proved by Euclid. Also, the classification of the postulates into sets helps to characterize the geometries. It has been found helpful to describe Euclidean geometry as the *COCPC* geometry, each letter representing a division of the Hilbert postulates. The postulational beginnings of other geometries can be indicated roughly in this manner:

Euclidean geometry	*COCPC*
Absolute geometry	*COC__C*
Lobachevskian geometry	*COC~PC*

The kind of congruence, for example, may be different in different geometries, but the congruence idea is present. The blank in the position of *P* in the absolute geometry means that the "uniqueness of parallels" postulate is missing. ~*P* means a denial of the parallel postulate; namely, in this case, that through the point *Q* there can be more than one line parallel to a given line ℓ. Riemannian geometry has, of course, incidence postulates, congruence properties, and a denial of the parallel postulate, but it is not an ordered geometry; its postulational origin might be represented roughly by *C__C~PC.*

Exercises

1. In Euclid's *Elements* the first proposition (theorem) is "On a given straight line to construct an equilateral triangle." Indeed every step in the proof can be well supported by Euclid's primitive statements except one. Which step is this? Formulate a postulate which Euclid needs.
2. It is interesting to note that although we seem to use Euclid's compass freely as a "line transferer," the compass was postulated to construct circles only, and if the compass were lifted from the paper, postulationally it col-

lasped. His Proposition 2 authorizes the use of the compass to transfer lines. His proposition states, "To place at a given point (as an extremity) a straight line equal to a given straight line." Prove that this can be done with a compass used only as a circle constructer, in the spirit of Euclid's Postulate 3. See Wolff's *Breakthroughs in Mathematics*[49] or other sources. This proof can be done by high school students.

3. Euclid's Proposition 16 is familiar to all geometry students: "In any triangle, if one of the sides be produced, the exterior angle is greater than either of the interior and opposite angles." Euclid's proof of this theorem does not use the fact that the sum of the interior angles of a triangle is that of two right angles. Euclid lets D be the midpoint of $\overline{BC}$, produces $\overline{AD}$ to F along $\overline{AD}$ so that $\overline{AD} \cong \overline{DF}$. Prove that $m\angle 3 = \angle 1$. Show that Euclid cannot prove this, unless in his primitive statements, he postulates that if AD is produced to F then D will remain between A and F. Discuss the result, if any, if it should happen that F lies between A and D. What primitive properties might Euclid have given to his "line" to strengthen his proof?

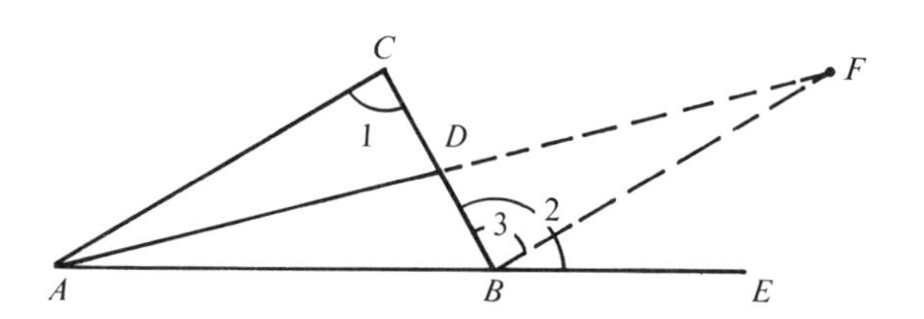

4. Consult geometry texts to find a (fallacious) proof that "every triangle is is isosceles." See, for example, Wentworth and Smith's *Plane Geometry*[50] and E. A. Maxwell's *Fallacies in Mathematics*.[51] What is wrong with the proofs presented? Is the error related to the idea of betweenness?

5. Let the notation [*ABC*] mean that *B* is between *A* and *C*. Let us postulate that "if [*ABC*] then [*CBA*] but not [*BAC*]." To practice drawing conclusions by the use of this postulate, what would be the consequence if it is given that points *P*, *Q*, *R* are on a line and we have (a)[*PQR*]; (b) [*QRP*]; (c) [*RQP*]; (d) [*RPQ*]? The purpose of this exercise is to show that it is helpful to students to have practice in interpreting the meaning of definitions and postulates before attempting to use them in proofs.

---

[49]Peter Wolff, *Breakthroughs in Mathematics* [New York: The New American Library of World Literature, Inc. (Signet Science Library Books), 1963], p. 54.

[50]George Wentworth and David Eugene Smith, *Plane Geometry* (New York: Ginn and Company, 1913), p. 273.

[51]E. A. Maxwell, *Fallacies in Mathematics* (Cambridge: University Press, 1959).

6. From the postulate stated in Exercise 5, prove that if $[ACB]$ then not $[BCA]$.

Problems as these help the student to explore postulational beginnings and bring to his attention the necessity of examining foundations of mathematics.

In 1932, George D. Birkhoff presented a set of postulates[52] for Euclidean geometry which made immediate use of the real number system; he employed the scaled line and protractor. In the hands of other writers his ideas have become the basis for a ruler-and-protractor geometry. In its beginnings the geometry is coordinate in character and not synthetic like the Hilbert-based geometries or like the beginnings of Euclid. Order and continuity come initially from these same properties of the number line.

It is not meant to suggest that there have been only two contributions to eliminate the logical shortcomings of Euclid, namely those of Hilbert and those of Birkhoff, for, indeed, there have been many. The two mentioned above are those who have affected the programs in secondary school geometry most recently. Some have adapted the Birkhoff postulates to secondary school use. Others have based their geometry for secondary schools on the Hilbert postulates. In the logical developments of the Hilbert-flavored texts one sees no coordinate-based beginnings; it is synthetic in treatment. The emphasis in the new programs is clarity, precision, and better logical undergirding.

The improved programs in geometry provide just as much opportunity for the use of student-centered approaches in teaching as the traditional courses. It does not change the learning process and the nature of the student to have newer concepts included. There is even more need for creativity. In the Birkhoff-based geometry, after the distance between points A and B has been defined as

$$d(AB) = |y - x|$$

then the student might be led to think about "betweenness."

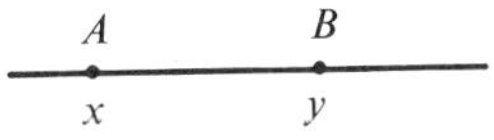

A B

x y

1. What relations do you suggest to define the property of "point B being *between* points A and C?"

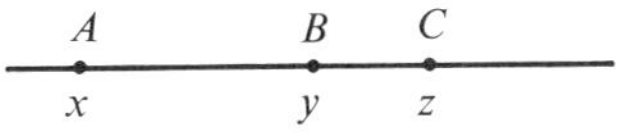

A B C

x y z

---

[52]George D. Birkhoff, "A Set of Postulates for Plane Geometry, Based on Scale and Protractor," *Annals of Mathematics*, 33 (1932), pp. 329-45.

Some students will suggest: "The coordinates must satisfy $x < y < z$" while others may say $(y - x) + (z - x)$ must equal $(z - x)$. Others will remark that $d(AB) + d(BC) = d(AC)$. Can one of these characterizations be proved from the other? Which will be adopted as the definition to be used in class? Which, therefore, will be regarded as theorems?

2. Given three points A, B, and C, on a line, is there always a betweenness relation. Can you argue why?
3. If A, B, and C have the betweenness relation $[ABC]$ can it have another? Can you support your answer by proof?

The challenge to the teacher is to have his subject so well at hand that he can plan experiences to help the students think about the subject actively rather than learning passively what others have done.

Such simple, yet fundamental, matters as what we mean when we say, "A is on one side of the line ℓ and B is on the other" or that "A and B are on the same side of the line ℓ" lead the student quite naturally to concepts of separation. Indeed, it is believed by some that the idea of separation is one of the earliest developed by children.[53] Such questions as these should be introduced:

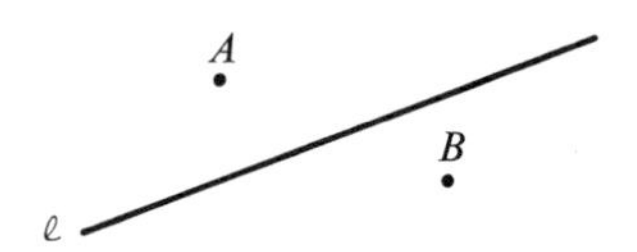

1. If A and B are on opposite sides of ℓ and A and C are on opposite sides of ℓ what conjecture can you make about B and C? Can you prove this?

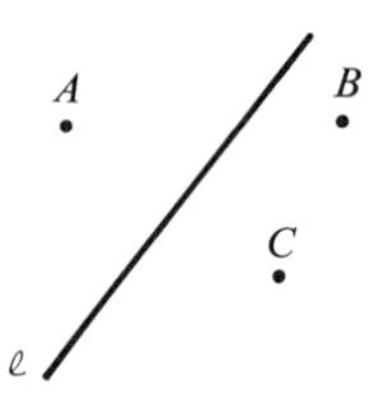

2. If we have the betweenness relation $[ABC]$ and the line ℓ is drawn through C, what can be said about points A and B relative to the line ℓ? Try to prove this.

---

[53]Nicholas J. Vigilante, "Geometry for Primary Children: Considerations, Considerations," *The Arithmetic Teacher,* 14 (October, 1967), 453-9.

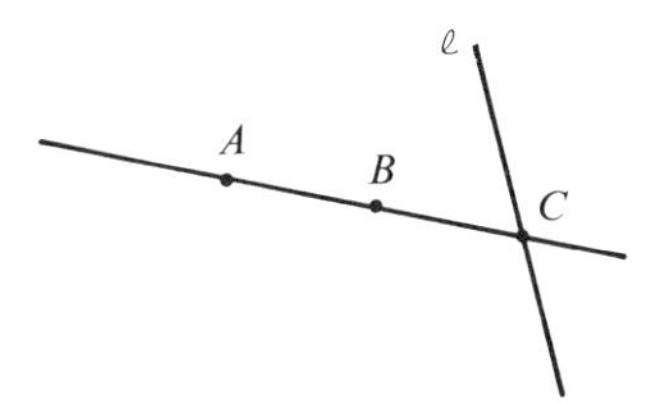

Students can be asked to try to explain what they think is meant to say that the point P is in the *interior* of $\angle AOB$ or what we mean by the interior of an angle. The requirement on P that the measures of angles $\angle BOP$ and $\angle AOP$ can be *added* is then an interesting topic. Finally, the student has enough background to prove the exterior angle theorem and might be helped in this way (the student has already shown in previous exercises that $\angle 4 \cong \angle 3$).

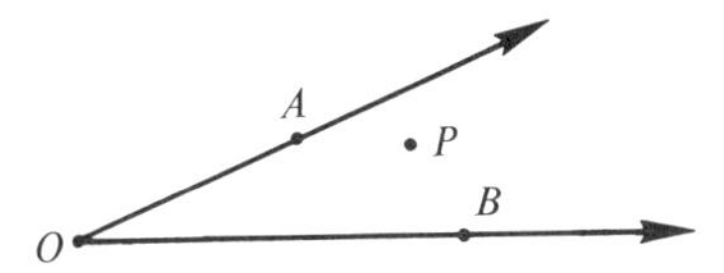

1. If we can show that $m\angle 4 + \angle 6 = m\angle 5$ will this prove that $m\angle 5 > m\angle 3$? How?

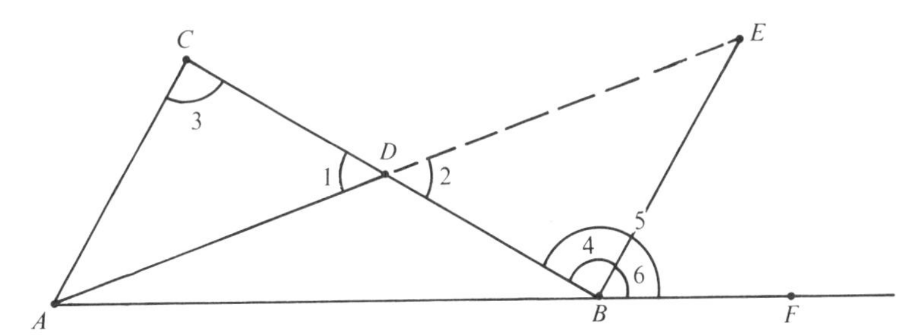

2. What must we prove about E before we can make the statement that $m\angle 4 + m\angle 6 = m\angle 5$?
3. See if you can prove on the basis of what we have done that E is in the *interior* of $\angle FBC$.

Such exercises guide the student, call for contribution on his part, and keep him close to the progress of the course.

After Hilbert-and Birkhoff-based beginnings are used in the more modern programs, the succeeding geometry in many cases is common to that found in older texts. Of course, it is Euclidean geometry. Better teaching and increased consciousness of student needs to facilitate the learning process are always in demand no matter what program is adopted. Sometimes improved instruction in mathematics is confused with increased rigor; these two facets are not the same.

6.12. Coordinate Geometry

In Chapters 2 and 3 there was some mention of coordinate geometry. Coordinate geometry may be Euclidean, affine, or projective, but it is distinguished from "synthetic geometry in that the primary elements in a coordinate geometry are assigned *numbers* and the secondary elements are indicated by *equations*. Usually the primary elements are points and the secondary elements are lines, but they need not be so chosen; *lines* may be considered as the primary elements in which case they have coordinates and the points will be expressed by equations. The former (usual) kind of geometry is called "point geometry" and the latter "line geometry"; this is another direction which students may take for individual study and exploration. Circles may even be used as primary elements and geometry studied from another point of view. A circle in point geometry is defined in the usual high school geometry manner but in line geometry it is considered as the envelope of the (primary) straight-line elements as shown in Fig. 6.40.

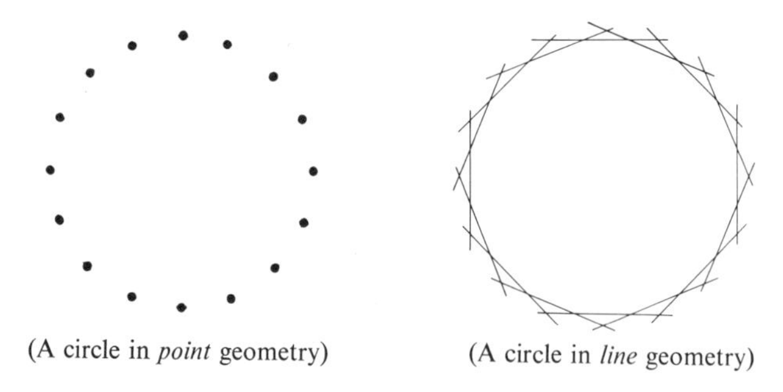

(A circle in *point* geometry) (A circle in *line* geometry)

Figure 6.40

The number of coordinates assigned to a primary element (think of a point) is dependent on the dimensionality of the space in which it is being considered and on the kind of geometry being studied. On a *line* in Euclidean geometry, for example, a point needs only one coordinate, as in the Birkhoff-based geometry beginnings. The "point at infinity" is not like the other (finite) points so it has no number, but usually only some symbol as P_∞ to indicate a point taking a position further and further from the origin. Similarly, on a Euclidean *plane*, two numbers are used to name a point and the points on the "line at infinity" are "different." In a Euclidean three-space, three numbers are used to represent the primary element (point). Coordinate geometry studies and results can be extended readily to higher dimensions becouse of the notation, one advantage in the use of coordinate methods.* If the setting is in *projective* geometry then *homogeneous coordinates* are used: a point on a line has two coordinates, as (3, 2); a point in a plane has three coordinates, as (1, 1, 4), and so on. The point (0, 1) represents the origin on a line, the point (1, 1) represents the unit point, and (1, 0) represents the point which, in Euclidean geometry, was called "the point at infinity." Now there is no distinction among any of the points as is the case on the projective line. Readers may be interested in pursuing further the use of these coordinates in geometry study and texts as Graustin's *Introduction to Higher Geometry*,[54] Milne's *Homogeneous Coordinates*,[55] and those on projective geometry may help. We see that projective geometry can be treated both synthetically and coordinatewise as well as Euclidean geometry.

In the current school programs coordinate geometry is introduced in the junior high school and certain facets are studied each year: the location of points, the equations of lines, the points of intersection of lines, the slope of a line, the length of a line segment, parallelism, some properties of the conics, and other figures. One of the precautions which should be taken is that the student may begin to think of coordinate geometry as an accumulation of skills or formulae by which to do certain things and may lose sight or never have realized that it is just as "logical" as synthetic geometry. Hence, the necessity and role of primitive terms, primitive statements, and necessary consequences should be emphasized at the proper time just as much as in the more usual geometry courses. Indeed, the student should help to develop these and at some point in his growing mathematical maturity he should look back into his junior high school years and decide what was assumed and what was deduced or proved. Cassius Jackson Keyser, in his *Thinking about Thinking*, says, "the task of postulate detection is urgent, immense, and omnipresent."[56]

There are many opportunities for student discovery and research in

---

*Professor Walter Prenowitz reported at the Toronto (August, 1967) Meeting of the Mathematical Association of American that he is working on a *synthetic* approach to geometry which will have the same advantage, that of extension to higher dimensional spaces.

[54]William C. Graustein, *Introduction to Higher Geometry* (New York: The Macmillan Company, 1937), Chapter XIII.

[55]W.P. Milne, *Homogeneous Coordinates for Use in Colleges and Schools* (London: Edward Arnold and Company, 1931).

[56]Cassius J. Keyser, *Thinking about Thinking* (New York : E.P. Dutton and Company, Inc., 1962), p. 88.

coordinate geometry. In the lower grades the results of throwing dice can be recorded on a lattice (coordinate system) constructed with one red and one blue axis which corresponds to the colors of the dice; or, if the rows and columns of the students' desks are straight, each student can be represented by a pair of numbers. Finally as the student moves into junior high school questions situations as these can appear on guide sheets or as parts of assignments:

1. Tommy and Johnny were camping in a field which had squares marked as in Fig. 6.41. Tommy was at position (1, 2) and Johnny at position

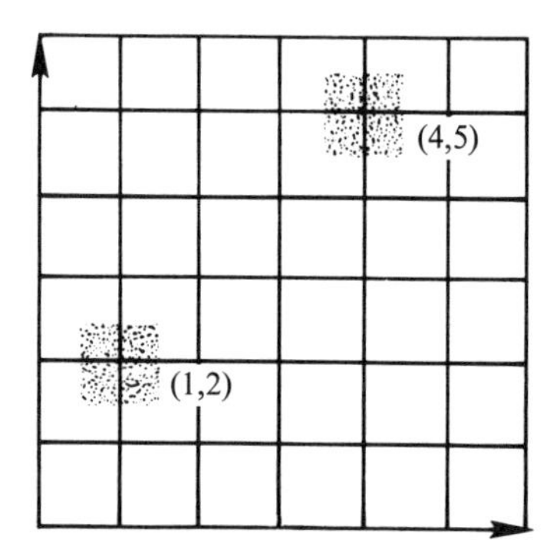

Figure 6.41

   (4, 5). How far must Tommy walk to go directly to Johnny's tent assuming that there are no tents in the way? (Some students will use a string applied to one of the scale lines to determine the distance; some will suggest the use of the Pythagorean Theorem. Merits of various methods are discussed; one never knows what suggestions he will receive.) The use of the Pythagorean Theorem is then encouraged in problems with physical methods used to check. Soon it is appropriate to use letters for coordinates and a theorem is born!

2. Joseph and Mark made themselves some ladders to go into trees by using heavy boards (Fig. 6.42). Joseph said that his ladder was surely steeper, but finally they wondered if one couldn't assign numbers to steepness. What suggestion do you have? (Some students have suggested the "height over the length" of the right triangle formed; others, the "hypotenuse" to yield a steepness number. One suggested once that if the height minus the length is zero the ladder is inclined at 45°—the greater this difference, the steeper the ladder. These definitions can be given the names of the students who suggest them and in subsequent work the "steepness numbers" can be compared. Some definitions seem to fit the picture; those which do not are discarded. For example, to

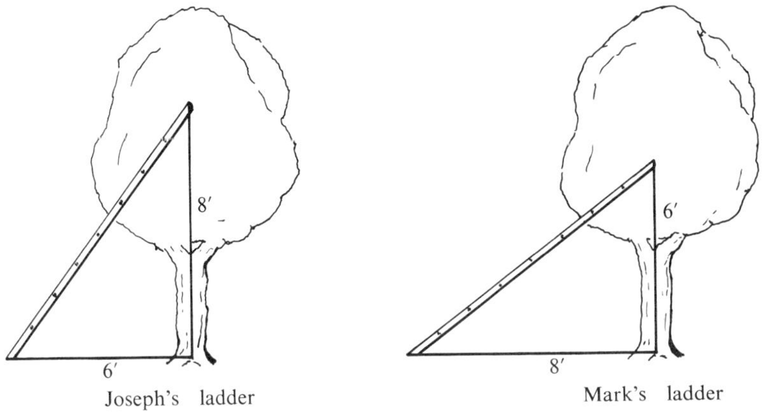

Figure 6.42

use "height minus length" as steepness does not seem natural, for when the steepness number is zero the line is not horizontal, but this helps the student to weigh and consider suggestions. Sometimes the teacher or the consulting of texts must eventually guide the students to what others

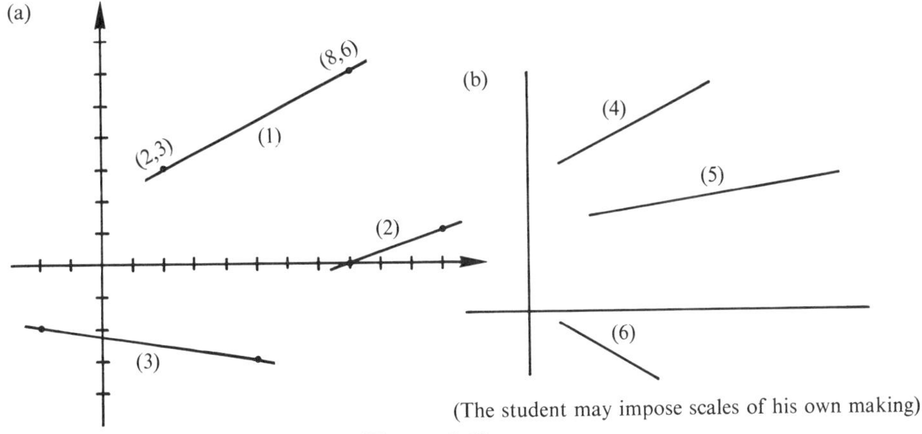

Figure 6.43

have done about steepness. Advanced classes, however, may well explore the consequences of different definitions. For example, if steepness is defined as the "height divided by the hypotenuse" what is the test for perpendicularity of two lines? The field is replete with problems for investigation at almost any level. Several types of problems can follow now to give practice for the student. Determine slopes in each of these cases using Fig. 6.43.

(c) if the line goes through $(-1, 3)$ and $(4, 7)$
(d) if the line goes through $(-1, -4)$ and $(0, -7)$

3. A laboratory approach to the test for the perpendicularity of two lines which has been found to be effective is as follows:

Note the line $\overleftrightarrow{\mathbf{AB}}$ of Fig. 6.44. What is the *slope* of this line? Now

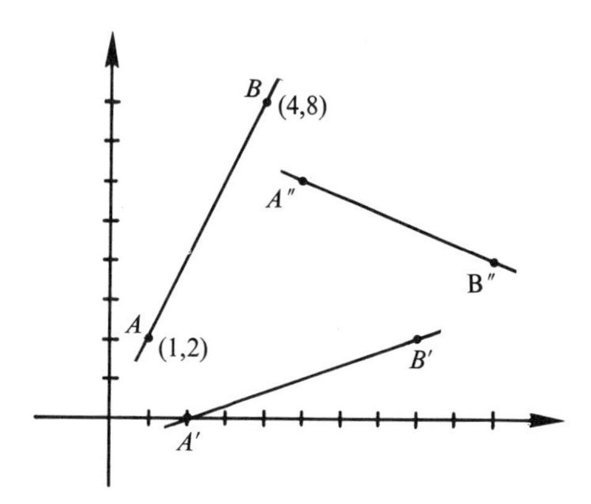

Figure 6.44

construct a line perpendicular to $\overleftrightarrow{\mathbf{AB}}$ at A. Determine from the coordinate system the slope of this perpendicular; keep answer in fractional form. Treat the other lines similarly. Does there seem to be a relation between the slope of a line and the slope of a line perpendicular to it?

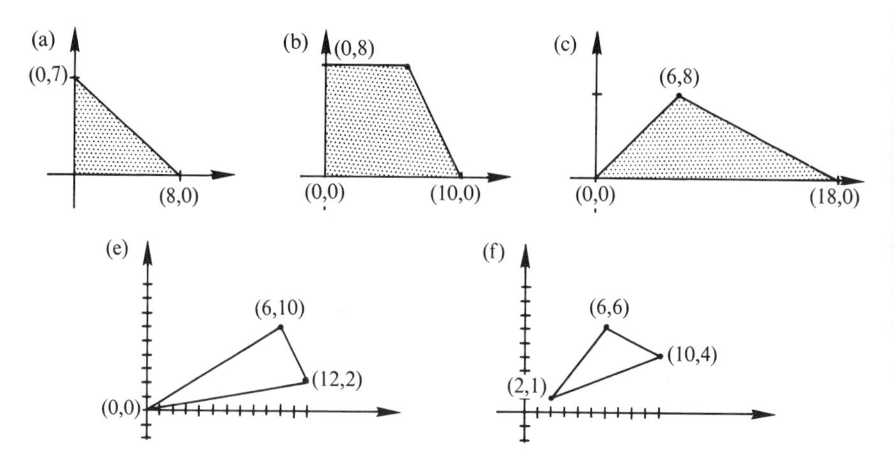

Figure 6.45

The "correct" conjecture usually emerges and, in some cases, remains a conjecture for a long time until eventually it is proved as a theorem.

4. It is interesting to develop an expression for the area of a triangle in coordinate geometry. Preliminary skills can be developed by the use of problems arising such as these (compute areas, Fig. 6.45). Even these will be looked upon differently by different students. Some will attempt complicated methods; others will surround the triangles in question with right triangles to make a rectangle and then determine the desired area by subtraction; others will use trapezoids; some have traced the figures on coordinate paper and have counted squares. Later the *numerical* exercise is changed to one more general to determine the area of the triangle with coordinates expressed in letters as shown in Fig. 6.46.

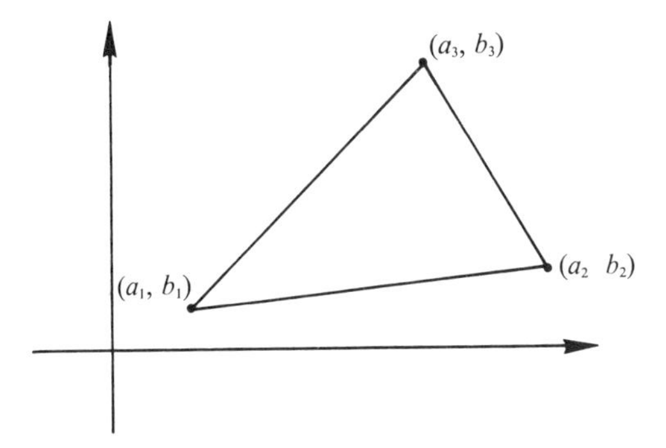

Figure 6.46

Students arrive at

$$A = \frac{a_1b_2 + a_2b_3 + a_3b_1 - a_1b_3 - a_2b_1}{2}$$

as one form of the answer. Students are asked to seek an easy way (a mnemonic) to remember this expression.* Usually some one in the class devises the mnemonic

$$\begin{pmatrix} a_1b_1 \\ a_2b_2 \\ a_3b_3 \\ a_1b_1 \end{pmatrix}$$

---

*The writer first did this with the hope that the determinant would be invented. It never was, but other suggestions were offered. $\begin{vmatrix} a_1 & b_1 & 1 \\ a_2 & b_2 & 1 \\ a_3 & b_3 & 1 \end{vmatrix}$

which is defined to represent

$$a_1b_2 + a_2b_3 + a_3b_1 - a_2b_1 - a_3b_2 - a_1b_3$$

This is named after the person who suggests it, as the "Georgian" or the "Hammerian." It may be up to the teacher or to a search of the literature to show the method of using determinants. Sometimes methods found in textbooks are not the most readily formulated or the ones most natural to the student.

Then can arise another test for the collinearity of three points and a method for determining conditions on the point (x, y) so that it lie on the line determined by two givenpoints, that is, another method for determining the equation of a line.

Advanced students may study the "Hammerian" or the "Georgian" (or the form invented to give the area of a triangle) to see if it has properties similar to those of determinants. Theorems can be formulated about this new form, too, mathematics in *statu nascendi*.

Although the above spirit and "engineering of experiences so mathematics emerges" may well permeate all coordinate geometry topics we close with the above with the hope they illustrate how this subject handled in a student-centered manner can also achieve some of the significant objectives of mathematics study.

Exercises

1. Review several properties of the determinant and see if they hold (or what parts of them hold) for the "Georgian" defined in this section. One can study mathematics with forms invented by the students themselves. (Already the property that two *rows* identical makes the determinant have a value of zero does not hold for the Georgian; examine this property for columns.)

2. Plan a series of exercises from which will emerge a method of computing the distance from a point to a line given the coordinates of the point and the equation of the line.

3. Plan a series of exercises from which will come the equation of a circle.

4. Let us take the *intercepts* of a line as its *coordinates*. Let the line $\overleftrightarrow{\mathbf{AB}}$ have the coordinates (3, 1) and the line $\overleftrightarrow{\mathbf{CD}}$ have the coordinates (5, 5). What conditions must exist on the line (u, v) so that it belongs to the "join" (or point of intersection) of the lines? Hence, we will have the *line equation* of the *point* of intersection. (Ans. u and v must satisfy $4u - 2v - 10 = 0$)

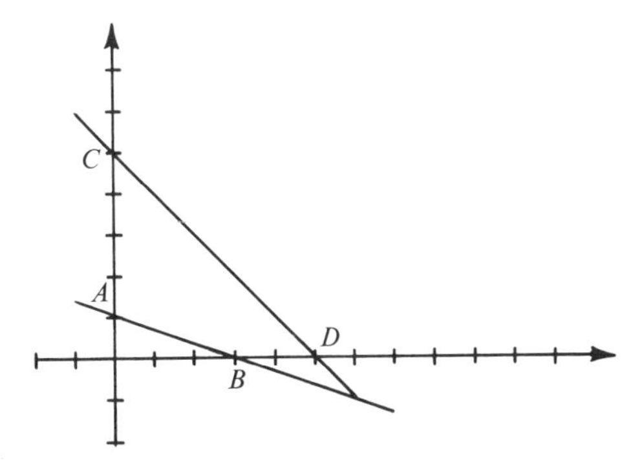

5. Find the (line) equation of the point determined by the lines whose (line) coordinates are $(-1, 3)$ and $(5, 8)$.

6. Suggest the plane dual of the problem in Exercise 5.

Coordinate methods sometimes enable one to perform proofs much more quickly than by the use of synthetic methods and sometimes the synthetic method is more elegant. After the student is skilled in the use of both methods, some teachers use them side by side and hence can point out similarities and advantages. Good reviews of algebra and coordinate methods can be accomplished by proving some of the projective geometry theorems by coordinate methods; the writers have done this in the upper level high school courses. A few suggested exercises are:

Exercises

1. Note the circle on which are placed six points with coordinates and numbers, Fig. 6.47. The point at which the line 12 meets 45 we shall call P; where 23 meets 56 we shall call Q; and where 34 meets 61 we shall call R. The Pascal Theorem states that the points P, Q, and R are collinear. Find equations of lines and points of intersections and confirm this.

2. It may be interesting to try to *prove* this by taking six points on a circle with letters as coordinates.

3. Use the ellipse $x^2/9 + y^2/4 = 1$ and confirm the Theorem of Pascal as in Exercise 1.

4. Use numbers as coordinates of points and confirm by coordinate methods the Theorem of Desargues.

5. Triangle ABC is inscribed in a circle. Fig. 6.48. Let P be any point on the circle and from P drop perpendiculars to the sides of the triangle. Confirm by methods of coordinate geometry, setting up a numerical case, that the feet of the perpendiculars are collinear. The line formed by the three feet is called the Simson line of the point P with respect to the triangle.

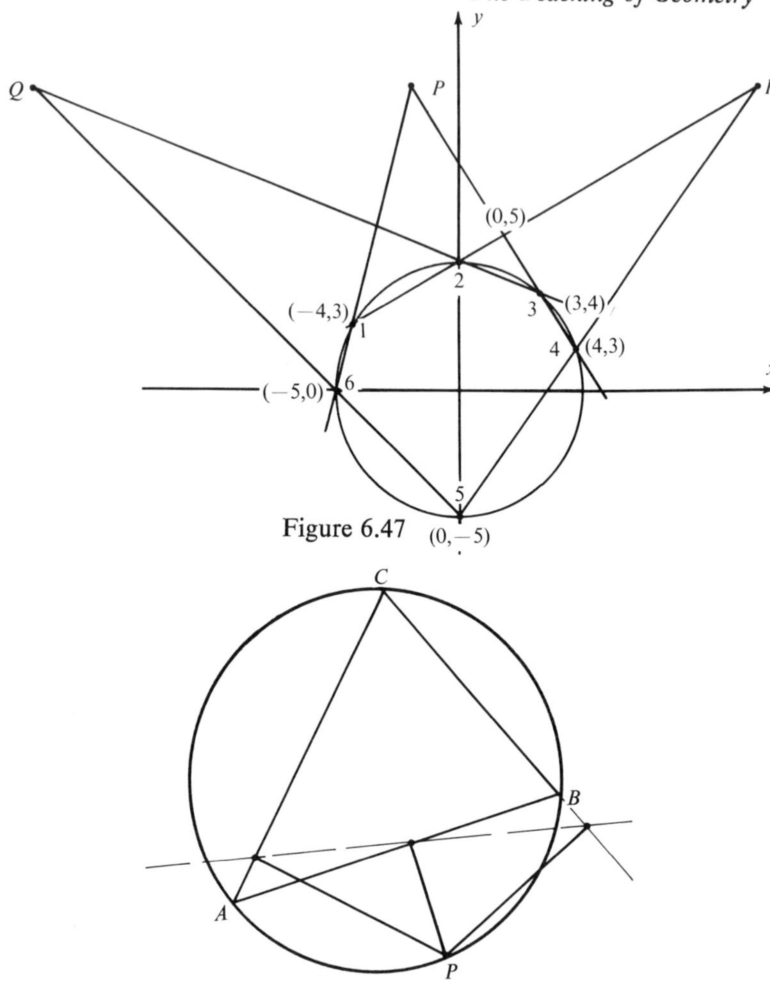

Figure 6.47

Figure 6.48

6. Could the theorem in Exercise 5 be considered in similarity geometry. Would such a theorem as stated ever appear in affine geometry? Why? In projective geometry? Why?

7. Students may be interested in investigating the kind of figure formed by the Simson lines as P takes different positions on the circle.

8. Assume a triangle with ordinary numbers (not letters) as vertices. Test deductively, by coordinate methods, the conjecture that the three medians are concurrent; that the three altitudes are concurrent; that the three perpendicular bisectors to the sides of the triangle are concurrent. Confirm that the three points of concurrency, namely the centroid, or the center, and the circum center, respectively, are collinear. The line on which these three points lie is called the *Euler line* of the triangle. Compute its equation.

6.13. Some Points of View on Geometry

Geometry is an old subject; it has been in existence as a formal discipline since the appearance of Euclid's *Elements* in 300 B. C. We have seen results of the continuing wrestle with Euclid's Postulate 5 until its independence from the other postulates of Euclid was established by Felix Klein and others in the latter part of the nineteenth century. Too, we have summarized some of the shortcomings of Euclid and have seen attempts to strengthen the logical beginnings of ordinary geometry. Then there is the rise of coordinate geometry, vector geometry, "line" geometry, and geometry from the standpoint of transformations. The geometry of Euclid became a model for logical organization and the transfer of this logical character to other fields and to thinking in nongeometric areas becomes inviting to some. Geometry, ever-present as a servant to man, looms overwhelmingly large as a part of that discipline which should be appreciated by the student. The fascination which geometry holds for many people, regardless of profession, and the challenges which it puts forth as its inner structure is being studied provide constantly new areas of inquiry and investigation. As proof of the latter statement, one need only to consider some of the problems introduced in this book and to browse through such works as Eves' *A Survey of Geometry*[57] and Coxeter's *Introduction to Geometry*[58] to catch a glimpse of topics which would be interesting in the secondary school if time permitted, especially if the teacher wishes to study the "traditional" Euclid also.

It is not surprising then that there are different points of view in the teaching of geometry. Teachers who must select texts and programs are or will be confronted with choices which we will attempt to describe here.

First of all the geometry of the future will start earlier and be studied at various times through the entire school program. It will begin with laboratory approaches in the elementary school and the postulational treatment will emerge and be developed as the student is able. Synthetic and coordinate methods will appear simultaneously. An important point of view, however, is that somewhere in the curriculum geometry should be studied as a course or as a unit of study. In *Goals for School Mathematics*, it is advised that if geometry is taught only "in little pieces" the meaning of geometry as a deductive sicence may be lost. "The meaning of geometry as a deductive science . . . depends on the fact that . . . (it) . . . is investigated at length and in depth by the application of the deductive method."[59]

There are points of view on the kind of logical undergirding the course

[57]Howard Eves, *op. cit.*

[58]H. M. S. Coxeter, *op. cit.*

[59]Cambridge Conference on School Mathematics, *Goals for School Mathematics, op. cit.*, p. 78.

in (Euclidean) geometry should have. Hilbert-based geometries do not use the number line to get started and the treatment is entirely synthetic. Coordinate methods are introduced later as an adjunct or as supplementary methods. "Betweenness" and "continuity" postulates are included. One of the earlier texts on the improvement of geometry to be Hilbert-based is that of *Geometry* by Brumfiel, Eicholz, and Shanks.[60] The Birkhoff-based, or the ruler and protractor geometry, was introduced in Section 6.11 and the *School Mathematics Study Group* texts and those of writers who have followed their outline are examples.[61] The aim of these programs is to improve ideas in geometry and to make the language more precise. In older geometry, for example, the notation *AB* had several meanings as the "line *AB*," the "segment *AB*," the "length of the segment *AB*." Both of the above programs have as their purpose to present more correct geometry in a careful manner, though the teacher is still the key for the wise use of texts. Indeed, the writers of *Goals in Mathematics* emphasize

> It is essential to show the greatest of pedagogical skill and insight in helping the student to decide the extent to which he should require proof of validity . . . it is important not to inhibit the student's enthusiasm and facility for solving problems by preoccupying him excessively with scruples about rigor.

The writers continue "The essential point here seems to be to develop intellectual honesty, so the student knows what is being assumed and what is being proved . . . "[62]

Another point of view concerning content and approach to geometry is that of using transformations as elementary operations. This does not mean that one is seeking to learn the properties invariant in different geometries by the study of these properties under various kinds of transformations as suggested in Section 6.10. Rather it means that early in school the student experiments with certain geometric figures to see if he can reproduce them or to see what changes take place by operations of translation, rotation, and other kinds of transformations. He also studies relations between these transformation; at an early age the "group concept" emerges. Some transformations leave lengths of line segments invariant and these are called *isometries;* with lights and shadows the student can learn that some transformations are *not isometric*. If figures are congruent, then they can be carried into one another by any of the transformations which are isometries. Students can study these matters by physical means.

Many interesting questions about transformations arise. If $T(AB)$

---

[60]Charles F. Brumfiel, Robert E. Eicholz, and Merrill E. Shanks, *Geometry* (Reading, Mass.: Addison-Wesley Publishing Company, Inc., 1960).

[61]School Mathematics Study Group, *Mathematics for High School : Geometry* (New Haven : Yale University Press, 1961).

[62]*Op. cit.*, p. 84.

(transformation of point A to point B) is followed by $T(BC)$ then what is the result when the first named transformation is followed by the second? (Ans. $T(BC)\ T(AB) = T(AC)$.) Young students can learn this with concrete materials and gain fundamental ideas without being burdened by notation. *Reflections* about a line are natural and intuitively clear, this is why geometry is so interesting. If $\overline{AB}$ is reflected about ℓ one gets $\overline{A'B'}$. Hence $\ell(\overline{AB}) = \overline{A'B'}$ where $\ell(\overline{AB})$ means *reflection* $\overline{AB}$ about the line ℓ.

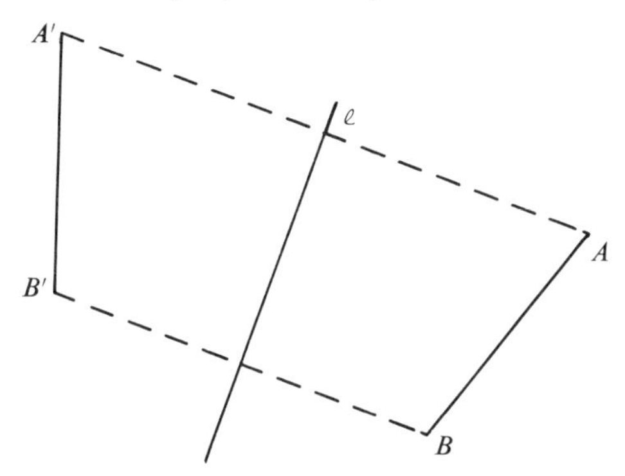

What is the meaning and result of $\ell(\overline{A'B'})$? Students in the elementary school can do this and arrive at the *idea* of an *involution.* Is there a relation between rotations, translations? And reflections? Many of these questions can be explored by the use of homemade and round-about-us things. These matters concern the world we live in, too. Synthetic methods can be used in argument later whereas, in the beginning, one is convinced simply by observing and feeling. The proof at this level is simply, "Behold!" Indeed, legend has it that Euclid has a habit of drawing a diagram on the board, retreating from it, pointing toward it, saying to his students, "Behold!"

Exercises

1. Prove that if two triangles ABC and $A'B'C'$ are congruent there is a unique isometry (1-1 correspondence which preserves lengths) so that the one triangle is carried into the other. Recall that an isometry is either a translation, a rotation, a reflection, or a combination of two or all of these.

2. Prove that if two congruent triangles ABC and $A'B'C'$ have one pair of corresponding vertices in common, then the one triangle can be carried into the other by at most two reflections about lines. Let B and B' be the common points. [The triangles may coincide, or appear as in (a), (b), or (c) (Fig. 6.49); are there other possible arrangements?] This problem is stated in more succinct language as, "If an isometry contains an invariant point, then it is the product of at most two reflections." It would be interesting to prepare a study guide to lead students to invent the idea of proof before recording it.

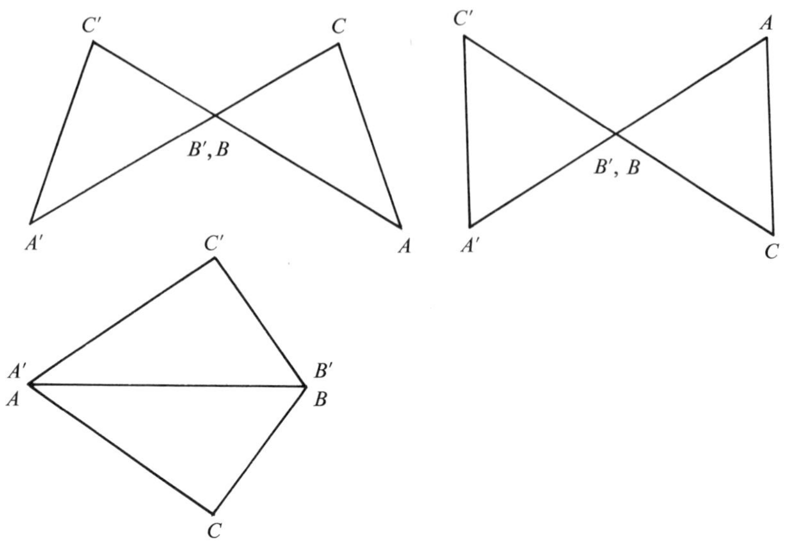

Figure 6.49

3. Prove that every isometry (in the plane) is the product of at most three reflections. Use congruent triangles which have no vertices in common.

Note that the transformation known as a *reflection* about a line, the simplest of which is given analytically by the equation

$$\begin{cases} \bar{x} = -x \\ \bar{y} = y \end{cases}$$

is not a Euclidean transformation at all as defined by Meserve and as discussed earlier. Hence when it is said in some circles that "Euclid must go," there is a grain of truth, if in the remark there is thought to make the basic element in the logical machinery the reflection transformation. Indeed, what is suggested in the remark is that perhaps there is a richer way for the students to begin geometry than to use the approach of Euclid at the outset. Danish, German, Belgian, and English schools are experimenting with programs in geometry based on reflections. A brief summary of some of the recent thinking on possible geometry curricula of the future is found in the booklet *Geometry in the Secondary School* which reports papers given at a joint meeting of The Mathematical Association of America and The National Council of Teachers of Mathematics.[63] For detailed programs one

---

[63]W.K. McNabb, ed., *Geometry in the Secondary School* (Washington, D.C.: The National Council of Teachers of Mathematics, 1967).

might consult F. Papy's *Mathematique Moderne 6-Geometrie Plane*[64] and Kelly and Ladd's *Geometry.*[65]

Another point of view would emphasize *applications* in geometry, and, indeed, would organize the material around some of the important applications. In such a program, of course, the logic of geometry would not be forfeited. The opposing viewpoint is the organization of mathematics as studied for mathematic's sake. Proponents of such programs claim that geometry is intrinsically interesting and need not depend upon motivation by practical application. Most courses of study combine these viewpoints.

So many significant, challenging, and inviting facets of geometry have been invented, discovered, or revived that the field is replete with subjects and problems which furnish opportunities to realize the very same objectives which the study of the traditional Euclid emphasized. Under some of the proposed radically new organization (for example, from the standpoint of *relection transformations*) the student sees Euclidean geometry as a special kind of geometry and much of the traditional geometry of Euclid is learned earlier in an appropriate manner. The use of transformations or projections makes some problems easier and is emphasized. In this mathematics revival, problems which were once studied only by advanced students now become interesting to secondary school students; as an example, one can mention the study of curves of constant width (Euclid had one of these in his construction for the proof of his first proposition) and perhaps the problem of Pick. There is a growing inclination to explore not only works of the past (even some of the "unsolved problems), but also mathematics which is being created now to provide topics in geometry to be used in the secondary school. Films and other materials from *The College Geometry Project of the University of Minnesota* provide hints relative to some of the ideas which may occur in the geometry of the future.[66]

All those working with materials for geometry, however, seem to agree that "in the first course, geometry (should) be formulated so that the student will feel that he is using natural descriptions of objects of experience"[67] and that both "excessive looseness and inaccuracy" and "excessive delicacy and austerity" can equally prevent objectives of the study of geometry from being realized. "Naturalness should be plain at the outset; the student should feel that he engaged in a careful analysis of things that he sees."[68]

---

[64]F. Papy, *Mathematique Moderne* 6—*Geometrie Plane* [Bruxelles 3, 342 rue Royale : Labor (or Montreal, 1029 Bever Hill : Didier)].

[65]Paul J. Kelly and Norman E. Ladd, *Geometry* (Chicago: Scott, Foresman and Company, 1965).

[66]The College Geometry Project of the University of Minnesota, University of Minnesota, Minneapolis, Minnesota 55455.

[67]*Op. cit.*, p. 79.

[68]*Ibid.*

Exercises

1. A circle is a curve of "constant width"; the width is the diameter. The construction with a compass of an equilateral triangle on a given line segment produces, in addition to the triangle, a curve of constant width. Are there others? Have the students make some of these and demonstrate physically that when these curves of constant width are used as rollers they produce the same effect as circular rollers.

2. Which of these rectangles seems most pleasing to you? This is a good way to introduce the "golden section"; it begins with ratio and then leads to algebra, sequences, to curves, to the arrangement of leaves on stems, to creation in nature. See Coxeter's *Introduction to Geometry.*[69]

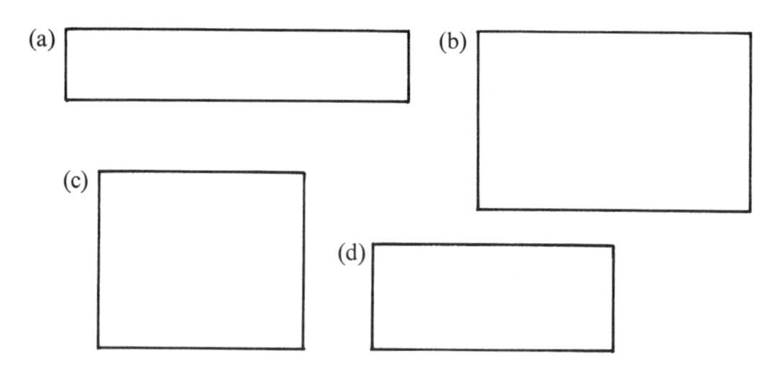

3. Let us consider polygons whose vertices are at lattice points. What is the area of A? of B? C? D? Devise more polygons to confirm the theorem of Pick (1899):

   If a polygon has its vertice at lattice points, then the area is given by $\frac{1}{2}b + i - 1$ where b is the number of lattice points on the

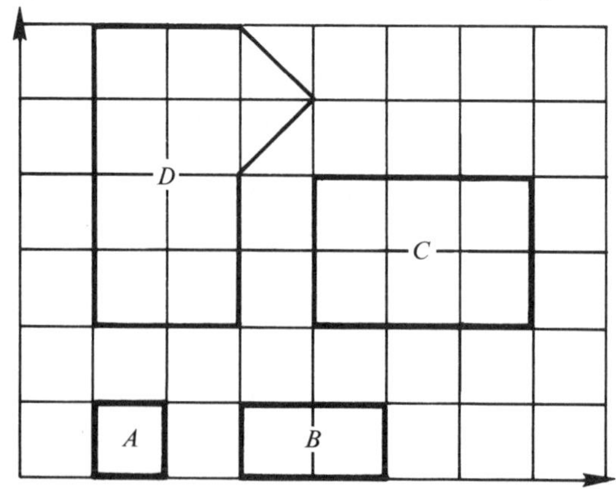

[69]H. M. S. Coxeter, *op. cit.*, Chapter 11.

boundary, and i is the number of lattice points in the interior of the polygon. Laboratory exercises can help students devise a formula for area.

4. If a circle is inscribed in an equilateral triangle the circle is tangent to the triangle at the midpoints of its sides and the medians are concurrent.

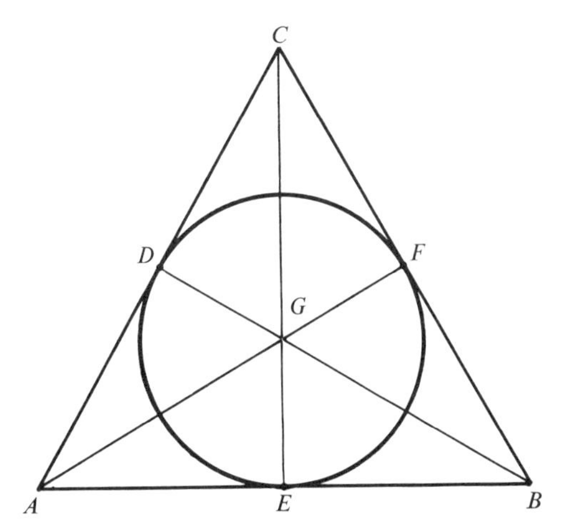

Now under an affine transformation (or under projection from a light source a finite distance away where the "screen" for the image need not be parallel to the plane of the object) the circle may project into what kind of figure? Also since "midpointness" and "concurrency" are invariant, what new theorem about ellipses being inscribed in a triangle will result?

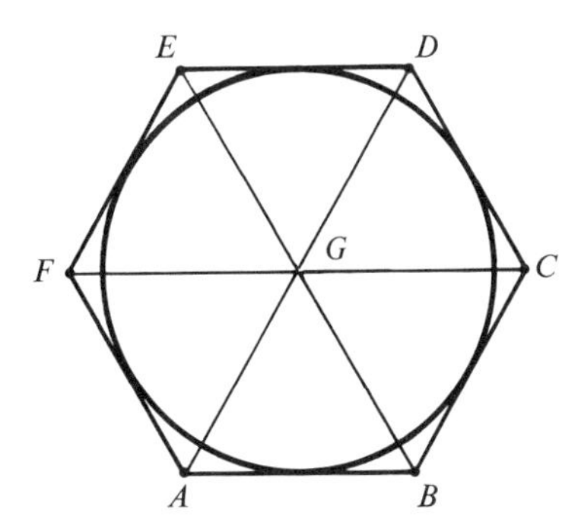

5. It can be proved that if a regular hexagon be circumscribed about a circle, then the diagonals are concurrent. But under a central (finite) point projection the circle may become an ellipse. What new theorem results? Suggest some theorems concerning the original figure which are not now valid?

6. In a circle if a diameter bisects a chord it also bisects the subtended arc. Would a similar theorem hold for an ellipse? What? Why?

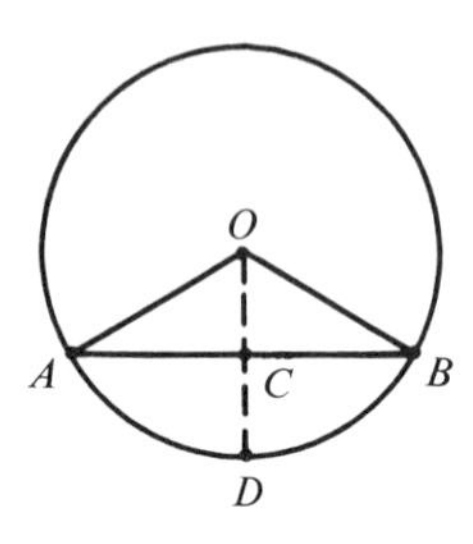

7. A reflection is an example of an isometry. As an example of the use of the reflection transformation to deduce a problem in geometry, let us consider the following:

(a) C is a circle with center O and radius r;

(b) P is a point given external to C;

(c) the problem is the construct a line tangent to the circle C from the point P.

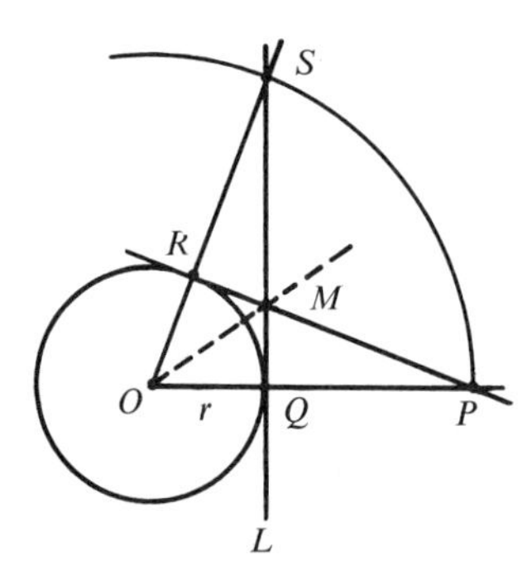

Reflection geometry suggests constructing $\overline{OP}$ with the intersection with circle C called Q; now construct the line L perpendicular to $\overleftrightarrow{\mathbf{OP}}$ at Q. With O as the center construct the circle with radius having the length of $\overline{OP}$. Let the point of interesction of this new circle with L be called S. Draw $\overleftrightarrow{\mathbf{OS}}$ intersecting C at R. $\overleftrightarrow{\mathbf{RP}}$ is the required tangent line. Now you argue that R is the reflection of Q and that P can be carried by reflection to S, and that the circle C remains invariant, all about the line $\overleftrightarrow{\mathbf{OM}}$. Since the property of two lines being perpendicular is invariant under reflection and since L is perpendicular to $\overleftrightarrow{\mathbf{OP}}$ then $\overleftrightarrow{\mathbf{RP}}$ is perpendicular to $\overleftrightarrow{\mathbf{OS}}$. You may be interested in this approach to geometry as discussed in the *Report* (of the) *Commitee on the Undergraduate Program in Mathematics* (*CUPM*) *Geometry Conference*, October, 1967, Number 18.[70]

6.14. Some Sample Study Guides

Sum of the Measure of the Interior Angle of a Triangle

1. *Review*. Evaluate the angles marked with a "?"

(a)

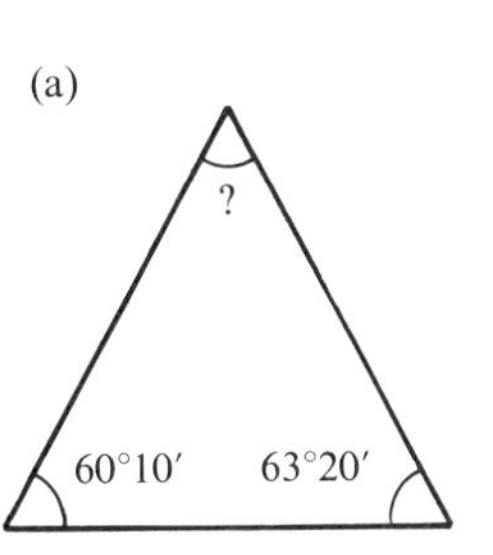

(b)

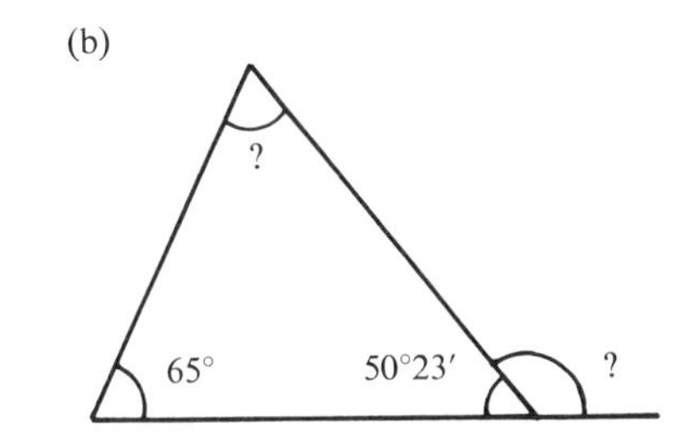

(c)

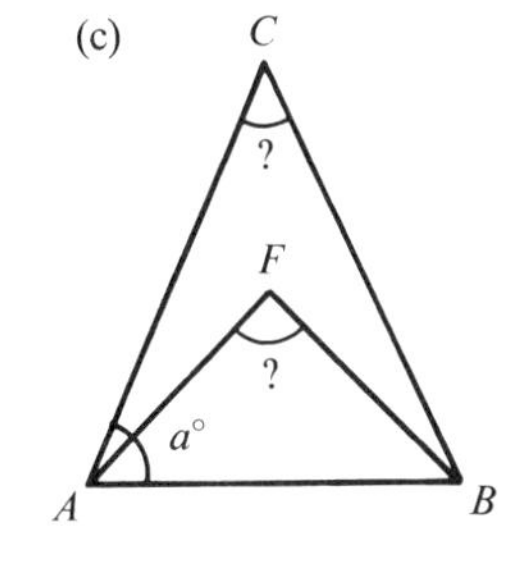

2. *Review*. Argue that the point E is in the interior of DBC.

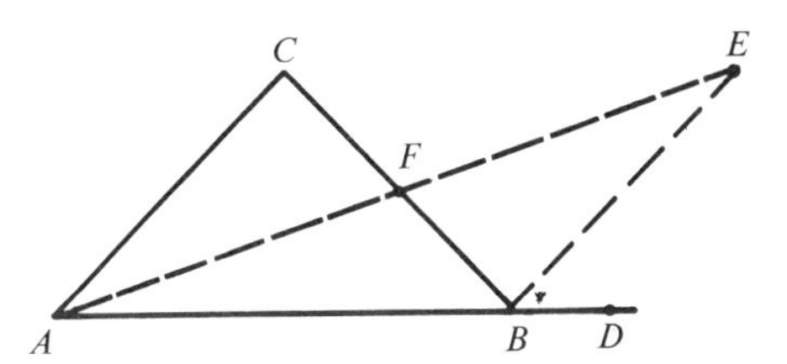

3. *Review*. Prove that the sum of the measures of the interior angles of a triangle is 180°. Write several of the syllogisms (in the form of major and minor premises and the conclusions).
4. *Text* (selected problems)
5. Given that the sum of the measures of the interior angles of the quadrilateral is 360°, find the sum of the measure of the *exterior* angles.
6. Do the same for the *pentagon*. Compute first the sum of the measures of the *interior* angles.
7. Do you think that you might obtain the same result for the sum of the measures of the exterior angles of a *hexagon*? Try it.

---

[70]CUPM Geometry Conference Proceedings, Part III: *Geometric Transformation Groups and other Topics*. Lectures by H.S.M. Coxeter and others. CUPM Report (October 1967), 18. Mathematical Association of America, Inc., 1968.

Secants and Tangents

1. *Review.*

   (a) An angle inscribed in a circle is measured by _______?

   (b) The angle between two chords is measured by _______?

   (c) The angle between two secants is measured by _______?

2. *Review.* Prove that $m\angle E$ is determined by $m(\widehat{AC}) - m(\widehat{BD})/2$.

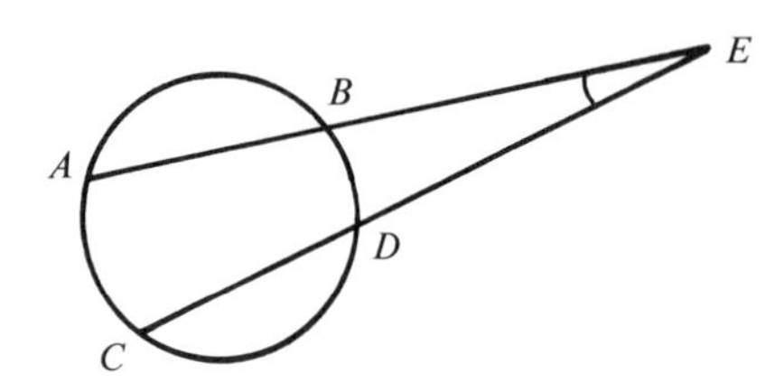

3. *Text* (selected problems)
4. In each of these cases we note the tangent $\overline{AB}$ and the secant $\overline{DB}$. Measure carefully the lengths of the segments AB, BC, and CD and see if there appears any promise of a relation among these three quantities.

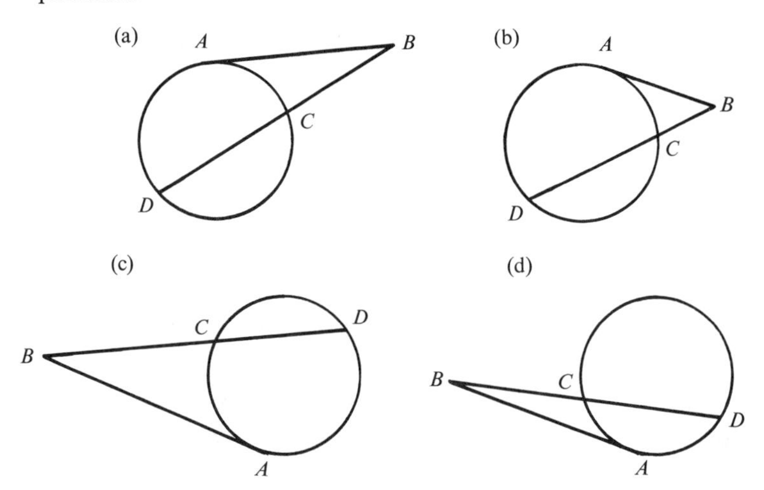

5. Construct several more examples and make similar measurements. Do the results seem to make your conjecture stronger?
6. *Review.* Give an example of

   (a) a dilemma

   (b) an *a fortiori* argument

   (c) a nonmathematical *reductio ad absurdum* argument

Parallelograms and Vectors

1. *Review.* What is the net effect of the "pulls" in the directions shown?

2. *Review.*

   Assumed Data: *ABCD* is a parallelogram

   $\overleftrightarrow{EF}$ is any line through *G*

   Conjecture to Test: $\overline{EG} \cong \overline{FG}$

   Write your examination (proof) of this conjecture.

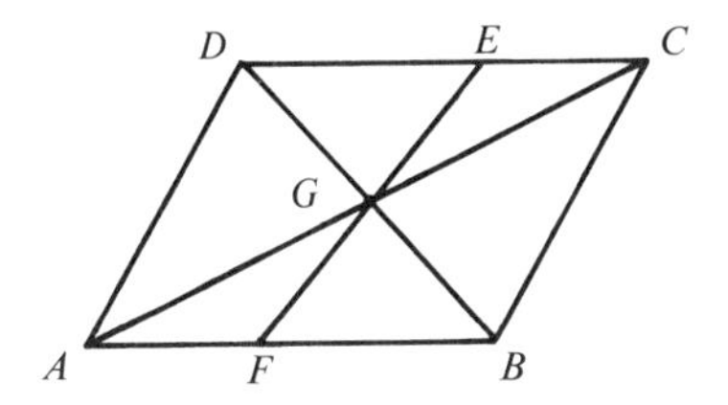

3. *Text* (selected problems)
4. A ball is kicked by two boys at the same time. The thrusts on the ball are summarized by the figure. Which way will the ball go? What *one* force would produce the same effect as the two given ones? What ways might you use to determine this?

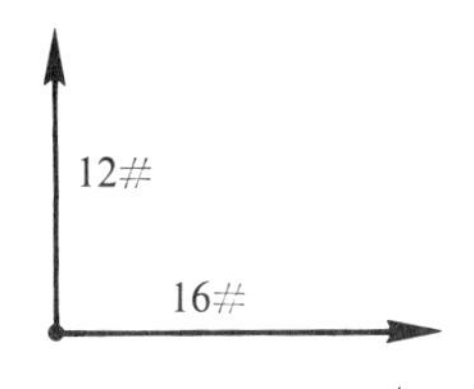

Bibliography

Albaugh, A. H., "Game of Euclid," *The Mathematics Teacher*, LIV (October, 1961), pp. 436–9.

Altshiller, Court N., "Dawn of Demonstrative Geometry," *The Mathematics Teacher*, LVII (March, 1964), pp. 163-6.

Avers, P. W., "Unit in High School Geometry Without the Textbook," *The Mathematics Teacher*, LVII (March, 1964), pp. 139-42.

Banks, M. S., "Flannelboard Geometry," *School Science and Mathematics*, LXIV (February, 1964), pp. 129-30.

Bassler, O. C., "Investigation of the Effect of Types of Exercises on Mathematics Learning," *The Mathematics Teacher*, LIX (March, 1966), pp. 266-73.

Black, J. M., "Geometry Alive in Primary Classrooms," *The Arithmetic Teacher*, XIV (February, 1967), pp. 90-3.

Brinkmann, E. H., "Programmed Instruction as a Technique for Improving Spatial Visualization," *Journal of Applied Psychology*, L (April, 1966), pp. 179-84.

Brown, O. R., "Using a Programmed Text to Provide an Efficient and Thorough Treatment of Solid Geometry Under Flexible Classroom Procedures," *The Mathematics Teacher*, LX (May, 1967), pp. 492-503.

Butler, Charles H., and F. Lynwood Wren, *The Teaching of Secondary Mathematics* (New York: McGraw-Hill Book Company, Inc., 1960), Chapters 16, 17.

Byrne, M. M., "Geometric Approach to the Conic Sections," *The Mathematics Teacher*, LIX (April, 1966), pp. 348-50.

Carrol, E. C., "Creatamath, or Geometric Ideas Inspire Young Writers," *The Arithmetric Teacher*, XIV (May, 1967), pp. 391-3.

Chaney, J. M., "Christmas at Palm Beach High Beach High School: The Geometree," *The Mathematics Teacher*, LV (November, 1962), pp. 600-2.

Christofferson, Halbert C., *Geometry Professionalized for Teachers* (Menasha, Wisconsin: George Banta Publishing Company, 1933).

Clarkson, D. M., "Taxicab Geometry, Rabbits, & Pascal's Triangle: Discoveries in a Sixth-Grade," *The Arithmetic Teacher*, IX (October, 1962), pp. 308-13.

Coltharp, F. L., "Simple Constructions Introduce Geometry," *The Instructor*, LXXV (October, 1965), p. 42.

Condron, B. F., "Geometric Number Stories," *The Arithmetic Teacher*, XI (January, 1964), pp. 41-2.

Davis, David R., *The Teaching of Mathematics* (Cambridge, Mass: Addison-Wesley Press, Inc., 1951), Chapter 10.

Davis, E., "First Days of Geometry," *The Mathematics Teacher*, LVI (December, 1963), pp. 645-6.

Dunnrankin, P., and R. Sweet, "Enrichment: A Geometry Laboratory," *The Mathematics Teacher*, LVI (March, 1963), pp. 134-40.

Eves, Howard, *An Introduction to the History of Mathematics* (New York: Rinehart and Company, Inc., 1953).

Fawcett, Harold P., *The Nature of Proof*, National Council of Teachers of Mathematics, Thirteenth Yearbook (New York: Bureau of Publications, Teachers College, Columbia University, 1938).

Gemignani, M. C., "On the Geometry of Euclid," *The Mathematics Teacher*, LX (February, 1967), pp. 160-4.

Glicksman, A.M., and H. D. Ruderman, *Fundamentals for Advanced Mathematics* (New York: Holt, Rinehart and Winston, Inc., 1964), Chapter 3.

Guggenbuhl, Laura, "Mathematics in Ancient Egypt: a Checklist (1930-1965)," *The Mathematics Teacher*, LVIII (November, 1965), pp. 7, 630-4.

Gurau, P. K., "Discovering Precision," *The Arithmetic Teacher*, XIII (October, 1966), pp. 453-6.

Harris, E. M., "Geometric Intuition and $ab = a + b/2$," *The Mathematics Teacher*, LVII (February, 1964), pp. 84-5.

Hesser, F. M., "Land of the Gonks, an Original Postulational System for High School Students," *School Science and Mathematics*, LXVI (June, 1966), pp. 527-31.

Hewitt, F., "New Look at Some Old Geometry Problems," *National Council of Teachers of Mathematics Yearbook*, XXVIII (1963), pp. 65-75.

Hewitt, F., "Visual Aid for Geometry," *The Arithmetic Teacher*, XIII (March, 1966), pp. 237-8.

Hydrographic Office, *U. S. Navy Aircraft Navigation Manual*, H. O. No. 216 (Washington, D. C.: United States Government Printing Office).

Johnson, Donovan A., and Gerald R. Rising, *Guidelines for Teaching Mathematics* (Belmont, California: Wadsworth Publishing Company, Inc., 1967).

Kattsoff, L.O., "Saccheri Quadrilateral," *The Mathematics Teacher*, LV (December, 1962), pp. 630-6.

Klinkerman, G., and F. Bridges, "Team Teaching in Geometry," *The Mathematics Teacher*, LX (May, 1967), pp. 448-92.

Levin, G. R., and B. L. Baker, "Item Scrambling in a Self-Instructional Program," *Journal of Educational of Psychology*, LIV (June, 1963), pp. 138-43.

Loeb, A. L., "Remarks on Some Elementary Volume Relations Between Familiar Solids," *The Mathematics Teacher*, LVIII (May, 1967), pp. 417-19.

Love, M. G., "Planning a Geometry Program in Junior High School," *Catholic Educational Review*, LXIII (September, 1965), pp. 393-400.

MacDonald, I. D., "Abstract Algebra from Axiomatic Geometry," *The Mathematics Teacher*, LIX (February, 1966), pp. 98-106.

Major, J.E., "Rings and Strings," *The Arithmetic Teacher*, XIII (October, 1966), pp. 457-60.

Marks, J. L., and J. R. Smart, "Using the Analytic Method to Encourage Discovery," *The Mathematics Teacher*, LX (March, 1967), pp. 241-5.

May, L. J., "Introducing the Compass as a Tool in Geometric Construction," *The Grade Teacher*, LXXXIV (March, 1967), pp. 98-100.

May L. J., "String and Paper Teach Simple Geometry," *The Grade Teacher*, LXXIV (February, 1967), pp. 110-2.

Mills, C.N., "Radii of the Apollonius Contact Circles," *The Mathematics Teacher*, LIX (October, 1966), pp. 574-6.

Moser, J. M., "Geometric Approach to the Algebra of Solutions of Pairs of Equations," *School Science and Mathematics*, LXVII (March, 1967), pp. 217-20.

Nannini, A., "Geometric Solution of a Quadratic Equation," *The Mathematics Teacher*, LIX (November, 1966), pp. 647-9.

National Council of Teachers of Mathematics, *The Teaching of Geometry*, Fifth Yearbook (New York: Bureau of Publications, Teachers College, Columbia University, 1930).

National Council of Teachers of Mathematics, *Enrichment Mathematics for High School*, Twenty-Eighth Yearbook (Washington, D. C.: The National Council of Teachers of Mathematics, 1963).

National Council of Teachers of Mathematics, *A Sourcebook of Mathematical Applications*, Seventeenth Yearbook (New York: Bureau of Publications, Teachers College, Columbia University, 1942).

Neureiter, P. R., "What's the Point?" *The Instructor*, LXXV (January, 1966), p. 15.

Piwnicki, F., "Application of the Pythagorean Theorem in the Figure-Cutting Problem," *The Mathematics Teacher*, LV (January, 1962), pp. 44-51.

Rosenberg, H., "Use of Vectors to Eliminate Construction Lines in Proving Theorems in Geometry," *School Science and Mathematics*, LXVII (June, 1967), pp. 567-8.

Schaff, M. E., "Discovery-type Investigation for Coordinate Geometry Students," *The Mathematics Teacher*, LIX (May, 1966), pp. 458-60.

Schuster, Carl N., and Fred Bedford, L. *Field Work in Mathematics* (New York: American Book Company, 1935).

Smith, L. B., "Geometry, yes; but how?" *The Arithmetic Teacher*, XIV (February, 1967), pp. 84-9.

Sweet, R., and M. DeWitt, "Geometric Christmas Decorations," *School Science and Mathematics*, LXIII (December, 1963), pp. 701-4.

Szabo, S., "Approach to Euclidean Geometry Through Vectors," *The Mathematics Teacher*, LIX (March, 1966), pp. 118-35.

Todd, R. M., "Direct Proof of the Theorem: if a Line Divides Two Sides of a Triangle Proportionally, Then it is Parallel to the Third Side," *School Science and Mathematics*, LXVI (November, 1966), p. 700.

Trigg, C. W., "Collinearity and Concurrency," *School Science and Mathematics*, LXII (October, 1962), pp. 524-5.

Wahl, M. S., "Easy-to-Paste Solids," *The Arithmetic Teacher*, XII (October, 1965), pp. 468-71.

Walter, M., "Example of Informal Geometry: Mirror Cards," *The Arithmetic Teacher*, XIII (October, 1966), pp. 448-52.

Walter, M., "Some Mathematical Ideas Involved in the Mirror Cards," *The Arithmetic Teacher*, XIV (February, 1967), pp. 115-25.

Wernik, W., "List of Standard Corrections," *The Mathematics Teacher*, LVII (February, 1964), p. 107.

Whitman, N. C., and Linda Oda, "Developing Spatial Perception Via the Experimental Technique," *The Mathematics Teacher*, LIX (November, 1966), pp. 631-3.

Wright, A. L., "Application of Combinations and Mathematical Induction to a Geometry Lesson," *The Mathematics Teacher*, LVI (May, 1963), pp. 325-28.

7

Teaching Advanced Topics in the Secondary School

> . . . mathematics need be neither austere or remote, . . . it has relevance for almost all human activities, . . .[1]

7.1. Introduction

Teachers interested in student-centered approaches and in devising situations from which mathematics can emerge, find topics in upper courses in mathematics as fruitful and as effective as those in lower grades. By this time students have a deeper background of experience from which to draw and many suggestions of excellent quality are likely to come from which the teacher learns (and must think, too).

In this chapter are discussed some topics which help to emphasize the outward-looking face of mathematics and which students have found helpful and interesting, also inward-looking items which are ordinarily considered in upper level mathematics whether this be in high school or in college. Of course, under continuous reorganization of the mathematics curriculum and methods of teaching, it is difficult to classify a topic as belonging to one level or to another. At times the presentation in this chapter may appear to assume

[1] R. C. Buck, "Goals for Mathematics Instruction," *The American Mathematical Monthly*, 72 (November, 1965), p. 956.

narrative style as experiences in the classroom are related. There is no special purpose in discussing one topic before another.

7.2. Dividing a Line Segment in a Given Ratio

Given the line segment from (a_1, b_1) to (a_2, b_2), the student usually recalls that the coordinates of the midpoint are

$$\left(\frac{a_1 + a_2}{2}, \frac{b_1 + b_2}{2}\right)$$

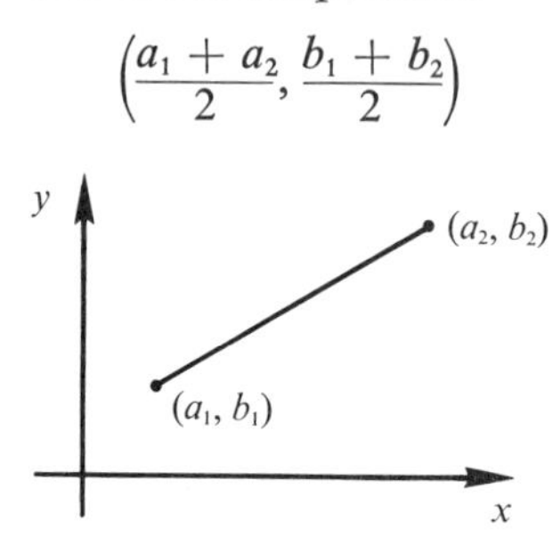

It is common experience, however, that on being asked for the coordinates of one of the points which *trisects* the line segment many students immediately suggest $[(a_1 + a_2)/3, (b_1 + b_2)/3]$. This response indicates that many students are not aware of a development for the midpoint formulae which may be generalized to include the derivation of the coordinates of a point which divides the line segment in any *given* ratio. Hence, for more advanced students, to reopen the question of deriving the coordinates of the midpoint of a line segment can provide for several different approaches by the students themselves. Many times, in teaching, all that is necessary is to ask students how *they* would attack a given problem. Some suggestions which have come in classroom practice are given here:

(a) Let C have the coordinates (x, y). If C is to be the midpoint then $\sqrt{(x - a_1)^2 + (y - b_1)^2}$ must equal $\sqrt{(x - a_2)^2 + (y - b_2)^2}$. Also the points A, C, B must be collinear so we now have a second relation

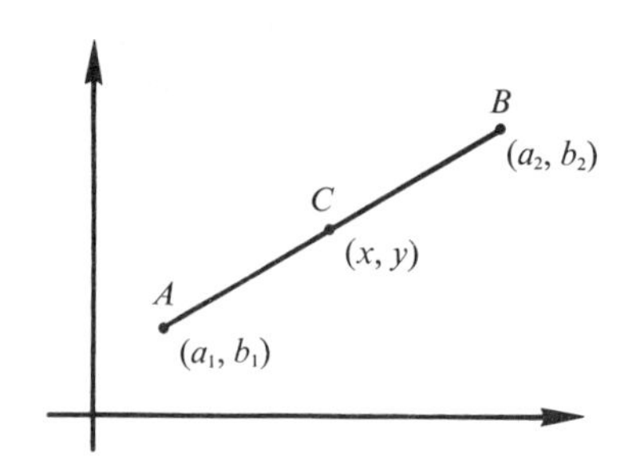

$$\frac{y - b_1}{x - a_1} = \frac{b_2 - y}{a_2 - y}$$

which must hold simultaneously with the first. The pursuing of this plan requires tedious but interesting work.

(b) If C is to be the midpoint, then $\ell(\overline{AC}) = \frac{1}{2}\ell(\overline{AB})$ and x and y must satisfy the relation

$$\sqrt{(x - a_1)^2 + (y - b_1)^2} = \tfrac{1}{2}\sqrt{(a_2 - a_1)^2 + (b_2 - b_1)^2}$$

and, of course, the points A, B, and C must be collinear.

(c) One student suggested that if C is the midpoint, then the area of the $\triangle ADC$ must be one fourth that of the $\triangle AEB$ and points A, C, B must be collinear.

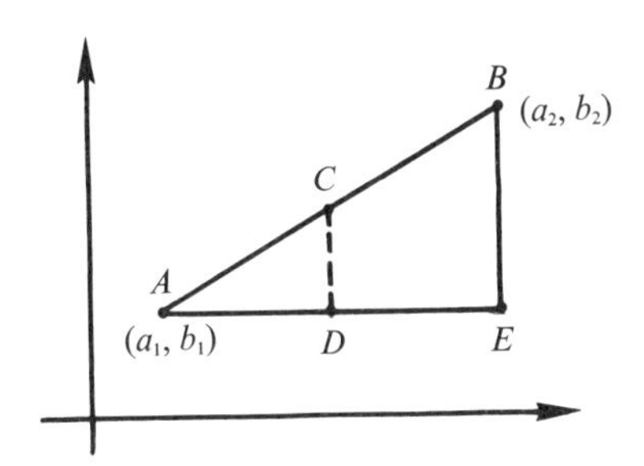

(d) Often several students reason that if C is the midpoint, then its x-coordinate is half the difference between the x-coordinates of A and E, or $(a_2 - a_1)/2$ Then momentary confusion, occurs for perhaps the conjecture or their memories are both poor! A little review and renewed insight into the meaning of a coordinate and the role of the *origin* corrects the error and the x-coordinate of C is made to read $a_1 + (a_2 - a_1)/2$ which, on simplifying, yields $(a_1 + a_2)/2$. Likewise the y-coordinate of the midpoint becomes $(b_1 + b_2)/2$.

The suggestions above involve the use of the distance formula, the computing of areas of right triangles (perhaps by the determinant or some other mnemonic), and computation as described in (d). In each case the student is asked to give logical support of his method. Do any of them involve "midpoint" in their own development? In (d) how can we prove that a perpendicular dropped from C to $\overleftrightarrow{\mathbf{AE}}$ will intersect $\overline{AE}$ at its midpoint? We see that we have put "old" geometry to use in a new setting. Using in this investigation what we learned causes the "old" to take on a new and significant role. Even in (d) students give logical support differently; some see similar triangles and say that $\overline{AD}:\overline{AE} = \overline{AC}:\overline{AB} = 2:1$, hence $\ell(\overline{AD}) =$

$\frac{1}{2}(a_2 - a_1)$. Others think of the theorem that "a line from the midpoint of one side of a triangle parallel to the third side intersects the other side at its midpoint" which is a *converse* of the standard theorem which begins "the line segment which joins the midpoints of two sides of a triangle . . ." Much more mathematics is really learned when students are encouraged to think about a problem in their own way than if the students are passively digesting mathematics developed for them, page by page, with no chance for mathematical exploration and intellectual adventure. Logical organization is emphasized as each student looks for support of his method and traces it back to more original beginnings.

Students soon see the advantage and economy in the "similar triangle" or "line parallel to the third side" approach and we are now ready with a *mathematical preparation*, which we would not have had otherwise, to find the coordinates of the point which is one-third of the way from A to B. This is now done easily. One-fourth of the way can be done by the midpoint formula or directly from similar triangles; one-fifth of the way by similar triangles; and so on. Finally comes $1/n$th of the way and also dividing the segments into lengths with the ratio $p:q$. It has been a fruitful and interesting journey, perhaps easy for the reader, but a genuine investigation for students encouraged by a classroom manner of the teacher which does not discourage any kind of contribution. Even the use of a piece of string which has the same length as the segment (with integers as coordinates) serves to examine conjectures at a physical level.

Exercises

1. Pursue the methods suggested in (a) or (b) to obtain the coordinates of the midpoints.
2. Use areas as in (c) to derive the midpoint expressions.
3. Use the "similar triangle" or "line parallel to a third side" method to compute the coordinates of the point C which divides the segment $\overline{AB}$ into two segments whose lengths have the ratio $p:q$. Does your answer give the midpoint formulae for $p = q$? It should.
4. Can you think of another way to approach this problem or any variation on what was done above? It is always interesting to try to proceed as if you didn't know.
5. Students have asked in class if the validity of the formulae for the coordinates of the point C $1/n$th of the way from A to B can be proved by *mathematical induction*. Investigate this. Already the validity has been substantiated with similar triangles, which corrects the sometimes mistaken notion that all theorems about natural numbers must be proved by the use of mathematical induction.

7.3. The Slope of a Line

On considering the "steepness" of a line, the student has a *picture*, as in Fig. 7.1, of a line not so steep and one very steep, which is good; mathematics

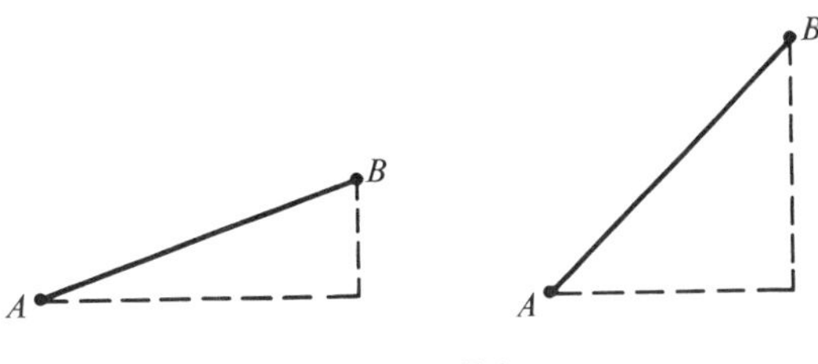

Figure 7.1

should give us practice in trying to express in number what we see. The two suggestions most often made to provide a number to express steepness are "height over base" and "height over hypotenuse," as were mentioned in the preceding chapter. In elementary work it is mentioned that although there could be several definitions, the one used by mathematicians is that of "height over base" (or the tangent of the angle of inclination of the line).

In upper level courses, however, with the student perhaps recalling the accepted definition of slope, it is interesting to ask if there could be other definitions, or other numbers to express steepness. The "height of the related triangle divided by the hypotenuse" usually comes again, and we now have two ways to indicate a measure of steepness (refer to Fig. 7.2). One of these

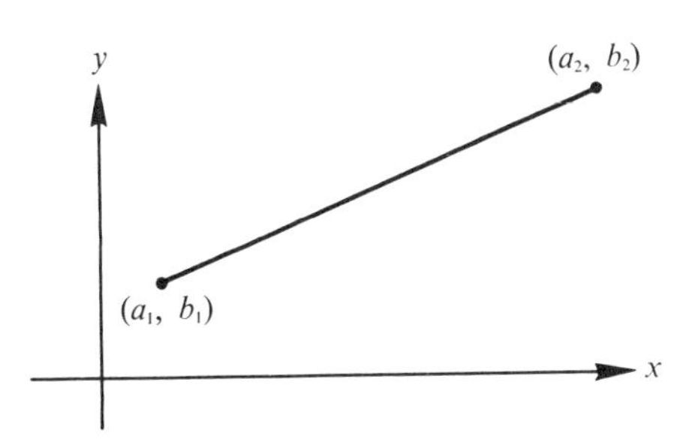

Figure 7.2

is $m=(b_2-b_1)/(a_2-a_1)$ and the other, $m=(b_2-b_1)/\sqrt{(a_2-a_1)^2+(b_2-b_1)^2}$.

Let us call the first the "Adams slope" (as a made-up student's name) and the other the "Birn slope." Laboratory work with graph paper suggests the conjecture that if two lines are perpendicular then the Adams slope of the one line is the *negative reciprocal* of the Adams slope of the other and conversely. Students are asked to prove this as a theorem. This is the ordinary definition of slope.

The Birn slope of a line is interesting in itself. What is the Birn slope of a line if it is horizontal? What is the Birn slope if it is vertical? Need there be any conventions on which point to call (a_2, b_2) and which to call (a_1, b_1) to have the sign of the Birn slope come out the same as the Adam slope? If two lines are perpendicular what, if any, is the relation between their respective Birn slopes?

To test the conjecture that if two lines are perpendicular, then the Adams slopes are negative reciprocals of each other, one can make use of similar triangles, for if two triangles have corresponding sides respectively perpendicular, then they are similar. We have the hypothesis satisfied (see Fig. 7.3) and hence, the triangles are similar. If a and b are the lengths of

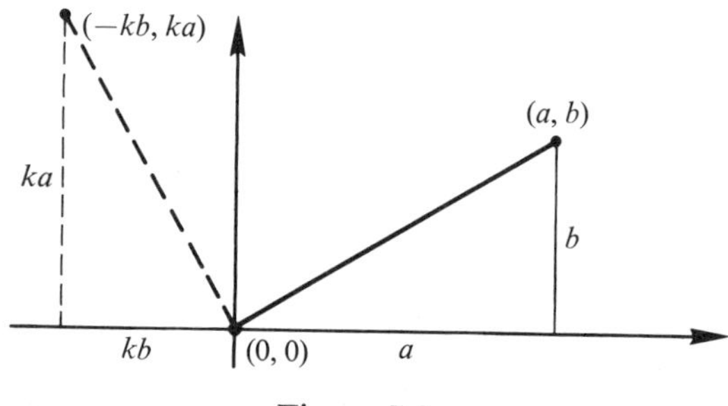

Figure 7.3

the sides of one right triangle, then ka and kb can represent the lengths of the sides of the other where k is the constant of proportionality. The relation between the Adams slopes follows with little effort.

Students have use of the same figure to develop a relation between the Birn slopes if the lines are given perpendicular. This we leave for the exercises.

It is just as proper to consider "two kinds of slopes" of a line as it is to study two kinds of curvature of surfaces or two kinds of logarithms or two kinds of "products" in vector mathematics. Indeed, direction cosines can be used instead of slope in the coordinate geometry of the *plane* as well as in *space*. Here the subject of the Birn slope can been pursued and mathematics is put to work investigating "new" mathematics (new to the student). The student is creating and is pursuing a problem just as real and challenging to him as more profound problems are to the researcher. It is urged that teachers carefully evaluate the dividends which come from investigating further a certain definition or idea; the time required for such study, may be worth the price of several theorems passively received.

Exercises

1. Complete the proof that if two lines are perpendicular, then the Adam slope of the one is the negative reciprocal of the Adam slope of the other.
2. Now prove the converse of the statement in Exercise 1.
3. One student conjectured that if two lines are perpendicular, then the sum of the squares of the Birn slopes of the two lines is 1. See if this conjecture can be made a *theorem* by proof.
4. If you wish the Birn slope to be positive when the Adams slope is positive, can the necessary conventions be made? If so, what will they be?
5. Can you think of another way to define a measure-number for slope of a line? If two lines are perpendicular, what relation, if any, can be determined?

7.4. Multiplication of Matrices

In the previous two sections, we have seen examples of the mathematics which can be developed and the deeper insights which can become a part of the student as he explores the consequences of different definitions or as he pursues further certain ideas which he himself has originated in the course of thinking about the subject in his own way. In this section we wish to continue in the same vein but to add a new facet to the situation approach used in the addition of matrices begun in Section 3.3.

Tommy had several uncles who were interested in his basketball success and they decided to encourage him by agreeing to pay him so much for each score. Uncle Tom held to the theory that Tommy should be urged to play hard at the beginning of a game so he agreed to pay 15¢ per score in the first quarter and 10¢ for each of the others. Uncle Tom's agreement looked like this:

	Quarter			
	1	2	3	4
Uncle Tom's payment	15¢	10¢	10¢	10¢

Uncle Charles planned to pay as follows:

5¢ 10¢ 10¢ 25¢

Uncle Edward made the proposal to pay 15¢ for each score without reference to the quarter. Well, this pleased Tommy, so he wondered what he would have earned in the 1958 season under this arrangement. His data appeared as follows:

	Quarter, payments			
Uncle	1	2	3	4
Tom	15	10	10	10
Charles	5	10	10	25
Edward	15	15	15	15

	1958 Season Scores in Games				
Quarters	1	2	3	4	5
1	3	2	1	0	1
2	4	1	3	2	1
3	1	3	2	4	5
4	3	0	0	2	1

How much would Uncle Tom have paid for Tommy's scores in the first game? (do you get $1.25?); for the second game? How much would Uncle Charles have paid for Tommy's scores in the fifth game?

Tommy thought he would make a chart showing what each uncle would have paid for each game. His chart looked like this:

	Game				
Uncle	1	2	3	4	5
Tom	$1.25	0.70	—	—	—
Charles	—	—	—	—	—
Edward	—	—	—	—	0.90

Complete the table. Since Tommy was interested in the total amount he would have received, he would add all the entries but the mechanics of how the entries can be computed is interesting.

What would the entries be in an *uncle-game* arrangement where each entry is the amount the uncle will pay for each game if the data for each quarter and each game is as shown?

	Cents Payments in Quarter			
Uncle	1	2	3	4
Tom	15	15	5	2
Charles	10	10	10	25
Edward	25	5	0	10

Scores in Games					
1	2	3	4	5	Quarters
1	0	3	4	1	1
1	0	0	2	1	2
1	3	2	4	6	3
3	0	2	0	3	4

	Games				
Uncle	1	2	3	4	5
Tom	—	—	—	—	—
Charles	—	—	100¢	—	—
Edward	—	—	—	—	—

Exercises

1. Use the same mechanics as you did above on these two arrangements to see what new arrangement you get; a few are furnished.

$$\begin{pmatrix} 1 & 3 & 2 & 1 \\ 4 & 0 & 1 & 1 \\ 1 & 4 & 2 & 1 \end{pmatrix}. \quad \begin{pmatrix} 1 & 3 & 2 & 0 & 1 \\ 0 & 1 & 2 & 0 & 1 \\ 1 & 2 & 0 & 1 & 1 \\ 1 & 1 & 1 & 1 & 1 \end{pmatrix} = \begin{pmatrix} 4 & - & - & - & 7 \\ - & 15 & - & - & - \\ - & - & 11 & - & - \end{pmatrix}$$

2. Do the same for the following arrangement.

(a) $\begin{pmatrix} 1 & 1 & 3 \\ 2 & 1 & 4 \\ 1 & -1 & 1 \end{pmatrix}, \quad \begin{pmatrix} 1 & 1 & 1 \\ 0 & 3 & -1 \\ 1 & 2 & -1 \end{pmatrix} = \begin{pmatrix} 4 & - & - \\ - & - & -3 \\ - & 0 & - \end{pmatrix}$

(b) $\begin{pmatrix} 1 & 3 & 1 \\ 2 & 4 & 6 \end{pmatrix}, \quad \begin{pmatrix} 1 & 4 \\ 3 & \frac{1}{2} \\ -1 & 0 \end{pmatrix} = \begin{pmatrix} 9 & - \\ - & 10 \end{pmatrix}$

(c) $\begin{pmatrix} 1 & 4 & 1 \\ 3 & 6 & 2 \end{pmatrix}, \quad \begin{pmatrix} 1 & 3 & -4 \\ 0 & 0 & 0 \\ 1 & 4 & -2 \end{pmatrix} =$

(d) $(-1 \quad 3 \quad 1), \begin{pmatrix} 4 \\ 0 \\ 2 \end{pmatrix} =$

(e) $(-1 \quad 3 \quad 1 \quad 2), \begin{pmatrix} 4 & 1 \\ 0 & 1 \\ 2 & 1 \\ 1 & -1 \end{pmatrix} =$

3. In Section 3.3 we learned to call these arrangement *matrices* and there we learned to *add* matrices. Is the operation being performed in Exercises 2 and 3 above adding? What name and symbol might be given to this new operation?

4. It is not known what names will be suggested for this operation, but others call this "row-by-column multiplication of matrices" and this is the name which should finally be adopted. Realizing that rectangular arrangements or matrices may have any number of rows and columns study the following pairs of matrices to see on which of these this operation of multiplication can be performed. Write answers to these which can be multiplied.

(a) $\begin{pmatrix} 1 & 2 & 4 \\ 1 & 6 & 3 \end{pmatrix} \times \begin{pmatrix} 1 & 4 \\ 8 & 6 \\ 7 & 3 \end{pmatrix}$

(b) $\begin{pmatrix} 1 & 2 & 4 \\ 1 & 6 & 3 \end{pmatrix} \times \begin{pmatrix} 4 & 6 & 3 \\ 1 & 8 & 7 \end{pmatrix}$
(Ans. Cannot be multiplied)

(c) $(1 \quad 2 \quad 4) \times (4 \quad 6 \quad 3)$
(Ans. Cannot be multiplied,

(d) $(1 \quad 2 \quad 4) \times \begin{pmatrix} 4 \\ 6 \\ 3 \end{pmatrix}$

(e) $\begin{pmatrix} 1 & 2 & 4 \\ 1 & 6 & 3 \end{pmatrix} \times \begin{pmatrix} 1 & 4 & 1 \\ 8 & 6 & 0 \\ 7 & 3 & 1 \end{pmatrix}$

(f) $\begin{pmatrix} 1 & 2 & 4 \\ 1 & 6 & 3 \end{pmatrix} \times \begin{pmatrix} 1 & 4 & 1 & 2 \\ 8 & 6 & 0 & 4 \\ 7 & 3 & 1 & 1 \end{pmatrix}$

(g) $\begin{pmatrix} 1 & 2 & 4 \\ 1 & 6 & 3 \end{pmatrix} \times \begin{pmatrix} 1 & 4 & 1 & 2 \\ 8 & 6 & 0 & 4 \end{pmatrix}$

(h) $\begin{pmatrix} 1 & 2 & 4 \\ 1 & 6 & 3 \end{pmatrix} \times \begin{pmatrix} 1 & 4 & 1 & 2 \\ 8 & 6 & 0 & 4 \\ 7 & 3 & 1 & 1 \\ 1 & 0 & 1 & 1 \end{pmatrix}$

(i) $\begin{pmatrix} 1 & 2 & 4 \\ 1 & 6 & 3 \end{pmatrix} \times \begin{pmatrix} 1 & 4 & 1 & 2 \\ 8 & 6 & 0 & 4 \\ 7 & 3 & 1 & 1 \\ 0 & 0 & 0 & 0 \end{pmatrix}$

5. What condition(s) must be met concerning rows and columns before two matrices are considered "multipliable." What condition(s) must be met before two matrices can be considered "addable?"

6. In multiplying two matrices, one gets the *product matrix.* Assuming the matrices are large enough, describe how one forms the entry for the second row and third column of the product. How does one form the entry for the second row and fourth column? The third row and second column? The fourth row and first column? The rth row and the cth column? The ith row and the jth column?

7. If the matrix multiplication involves

$$\begin{pmatrix} a_{11} & a_{12} & a_{13} \\ a_{21} & a_{22} & a_{23} \\ a_{31} & a_{32} & a_{33} \end{pmatrix} \times \begin{pmatrix} b_{11} & b_{12} & b_{13} & b_{14} \\ b_{21} & b_{22} & b_{23} & b_{24} \\ b_{31} & b_{32} & b_{33} & b_{34} \end{pmatrix}$$

What is the entry in the third row and second column of the product matrix? The second row and fourth column? The ith row and the jth column? (Ans. $a_{i1}b_{1j} + a_{i2}b_{2j} + a_{i3}b_{3j}$)

8. Can you recall or invent a notation which shortens the answer in the last part of Exercise 7? Suggest some advantages to this notation.

9. Now in previous contacts with *operations* in different domains of elements (natural numbers, integers, ..., matrices) we asked certain questions about the operation. What questions should be asked about the multiplication of matrices? Investigate using two (rows)-by-two (columns) or three-by-three matrices. State results as theorems.

10. Is there a matrix which when multiplied by any other matrix A gives the matrix A as the product? See if you can devise one. If successful what might such a matrix be called? Is the multiplication with this new matrix commutative?

11. Try to arrange a logical organization of what you have done. Taking the real numbers and operations on real numbers and their properties as primitive, start with the definition of a matrix; then define multiplication, then add other features and theorems you have formulated above. Proofs need not be included; just make statements.

12. Consult a text which contains work on matrices. Do you see improvements you might make?

The above treatment, or sample guides, have been used in classes and it is rewarding to see the rapidity with which the technique of multiplication comes from stories like the situation of Tommy's uncles and the basketball scores. Students and others can learn to do this easily, although had they been told they were going to "multiply matrices" a wall of mystery and fear would have caused closed minds in many cases. Compare this approach in meaning for the student with the sudden appearance in a text of

Definitions: The product of two matrices (a_{ij}) and (b_{jk}) is another matrix (c_{ik}) where each entry in (c_{ik}) is given by

$$c_{ik} = \sum_{n=1}^{j} a_{in}b_{nk}$$

in which the student may be asked immediately to multiply several pairs of matrices or to prove that the operation is commutative! Indeed, the definition is beautiful and compact and it commands the respect of mathematical readers but, in teaching, the operation should be taught so that the student can formulate or have a part in framing a definition as a climax to his experience. He now reads the definition made by other writers and coworkers with understanding that usually results in purposeful questions. A text's

$$c_{ik} = \sum_{u=1}^{j} a_{in}b_{nk}$$

may be better than his definition, but he has deeper understanding after he has first formulated a definition by himself (Exercises 7 and 8). The student can learn to *read* mathematics by comparing the work of others with his own and by learning how the author of a text extends work which he has begun. This kind of methodology helps to attain the important Goal 2 in Buck's "Goals for Mathematics Instruction" which urges us "to convey the fact that mathematics, like everything else, is built on intuitive understandings and agreed conventions, and that these are not eternally fixed."[2] Buck concludes with the thought that "... mathematics need be neither austere or remote,... it has relevance for almost all human activities,... it can be of value to any sincerely interested person."[3]

---

[2]*Ibid.*, p. 950.
[3]*Ibid.*, p. 956.

7.5. Fitting Curves to Experimental Data

Sometimes the teacher has the opportunity to teach both science and mathematics to the same students or has made arrangements to work co-operatively with science teachers in certain facets of instruction. Much interesting and horizon-widening mathematics can arise from the simple problem of trying to find the best straight line which describes a seemingly linear relation. Such a discussion can begin with an experimental study of elasticity (stretching a spring, bending a wooden lattice strip, twisting a wooden rod) whereby the class secures data on weights applied and corresponding displacements. We assume for the work of this section that the student knows that the quickest way to search for a possible relation between two sets of data is to construct a graph. Also we expect that in his previous mathematical study he has learned that he can "complete the square" on the quadratic expression

$$y = ax^2 + bx + c$$

to obtain

$$\frac{y}{a} = x^2 + \frac{b}{a}x + \frac{c}{a}$$

and

$$\frac{y}{a} = x^2 + \frac{b}{a}x + \frac{b^2}{4a^2} + \frac{c}{a} - \frac{b^2}{4a^2}$$

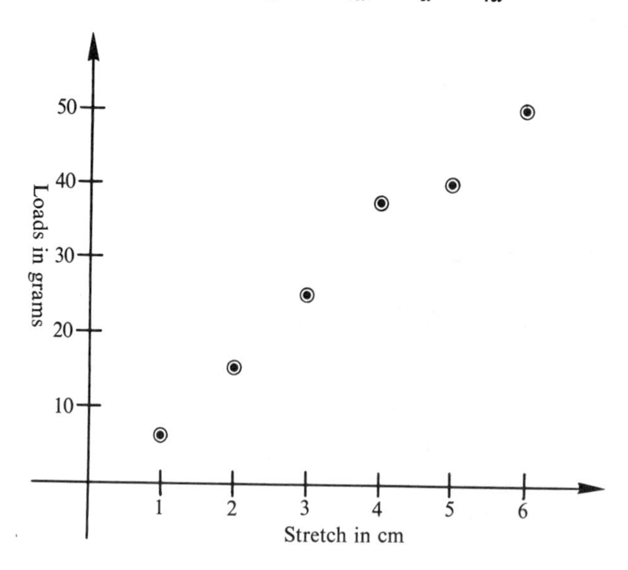

Figure 7.4.

and finally

$$\frac{y}{a} = \left(x + \frac{b}{2a}\right)^2 + \left(\frac{c}{a} - \frac{b^2}{4a^2}\right)$$

and from this see that at $x = -b/2a$ y has minimum or maximum value. Guides for study and development might include a review of determining maximum or minimum on quadratic expressions *ab initio* along with current work and then an exercise like the above to set the stage for discussion. The data which we obtained "yesterday" appear on cross section paper as in Fig. 7.4. Draw the line which you think best fits these data. Of course different students will have different lines and for different reasons. Some will suggest that some points shoud be below the line and some above. It is then time to stop and discuss this entire problem.

All points of view deserve attention. Some students will suggest that if vertical distances above the proposed line be called positive and those below the proposed line be called negative, then the line of best fit will be that line such that these "signed" distances add to zero. The teacher or others might show that the line ℓ in Fig. 7.5. obeys that demand exactly, yet surely it is not the line whose mathematical description best describes the behavior represented by the points on the graph.

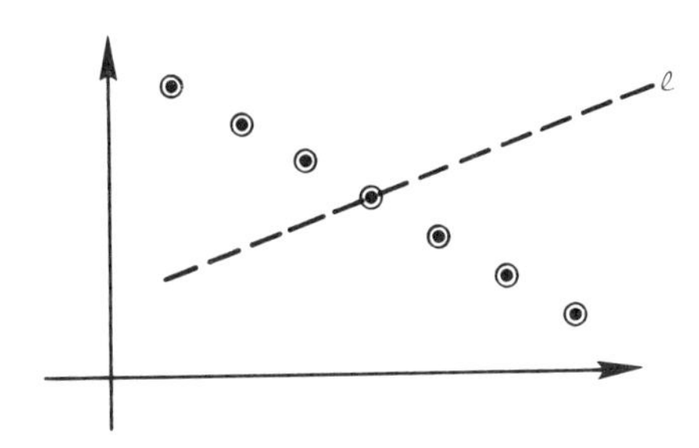

Figure 7.5.

The reader may know that a much used definition for the line of best fit is that line $y = mx + b$, which makes the sum of the squares of the ordinate distances a minimum. That is, the coefficients m and b in the equation of the line $y = mx + b$ are determined so that the sum $a^2 + c^2 + d^2 + e^2 + f^2$ has minimum value (Fig. 7.6). The points on the cross-section paper come from experimental data. Whether this is defined by the students or not, the discussion of their own proposals for the line of best fit is always beneficial.

With this definition we are now ready to ask the student to compute ordinate differences and see if he can complete the solution of the problem. As an exercise we might devise Fig. 7.7.

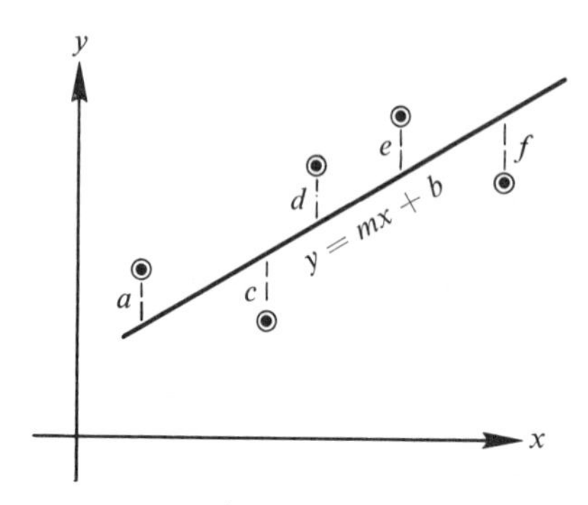

Figure 7.6

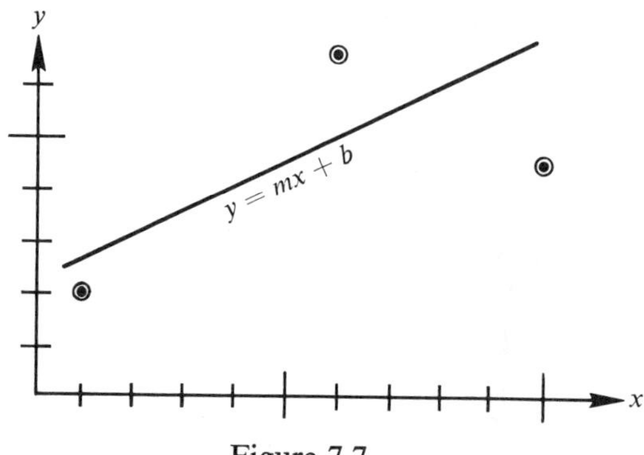

Figure 7.7

Let $y = mx + b$ be the equation of the line of "best fit."

(a) What is the ordinate distance from the point (1, 2) to this line? (Ans. $m + b - 2$)

(b) What is the ordinate distance from the point (6, 7) to the line? from (10, 5) to the line?

(c) Hence, compute the sum of the squares of these distances.

The answer should be $137m^2 + 34mb + 3b^2 + 78 - 188m - 28b$ which we might label as D^2, but now the student needs some help. Some helpful questions are "How many variables do we have? If we think momentarily of b as being a constant then what letter is considered as the variable? Is the expression a quadratic expression? In what variable if b is fixed? Could we find the value of m in terms of b which makes D^2 a minimum? See what you can do with this! This is a review of coordinate geometry and algebra and the student sees mathematics in use. The student "completes the square," regarding m as the variable and b fixed, then again regarding b as the variable and m fixed. Hence there are obtained two relations in m and b to be satisfied simultaneously. What began as a simple question in moving a ruler on a graph to obtain what the eye considered as a best fit has now come under mathematical scrutiny and a best-fit condition defined mathematically in terms of the given points.

For the teacher, two comments are in order: the simultaneous conditions in m and b can be obtained by the use of the calculus; also the discussion concerns determining the line of best fit by the method of *least squares*. There are other definitions and methods for "best fit."

Exercises

1. Determine the two conditions on m and b mentioned in the text by "completing the square." (Ans. $137m + 17b = 94$; $17m + 3b = 14$)

2. Now determine values of m and b which satisfy the conditions in Exercise 1 and, therefore, make the line whose equation was taken as $y = mx + b$ the line of best fit according to the "least square" definition. (Ans. $m = 44/123$, or approximately 0.36, $b \doteq 2.6$)

3. In a similar manner, determine the equation of the line of best fit if the three points on the graph have names $(x_1, y_1)(x_2, y_2)(x_3, y_3)$. Here is a real challenge to students from which they may be able to conjecture relations between m and b which must be obeyed simultaneously for any number of points. Ans. the relations between m and b for n points are

$$\sum_{i=1}^{n} y_i = m \sum_{i=1}^{n} x_i + nb$$

$$\sum_{i=1}^{n} y_i x_i = m \sum_{i=1}^{n} x_i^2 + b \sum_{i=1}^{n} x_i$$

4. Now one can return to the original problem of the line of best fit for the data from the spring and weights. Students will see the advantage of a desk calculator and the advantage of neat arrangement of data.

Students are really doing something here and classroom experience reflects their enthusiasm for the fact that many different facets in their mathematical experience have been brought to bear effectively on a very interesting problem.

Various natural phenomena can be studied. Is the relation between the *frequency* and the *tension* on a vibrating string a linear relation? With help from the physics students one can obtain a set of data. Graphing yields a suggested curve which is not a straight line. The distance a marble has rolled down an incline by the end of each second can furnish another set of data (one may use a seconds pendulum to indicate time intervals); the graph is not a straight line. By graphing known relations as $y = x^2$, to illustrate the "square relation" $y_1 : y_2 = x_1^2 : x_2^2$, or the equation $xy = k$ to illustrate the type of graph which describes the inverse relation $y_1 : y_2 = x_2 : x_1$, and others, the student creates a reservoir of curves which he can use for comparing a graph which results from the use of given data. If a graph resembles a parabola, then the proper elements can be squared and a new graph made; if the new graph suggests a straight line, then the conjecture that the original relation was a square relation is strengthened. Students might be directed in such cases as follows:

(a) Below is the set of data obtained from the study of the marble rolling down the trough. Construct a graph to see what kind of relation is suggested.
(The student derives a curve which looks like a parabola.)

Time Ball Has Been Rolling	Distance Traveled
0 sec	0.0 cm
1 sec	4.0 cm
2 sec	15.0 cm
3 sec	36.3 cm
4 sec	65.0 cm
5 sec	98.2 cm

(b) Now change the data, if necessary, to study further for a possible relation. Is your conjecture strengthened? (The student has learned now to square the times and graph again. This time he graphs the set: (0, 0. 0), (1, 4. 0), (4, 15. 0), (9, 36. 3), (16, 65. 0), (25, 98. 2) and the graph is nearly a straight line.

(c) It the new graph seems approximately to fit a straight line, compute the equation of the "line of best fit." It will have the form $d=mt^2+b$.

There is no end to laboratory studies which can be made and followed by mathematical considerations as those above although some may involve logarithmic, trigonometric, and other relations. A clarinetist may study the frequency of tone versus the distances from the mouthpiece to the first open hole. This gives a curve resembling one branch of a rectangular hyperbola. Then one set of data must be graphed against the inverts, or the reciprocals, of the measurements in the other set to see if one gets a straight line, and the least squares method may be used. The xylophonist may simply measure the lengths of the bars and graph against frequency of the tone; the advanced biology student may be interested in the height of a plant versus time; a boy interested in auto mechanics may study the relative viscosity of oils versus temperature; the chemistry student might secure and study data on the acidity of milk versus time as it sours. Some "relations" may not belong to any of the standard well-known behaviors and hence cannot be expressed in simple mathematical terms, but this helps the student to keep searching. Students find this topic quite fascinating and it convinces them again of the utility of mathematics in many fields. The outward-looking face is surely emphasized here.

Exercises

1. Construct a graph of the data in (a) above and see if a parabola or square relation is suggested.
2. Now try to confirm your conjecture by "squaring the times" and graphing again.
3. Determine m and b in $d = mt^2 + b$ from the data in (b) for the line of best fit by the method of least squares.
4. Suggest several studies which you could encourage upper level high school students to make with the limitations of equipment and experience which they have.

7.6. Developing Formulae for Transformations

In developing the formulae,

$$\bar{x} = x + h$$

$$\bar{y} = y + k$$

for the translation of the axis system one can simply move the axis system to the right or left or up or down and ask what new names certain given points have and ask further what relations exist between $\bar{x}$ and x and $\bar{y}$ and y where $(\bar{x}, \bar{y})$ is the new name which the point with the old name (x, y) has acquired during the translation. Also translation of axes and subsequent relations can be shown by the use of the overhead projector with sliding transparencies.

A story which has been found interesting, however, is this:

(a) Charles lived in a residential housing development in a city and he and his boy friends had their houses marked by pairs of numbers (Fig. 7.8). The pair (3, 4) meant three houses east of Center Street and four houses north of Main Street. Johnny's house was marked __; Henry's; house had the mark __; Tommy's house had the mark__.

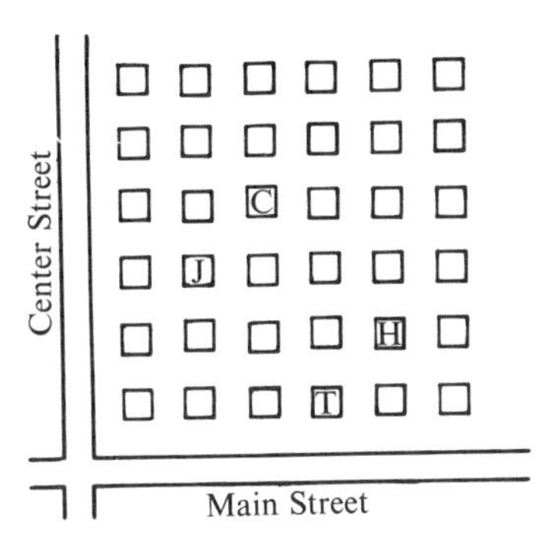

Figure 7.8

(b) One summer Charles went to visit his grandmother and he was gone for several months. When he returned there were many more houses and the *streets had been moved!* Houses were found where the streets had been. The plot looked like Fig. 7.9.

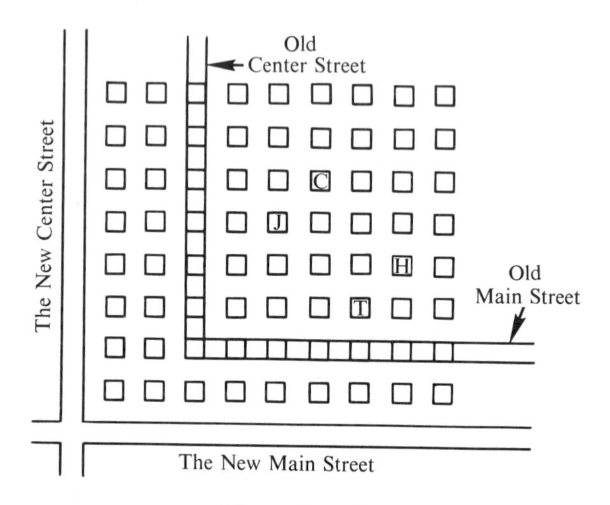

Figure 7.9

Charles and his friends had to assign new pairs of numbers. How was Charles' house named now? Johnny's house? Henry's house? Tommy's house?

(c) What must be done to each old name to get the new one? If we let (f, s) be the old name and $(\bar{f}, \bar{s})$ be the new name, then, after the streets are moved, what relations exist between f and $\bar{f}$ and s and $\bar{s}$? We can say $\bar{f} =$ __; $\bar{s} =$ __.

(d) Now had the streets been moved 10 to the west (or left) and 3 to the south (or down) then we would have $\bar{f} =$ ______ and $\bar{s} =$ ______.

(e) Had the streets been moved 1 to the east and 0 to the north, then we would have $\bar{f} =$ ______ and $\bar{s} =$ ______.

(f) Tell what movement of the streets must have taken place had Charles figured that

(i) $\begin{cases} \bar{f} = f + 2 \\ \bar{s} = s + 7 \end{cases}$ (ii) $\begin{cases} \bar{f} = f \\ \bar{s} = s + 8 \end{cases}$ (iii) $\begin{cases} \bar{f} = f - 3 \\ \bar{s} = s - 4 \end{cases}$

(g) What would be a good name for the expressions like $\bar{f} = f - 4$, $\bar{s} = s + 7$?

For (g) various names are suggested, perhaps "movement formulae," but whatever they are they have come *from the student*; he became involved in a story, there was no fear, some mathematics emerged, and he understands because he helped to formulate the relations himself. The teacher will wish to explain that others call this *translation* and, to be conversant with others, we should adopt this term. Next comes *practice* in interpretations.

(h) We have seen that $\bar{f} = f + 4$, $\bar{s} = s - 2$ or, by this time, $\bar{x} = x + 4$, $\bar{y} = y - 2$, can be interpreted as a translation of axes while the point remains fixed, but acquires a new name. Is there any other interpretation which can be given to the translation equations

$$T: \begin{cases} \bar{x} = x + 4 \\ \bar{y} = y - 2 \end{cases}$$

This introduces the active and passive points of view of translations in particular and of transformations in general. Another challenge can be the query, "Can we do anything to the coordinate axes other than translate?" Yes!—the students will suggest other things.

Rotation about the origin is interesting and the following approach has been found effective. Starting with simple laboratory beginnings the student culminates his work on rotation by developing the formulae for rotation himself.

(i) Let us suppose that the point P takes on the name (6, 8) after a coordinate system has been imposed (Fig. 7.10). Now let us suppose further that some one rotates the axes about the origin by 30°. Note the $\bar{x}$- and $\bar{y}$- axes and estimate the new coordinates for the point P. (α) Tell

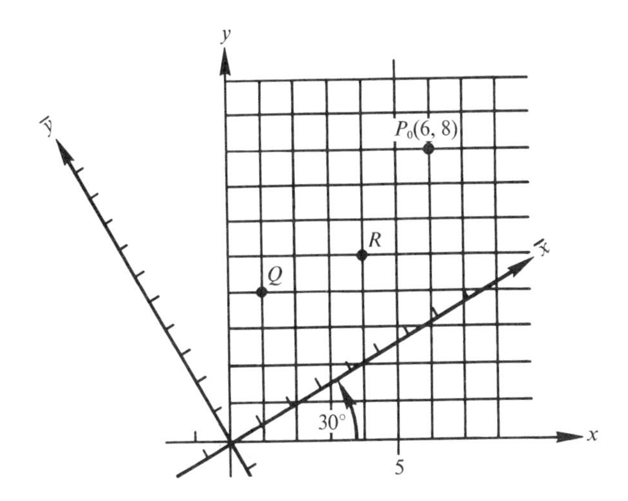

Figure 7. 10

both the old names and the new names for the points Q and R. (β) See if you can arrive at the new names for P by *computation.* The angle of rotation is 30°. Perhaps we can help by drawing a new figure, 7.11. We wish to learn the values of $\bar{x}$ and $\bar{y}$.

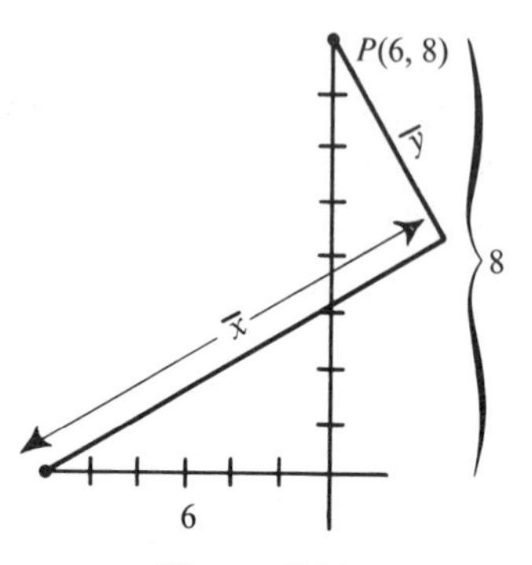

Figure 7.11

Some students have made use of the relations in the 30°–60° right triangle and similar triangles whereas others use trigonometric ratios. In either case the student is putting his mathematics to work for a very definite purpose and is reviewing and creating at the same time. One might consider a rotation of 25° as an additional problem if most of the group used the 30°–60° right-triangle relations.

(j) See if you can develop formulae for $\bar{x}$ and $\bar{y}$ if P has original coordinates (x, y) and the angle of rotation is θ as shown in Fig. 7.12. Here

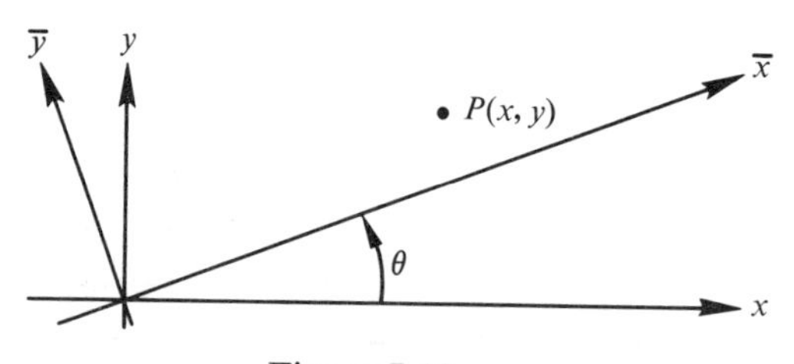

Figure 7.12

the student makes use of trigonometric ratios and identities and obtains, finally

$$R: \begin{cases} \bar{x} = x \cos \theta + y \cos \theta \\ \bar{y} = -x \sin \theta + y \cos \theta \end{cases}$$

It is interesting to note that in the writer's experience no student has ever developed this in the manner in which it is usually done in texts on the calculus. This may be due to the way the numerical approach was started, but it also suggest that some developments, though less elegant, may be *more natural* for the student. In teaching, the ways more natural for the student are the effective ways and, to learn what these are, we need to give the student a chance to contribute his method. Elegance may well come in a second or third "round."

The topic on rotation could start with observing by the eye what the new coordinates are (the student now knows what we are looking for), then a numerical computation, and finally a computation for the general case. The writers have found it useful sometimes to find the new name for (x, y) if the angle of rotation be $30°$. Sometimes a gradual approach to the general case is more successful and the student thinking in his own way can produce much. The teacher should think, "How would *I* approach this topic if I didn't know?" and perhaps become more sympathetic to student suggestions.

Exercises

1. Do the computations (i).
2. Derive the formulae in (j) by methods of your own devising.
3. Consult several texts in the calculus or in analytic geometry under the heading "Transformations of Coordinate Systems" and see how the formulae for rotation are developed.
4. We have *translated* and we have *rotated* axes about the origin. Can you think of anything else which might happen to axes (other than bending)?
5. What change-in-axis interpretation can be given to the transformation*

$$\begin{cases} \bar{x} = x \cos 30° + y \sin 30° + 4 \\ \bar{y} = x \sin 30° + y \cos 30° - 5 \end{cases}$$

Would a rotation followed by a translation yield the same change as a translation followed by a rotation? Investigate.

6. Make up a situation which would lead a student to the change-of-scale transformation, as

$$\begin{cases} \bar{x} = 3x \\ \bar{y} = 2y \end{cases}$$

---

*Note that if this transformation is thought of as

$$\begin{cases} \bar{x} = ax + by + c \\ \bar{y} = dx + ey + f \end{cases}$$

then $a = e$, $b = -d$, and $ae - bd = 1$. This the writer uses to introduce a test for a Euclidean transformation in the plane; for the transformation represents the combination of two rigid motions.

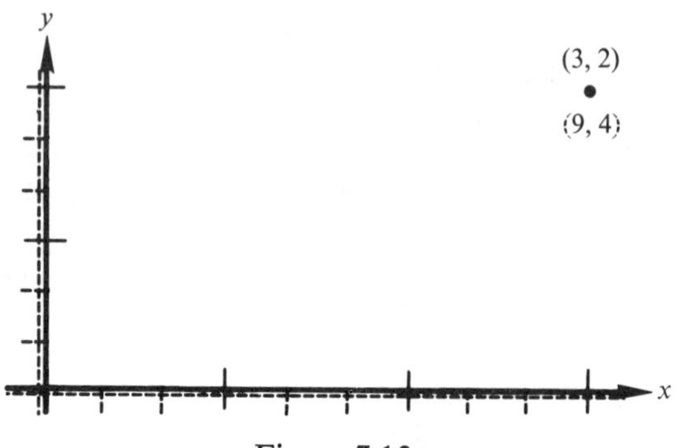

Figure 7.13

It is instructive to ask the student to show these changes in the axis system. Referring to Fig. 7.13, the point whose name was (3, 2) now assumes the name (9, 4) because there has been a change in the lengths of the units. Each (old) unit on the x-axis consists now of three (new and shorter) units; each (old) unit on the y-axis consists now of two (new and shorter) units. Some call these *dilations*, some *dilatations*. Of course, one can think also of the point being "carried" from the position (3, 2) to the position (9, 4) under the transformation Fig. (7.14).

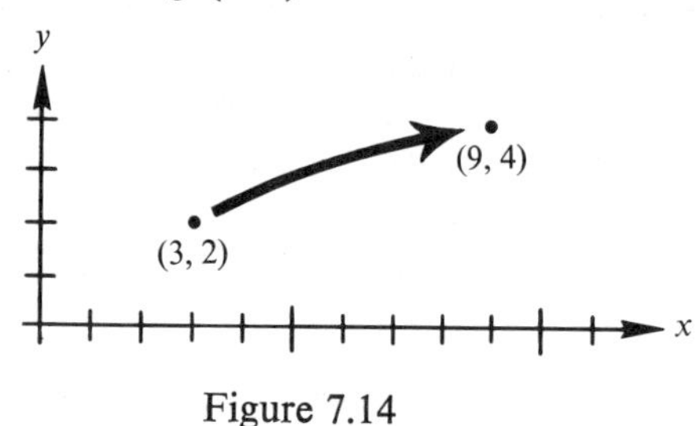

Figure 7.14

7. Devise a change-in-axis interpretation for the transformation

$$\begin{cases} \bar{x} = 3x + 4 \\ \bar{y} = 2y - 7 \end{cases}$$

Sketch a figure to show the changes and check your interpretation using the point whose original name is (2, 3). The changes you describe should yield (10, -1) for the new name.

8. Does the transformation in Exercise 7 describe a change in scale followed by a translation or a translation followed by a change of scale? Which Illustrate.

7.7. An Interpretation of the More General Linear Transformation

The change-of-axis pictures generated by transformations which have come to be called translations, rotations, and changes of scale give rise to devising an interpretation for the transformation of the type

$$\begin{cases} \bar{x} = 3x + 2y + 7 \\ \bar{y} = 5x + 4y - 6 \end{cases}$$

Does it represent a change of scale on each axis? Is there rotation? Is there translation? It is somewhat difficult to devise an interpretation for this transformation, but the student can be helped with some exploratory exercises like these:

(a) Let us consider the transformation

$$\begin{cases} \bar{x} = 3x + 2y + 7 \\ \bar{y} = 5y \end{cases}$$

If $y = 3$ then $\bar{x}$ becomes $\bar{x} =$ __ and this means that each unit on the x-axis is replaced by _______ and then the origin is moved _______. (Ans. 13 new units to the left.) If $y = 4$ then $\bar{x}$ becomes $\bar{x} =$ __ and this means that each unit on the x-axis is replaced by _______ and then the origin is moved _______. Hence along the line $y = 5$ the x-axis scale is having each unit replaced by _______ and the $x = 0$ point is being shifted _______. Along the line $y = 6$ the x-axis scale is having _______ and the $x = 0$ point is being shifted _______. Along the line $y = 2$ what happens to the x-axis scale? The $x = 0$ point on the line $y = 2$ is being shifted _______. Along the line $y = 1$ what is happening to the x-axis scale? The $x = 0$ point on the $y = 2$ line is being shifted _______. Along the line $y = 0$ what is happening to the x-axis scale? The $x = 0$ point on the line $y = 0$?

(b) In Fig. 7.15 is shown an interpretation of the transformation

$$\bar{x} = 3x + 2y + 7$$

$$\bar{y} = y$$

along $y = 0$ and $y = 1$. You show the transformations along the lines $y = 2, 3, 4, 5, 6, 7$. Is the scale-change the same along every line parallel to the x-axis? Is the translation the same? How can you describe the transformation

$$\begin{cases} \bar{x} = 3x + 2y + 7 \\ \bar{y} = y \end{cases}$$

in words?

(c) The point (3, 4) takes on what new name by substitution of 3 and 4 in the transformation? [Ans. (24, 4)]. Is this the new name in the figure you have constructed? The point (1, 3) takes on what new name under the transformation? Is this the new name on the figure you have constructed? Check in a similar way the names of some other points.

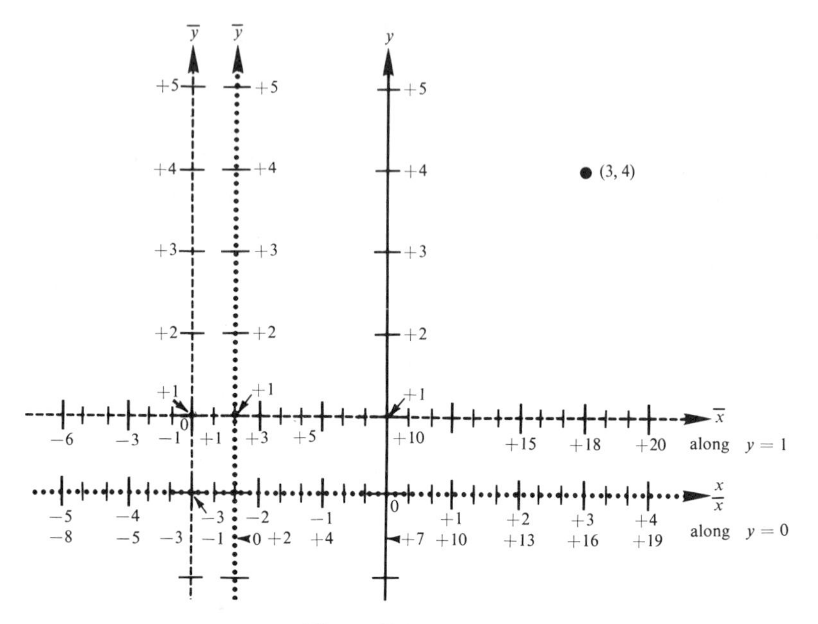

Figure 7.15

(d) Let us now look at

$$\begin{cases} \bar{x} = x \\ \bar{y} = 5x + 4y - 6 \end{cases}$$

along the line $x = 1$ we note that $\bar{y} =$ __. This means that each unit on the y-axis is replaced by __; is there any shift of the $y = 0$ point on the line $x = 1$? Along the line $x = 2$, $\bar{y}$ becomes __. Transformationwise, what does this mean? Along the line $x = 3$, $\bar{y}$ becomes $\bar{y} =$ __. What does this mean?

(e) Can you indicate on a figure of the axis system the effect of the transformation (Fig. 7.16).

$$\begin{cases} \bar{x} = x \\ \bar{y} = 5x + 4y - 6 \end{cases}$$

(f) Under the transformation

$$\begin{cases} \bar{x} = x \\ \bar{y} = 5x + 4y - 6 \end{cases}$$

the point (3, 2) takes on what new name? (Ans. 3, 17). Does this new name appear correctly on the figure?

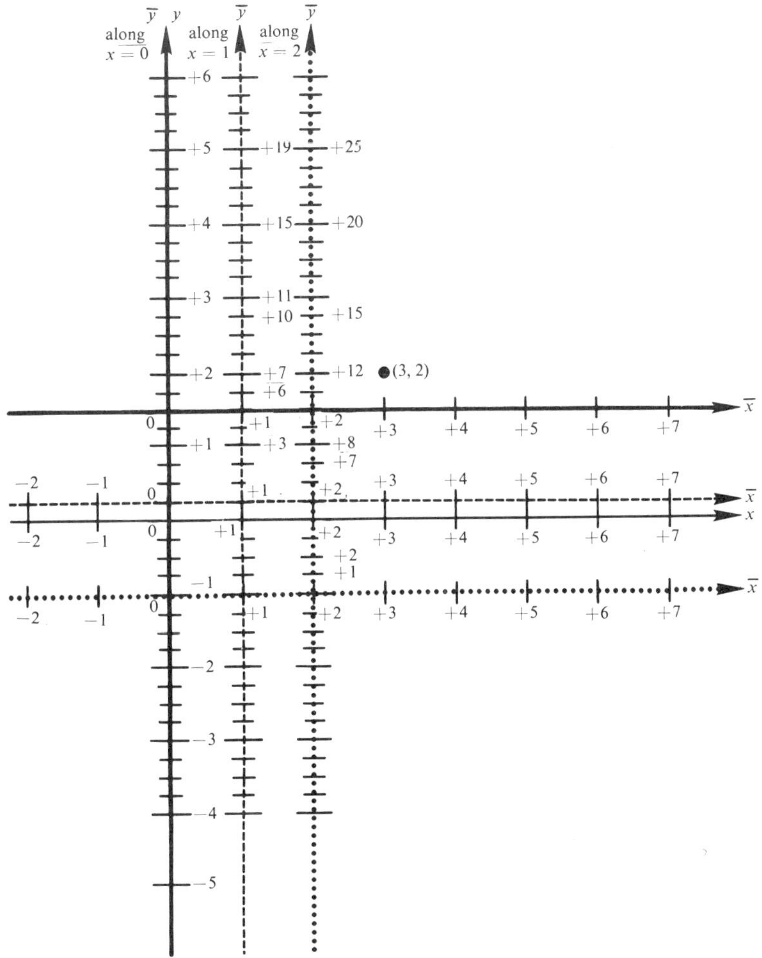

Figure 7.16

(g) How can you describe the transformation

$$\begin{cases} \bar{x} = x \\ \bar{y} = 5x + 4y - 6 \end{cases}$$

in words?

This guided development is rather long, but sometimes it is done in parts along with other current work. It *is* interesting to devise an interpretation for the more general transformation after the simpler ones are done. The challenging problem yet remains: to interpret geometrically the effect

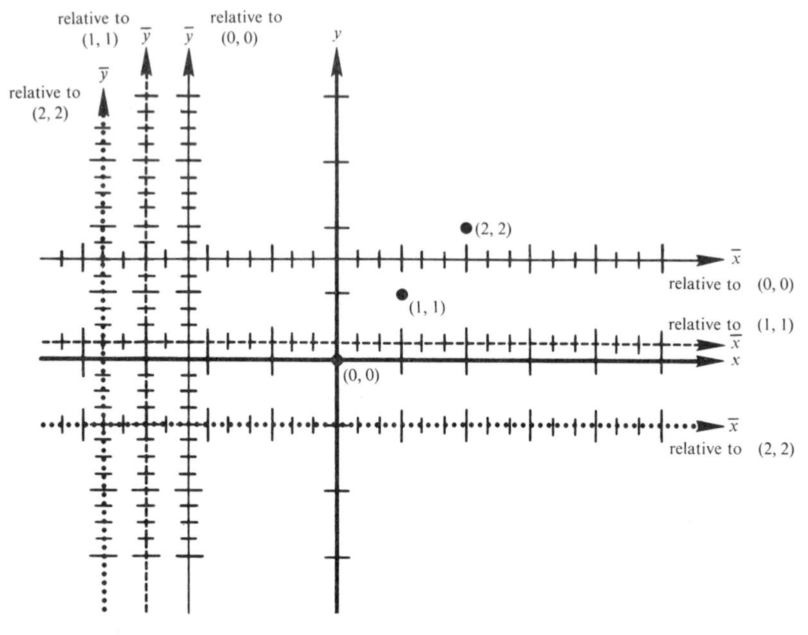

Figure 7.17

on the axis system of the transformation

$$\begin{cases} \bar{x} = 3x + 2y + 7 \\ \bar{y} = 5x + 4y - 6 \end{cases}$$

We have studied the effect of each part separately; now to put them together. We might help to develop a picture (Fig. 7.17) by questions as these:

(h) What does the given transformation do to the scale of the x-axis? the scale of the y-axis?

(i) Applied to the original point (0, 0) of the coordinate system, the transformation moves the axis system with its new scales __ units to the left and units up. Hence, there is a new origin.

(j) Applied to the original point (1. 1) the transformation moves the axis system with its new scales __ units to the __ and __ units to the __.

(k) Relative to the point (3, 3) the given transformation does what to the axis system?

(l) What part of the effect of the transformation remains the same relative to any point? Which part keeps changing? Can you describe in words what this transformation does?

This is a complicated picture. The scale changes are the same, but the translations are different at different places on the coordinate system. Students might like to study this further. The remarkable property about this intricate transformation, however, is that if the coefficients are "just right," that is, if in

$$\begin{cases} \bar{x} = ax + by + c \\ \bar{y} = dx + ey + f \end{cases}$$

$a = e$, $b = -d$, and $ae - bd = 1$ (and, for simplicity at the moment, $c = f = 0$) then this changing of scales and the shifting of axes as studied above produces the same effect as *if the axes had been rotated about the origin.*

From an active point of view, the transformation is interesting (Fig. 7.18). The point (0, 0) is carried into (7, −6); the point (1, 1) is mapped

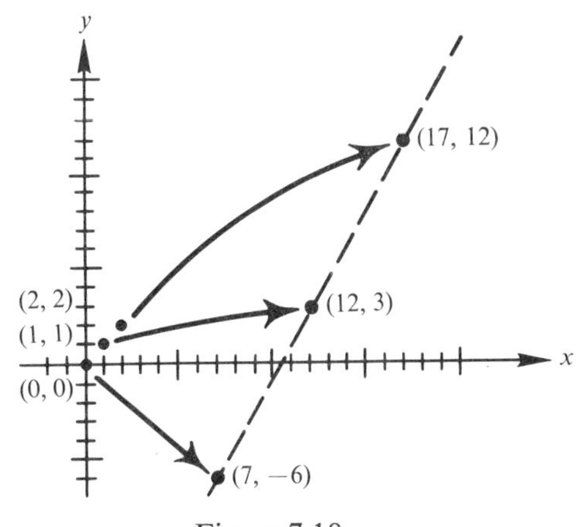

Figure 7.18

into (12, 3); the point (2, 2) takes the position (17, 12). The transformation has the effect of "displacing" the points by different amounts at different places. The effect of the transformation is not the same everywhere.

We have attempt to show the use of carefully prepared exercises in approaching an interesting but more complicated topic. The response to such guides is good, and, for many students, their use is much more effective than the mere introduction of the final question whose answer is sought. The use of guides educates; the final question proposed immediately often causes the student to be puzzled and frustrated.

7.8. Devising a Projection

A beginning to the geometry of projection can be made very readily with modern projection equipment: the overhead projector, the slide projector, or with an electric light bulb, along with wire and cardboard models

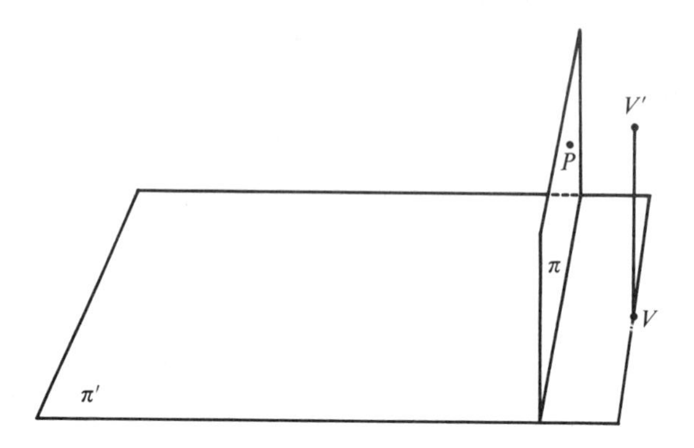

Figure 7.19

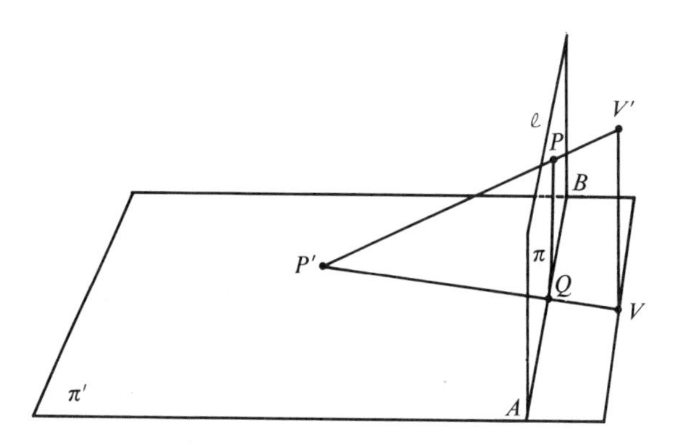

Figure 7.20

of figures and curves. One can demonstrative various kinds of projection and the student gains ideas of the invariance of certain geometrical properties under different kinds of projection which can be associated with different kinds of geometry. There comes a time, however, when one would like to have a projection device on paper and the discussion in this section begins here.

We have found it profitable to begin with an exercise like this (Fig. 7.19 and 7.20):

(a) A lamp is located at V' and P is a point in the plane π. Devise a way to construct the image of P on the plane π'.

It is interesting to observe what methods students devise. Some of them do not accomplish the desired projection, but usually someone originates an acceptable method and we can enter into an interesting and fascinating series of studies. One method to construct the image of P on π' is to drop a line from P parallel to $\overleftrightarrow{\mathbf{VV'}}$ and let it intersect $\overleftrightarrow{\mathbf{AB}}$ at Q. Where $\overleftrightarrow{\mathbf{VQ}}$ intersects $\overleftrightarrow{\mathbf{PV'}}$ yields P' the image. What term we apply to the image P' if $\overleftrightarrow{\mathbf{PV'}} \| \overleftrightarrow{\mathbf{QV}}$ depends on what kind of geometry we are in at the moment. In projective geometry we would rule such a physical situation not permissible; in affine geometry, P' would be an "ideal" point. If P on ℓ makes $\overleftrightarrow{\mathbf{V'P}} \| \overleftrightarrow{\mathbf{VQ}}$ then, in an affine projection, the line ℓ projects into the ideal line.

One can readily introduce more situations now.

(b) Suggest how you might construct the image (or projection) of a line segment. Is the length of the projection the same as that of the original segment?

(c) Begin with three points collinear on the plane π. Are the projections of these points collinear?

(d) Take four points on a line in the plane π. Use a centimeter scale to determine the cross-ratio of these four points. How does the cross-ratio of the projections of these points compare with that of the original points?

(e) Place a closed curve such as an ellipse on the plane π. What can you say about the projection?

(f) Can you place an ellipse on the plane π in such a way that the projection seems to be a *parabola?* How? Try it.

The above exercises lead the way to some ideas of invariance and noninvariance in projective geometry and set the stage for the idea that in projective geometry the parabola, the ellipse, and the hyperbola are simply all *conics*. It is not until *parallelism* enters in the postulational evolution of geometry that the distinction among conics is made. Next, one can alter the projection method and pursue the study further.

(g) Let us suppose now that we *tilt* the plane π, with everything on it, so it lies on the π' plane (Fig. 7.21).

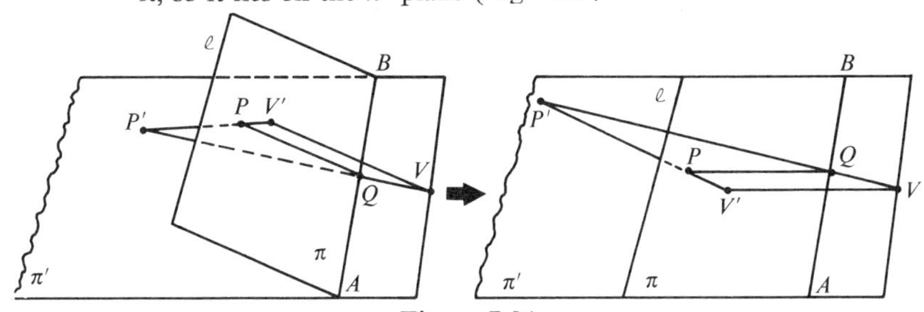

Figure 7.21

Using the same mechanics as before, construct the *image* of a point P.

(h) Using the same mechanics construct the image of this closed curve. (Fig. 7.22). We can repeat some of the studies done before and investigate new ones.

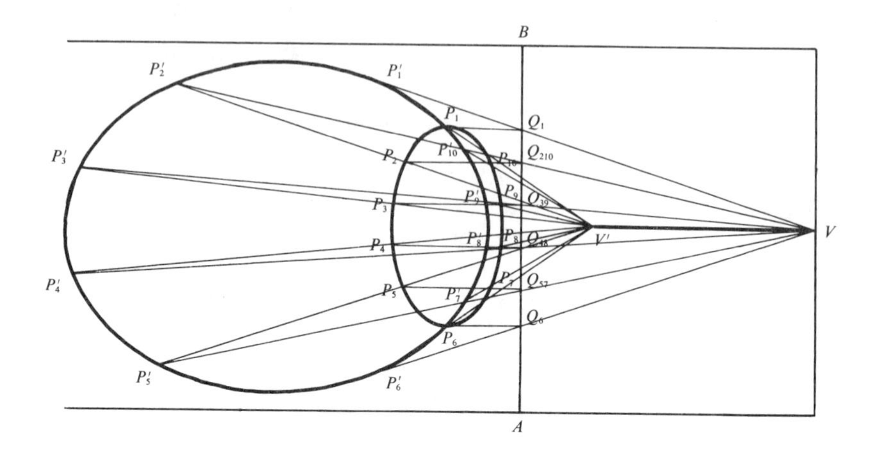

Figure 7.22

The introduction above is a rather natural one to the subject of projection and it illustrates well how a natural approach to a discipline can bring out some very significant properties in an experimental or laboratory level while the student has a physical model and method with which to work. Generalizations, abstractions, and conclusions can develop while the student works. Soon he will be ready for a postulational treatment. Then the more formal approach, which rises up from his previous experiences in handling parts of the subject, emerges as challenging and meaningful. The student is now trying to erect an axiomatic structure concerning something with which he is already acquainted. In his Goal 6, R. C. Buck writes "... it pays to subject a familiar thing to detailed study ..."[4] Approaches as the above help orient students to more abstract treatments. Out of the above physical introduction, for example, comes the idea of the formation of a conic by the intersection of two pencils of lines and the observation of the apparent invariance of certain properties, both significant concepts in projective geometry.

Exercises

1. In the last figure the conic does not intersect the line ℓ. Study its projection if (a) the conic is tangent to ℓ and (b) the conic intersects the line ℓ.

   (The distance between ℓ and AB should be equal to the length of $\overline{V'V}$.)

2. Why is the line ℓ called "the line at infinity?" It really isn't "at infinity" for it is now present on the plane π.

[4] *Ibid.*, p. 955.

3. The mechanics of projection we have employed does not belong to the "pure" projective geometry; our projections are in an "affine setting." Why? (Ans. In our construction we drew lines *parallel*, and in projective geometry, *all* lines meet.)

7.9. Approximations of the Zeros of a Polynominal

The purpose of this section is to show how closely the usual suggestions for methods to approximate roots agree with those which have come to us from the past. Indeed, when students are permitted to think about a problem in their own way, it is sometimes true that they suggest methods very close to those found in texts. The difference in the entire procedure, however, from the method of teaching wherein the teacher tells is that the student has *invented* the method himself and, therefore, has better understanding; he need not learn by rote for he can recreate the method when it is necessary.

Let us suppose that we are interested in solving cubic equations and the exercises have been progressing like this:

.

.

.

4. Solve these cubic equations:

(i) $x^3 - x = 0$

(ii) $x^3 - 3x^2 + 2x = 0$

(iii) $x^3 - 7x^2 + 14x - 8 = 0$

(iv) $x^3 + x^2 = 10x + 8 = 0$

5. Solve Exercises 4(i), 4(ii), 4(iii) by graphical methods. Do you obtain the same answers as you did by algebraic methods?

6. Consider the cubic equation $x^3 + x^2 - 4x - 10 = 0$. Can this equation be solved by "factoring"? Can you find a root by the use of its graph?

7. Is the answer from the graph in Exercises 6 the actual root or is it an approximation? What methods can you suggest to yield a closer approximation?

8. Try one of the methods you suggested in Exercise 7.

With the beginnings of solution by graphing, student already has a setting and a picture to assist in making suggestions. Using a table of values (see Exercise 6) the student sees 2.4 as an approximate value of the root between $x = 2$ and $x = 3$. Several different suggestions often come from the students.

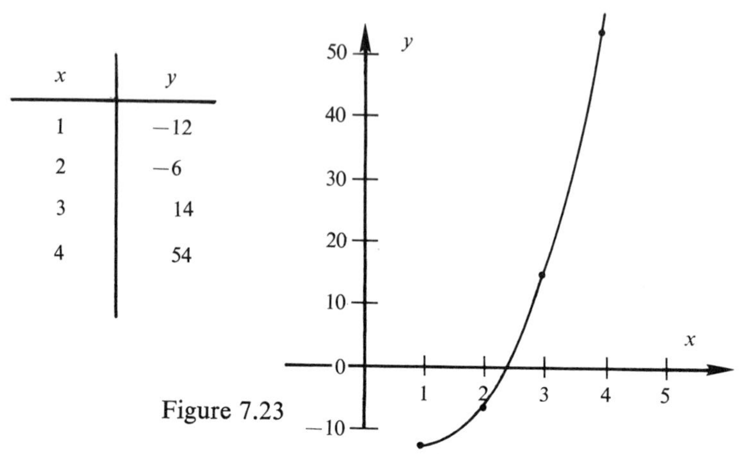

x	y
1	−12
2	−6
3	14
4	54

Figure 7.23

(a) One student outlines a plan like this: simply try 2.4 and 2.5 in the table of values and watch the value of y. If $x = 2.4$ makes y negative but $x = 2.5$ makes y positive, then there is a root between 2.4 and 2.5. If one wishes a closer approximation, he just tries, say, the values of

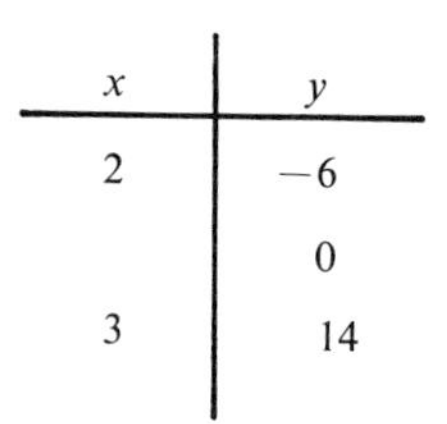

x	y
2	−6
	0
3	14

Figure 7.24

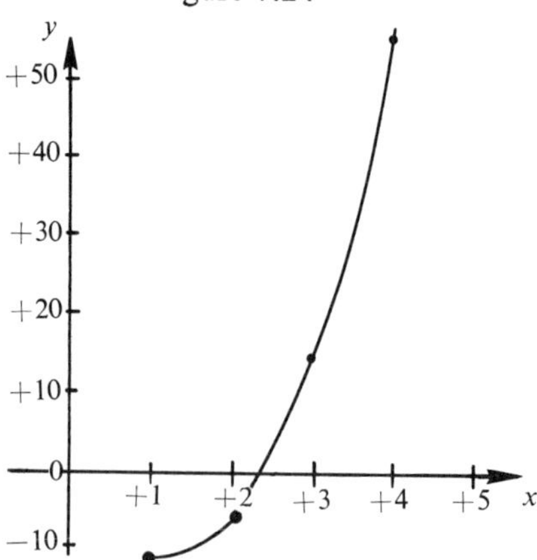

Figure 7.25

$x = 2.40$ and 2.45 to see if y changes sign. If it does, one just keeps "closing down" on the root by trial.

(b) Sometimes students suggest *interpolation* on the table of values. Here, they reason, we have and for $y = 0$ one can estimate the corresponding x by using 2 plus $\frac{6}{20}$ or x is approximately 2.30. In the class, if it were not a direct suggestion, the teacher might seek again the assumption behind interpolation and ask the class to *show* on the graph what one is doing when he interpolates (Fig. 7.25). Out of this comes the usual straight-line interpolation method.

Now can come the question on study guides:

9. Our interpolation method indicates 2.30 as an approximation to the root between 2 and 3. Yet other methods tell us it is between 2.4 and 2.5. What caused this approximation to be so much in error? Can we still use the method to get a better approximation? Some will say, "Yes! Compute y for $x = 2.3$ and use it, if y is negative, along with the $y = 14$ for $x = 3$ and interpolate again. Making another entry in the table of values, we approximate the root again:

x	y
2	$^-6$
2.3	$^-1.75$
2.0	14.00

$$x \doteq 2.3 + \frac{1.75}{15.75}$$

$$\doteq 2.3 + .11 = 2.41$$

Our students now feel better that there is close agreement between approximations made by the three methods: graphing, trial, and interpolation.

10. How could we make a still better approximation?

Students usually suggest repeating this process.

One is sometimes surprised at what the student will produce when invited to think more about this, and, not infrequently there are some proposals which are very refreshing. One such follows: This curve, I know, is a cubic, but it does remind me of a portion of a parabola so I shall compute the equation of the parabola, which is about like the cubic through the three points shown. Then I can use the quadratic formula to make an approximation (Fig. 7.26). Students had had practice in determing equations of curves through given points and this ability was called into use to attack the present problem. To find the equation of the parabola through the points $(1, -12)$, $(2, -6)$, $(3, 14)$ one has, from $y = Ax^2 + Bx + C$, the following relations

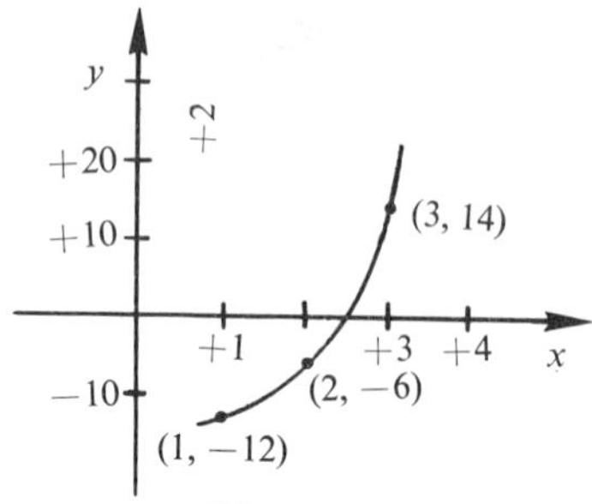

Figure 7.26

$$-12 = A + B + C$$
$$-6 = 4A + 2B + C$$
$$14 = 9A + 3B + C$$

to be satisfied simultaneously. These conditions yield $A = 7$, $B = -15$, $C = -4$, whence the parabola through the three given points of the cubic has the equation $y = 7x^2 - 15x - 4$. By the use of the quadratic formula one root is approximately 2.4.

A field is opened and, on reflection, an interesting observation. At first we replaced a portion of the cubic by a straight line and used interpolation; next we replaced a portion of the cubic by a parabola and used the quadratic formula. Not every day is as fruitful when student-centered and student-suggested approaches are used, but unless the teacher *invites* and *encourages* he will get practically none. Mathematics should not be handed down as ready-made—it should, as much as possible, be *created* by the student. Note in this section the simple beginnings of approximation by trial, by graphing, by interpolation (linear), and finally by a replacing a portion of the cubic by a parabola. Not everyone made all these suggestions, but the thinking of individuals brings to light several methods. All students, however, are expected to become proficient in all methods which are retained as the course moves forward.

Exercises

1. Work through the "replacing a portion of the cubic by a parabola" to confirm the results of the text.

2. Apply all four methods to determining an approximation to a real root of $x^3 - 6x^2 + 8x - 10$.

3. With your experience in mathematics, do *you* have any further suggestions on how to approximate the root between $x = 2$ and $x = 3$ for the cubic given in the text?

4. Could one replace a portion of a quartic curve by a parabola in an effort

to approximate a real root? Could one apply similar methods to a quintic? Are there reservations which should be made? Are there portions of cubics which might more effectively be replaced by straight lines? Illustrate.

7.10. Some Topics in Coordinate Geometry

Although there was some discussion of the method of coordinate geometry and its place in secondary school courses in the last chapter, we wish to suggest here some approaches in teaching parts of this subject which the writers have found effective. Most analytic geometry texts *tell* the student and write that "we shall now show" so that the student is almost always receiving his mathematics in a passive matter-of-fact way rather than being a part-creator. In many cases it would be very simple to structure problems and experiences so that some of the facts of analytic geometry could emerge as products of the students' suggestions.

One theorem most often presented by a synthetic type of methodology (see Fig. 7.27) is that of the "distance from a line to a point" or, as stated by others, "the distance from a point to a line." In some texts this is developed by placing the equation of the line in the normal form and identifying what part of this equation tells the distance; others compute the *equation*

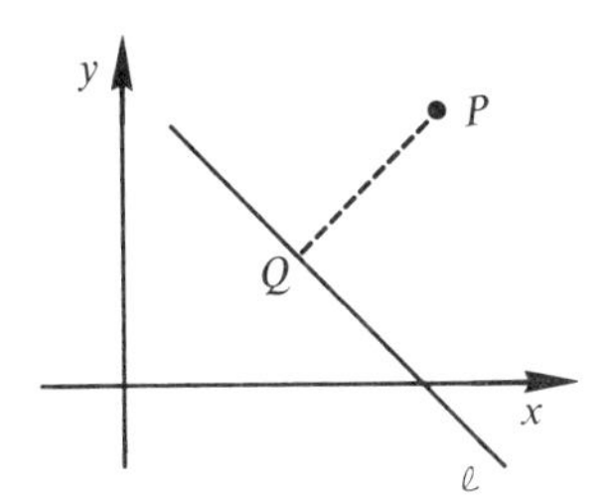

Figure 7.27

of the line $\overleftrightarrow{\mathbf{PQ}}$ which is perpendicular to ℓ, solve the equations of ℓ and $\overleftrightarrow{\mathbf{PQ}}$ simultaneously, and then use the resulting coordinates of Q and P to find the length of $\overline{PQ}$. All of this is figured out and *presented* to the student; he then reads some sample problems and is instructed "to do likewise." How uninteresting and nearly lifeless, although there may be some "different" and challenging problems later. Rather, *methods of discovery* can be used to help the course itself grow—discovery exercises need not be adjuncts to lecture and other synthetic methods of presentation. When a class is conducted with student suggestions taking a dominant role, there is discovery and creativity much of the time. A study guide sheet or an assignment may contain, in part, these problems (Fig. 7.28):

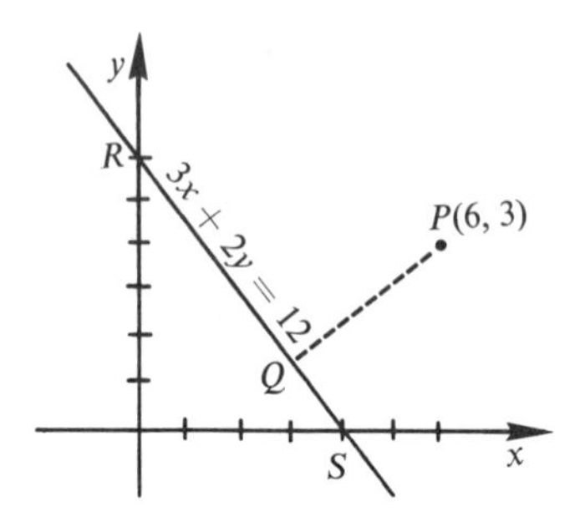

Figure 7.28

1. $\overleftrightarrow{\mathbf{RS}}$ has the equation $3x + 2y = 12$. If $\overleftrightarrow{\mathbf{PQ}} \perp \overleftrightarrow{\mathbf{RS}}$, then what is the *slope* of $\overleftrightarrow{\mathbf{PQ}}$?
2. Compute the *equation* of the line $\overleftrightarrow{\mathbf{PQ}}$.
3. How could you find the *length* of the segment $\overleftrightarrow{\mathbf{PQ}}$? What length do you obtain?

Of course, some teachers might wish to ask Problem 3 immediately, and students would be challenged to furnish more facets from their own experience at the outset. It *is* true that being asked to work Problem 1 and 2 first may guide the student into a direction he would not have gone had he been asked Problem 3 or the ultimate problem immediately. After several problems like this, however, the following might appear on a study sheet, or, if the class is oriented toward seeking generalizations, a student may suggest the investigation.

4. If P has the coordinates (a, b) and $\overleftrightarrow{\mathbf{RS}}$ the equation $3x + 2y = 12$, what is the length of $\overline{PQ}$?

The students will get $|3a + 2b - 12/(\sqrt{13})|$ as the length. This is especially interesting when the student reflects on his result, and is profitable too. An efficient method emerges for computing the length of $\overline{PQ}$. Also, it was invented by the student. Thus far the method applies only to finding distance from $P(a, b)$ to the line whose equation is $3x + 2y - 12 = 0$ but it is a start and now the student can suggest the next query. Through practice first with concrete numerical exercises and, later, with a partially generalized situation, he now has the knowledge and skill to suggest and investigate the problem of the distance from $P(a, b)$ to the line whose equation is the most general form $Ax + By + C = 0$. He arrives at

$$\ell(\overline{PQ}) = \frac{Aa + Bb - C}{\sqrt{A^2 + B^2}}$$

not as a reader, but as a *producer*, not as a follower, but as a *creator*. This is a good example of the use of successive stages of abstraction in approaching problems; such a plan is often very effective. A few students can consider general cases immediately, but most find the use of more concrete cases at the beginning most helpful. Indeed, Saxelby suggests that "The most natural method of advance is by a series of successive approximations... and, in fact, this is the way in which the subject has actually grown up."[5]

One might now devise and explore situations leading to expressions for the equation of a line in terms of "distance from the origin" and the angle ω which the normal makes with the x-axis (Figs. 7.29 and 7.30). Indeed, a beginning question might be

1. The line ℓ has the name $4x + 5y = 40$. Suppose that a perpendicular P is dropped from the origin to ℓ. Can you express the slope of the line p? Can you rewrite the equation of ℓ in terms of the cot ω? sin ω and cos ω? (Ans. $y = -\cot \omega + 8$; $x \cos \omega + y \sin \omega = 8 \cos \omega$).

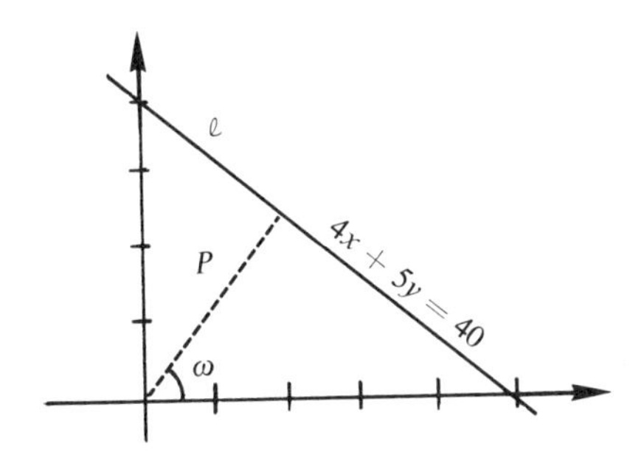

Figure 7.29

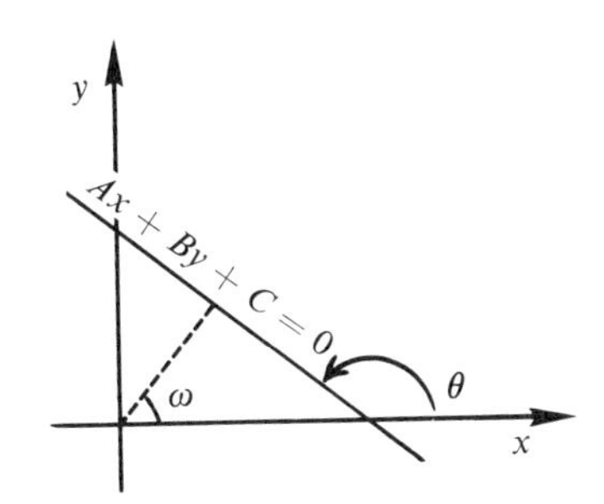

Figure 7.30

[5]F. M. Saxelby, *A Course in Practical Mathematics* (London: Longmans, Green, and Company, 1907), p. vi.

2. What geometric interpretation can you give to each coefficient and constant in $x \cos \omega + y \sin \omega = 8 \cos \omega$?

3. Rewrite $Ax + By + C = 0$ in terms of ω. (Ans. $x \cos \omega + y \sin \omega = -C/B \sin \omega$).

4. What interpretation can be given to $-C/B \sin \omega$? Hence what interesting facts about the line are revealed from the use of the coefficients A, B, C? (Ans. $-A/B$ is the slope of the line, $-C/B$ is the y-intercept, $B/A = \tan \omega$, $-C/B \sin \omega$ is the distance of the line from the origin.)

The above guided development helps the student to formulate another "equation" of a straight line; this is recognized by the reader as the *normal form* $x \cos \omega + y \sin \omega - p = 0$. What kind of problem can we now work? (A line is 10 units from the origin and the angle which the normal makes with the x-axis is $\pi/6$; what is the equation of the line?) Is the possible to start with $Ax + By + C = 0$ and obtain expressions for $\sin \omega$ and $\cos \omega$ and p so that one may move from one form to the other? Students can work on this question profitably.

In the study of conics "stages" can be set so that the student can devise certain forms (Fig. 7.31). He may begin by considering:

1. What algebraic condition must be placed on the point whose name is (x, y) so that it is always at distance 4 units from the point (3, 2)? [Ans. The algebraic conditions that $(x - 3)^2 + (y - 2)^2$ must equal 100].

2. All the points (x, y) which satisfy this condition will lie on what kind of curve?

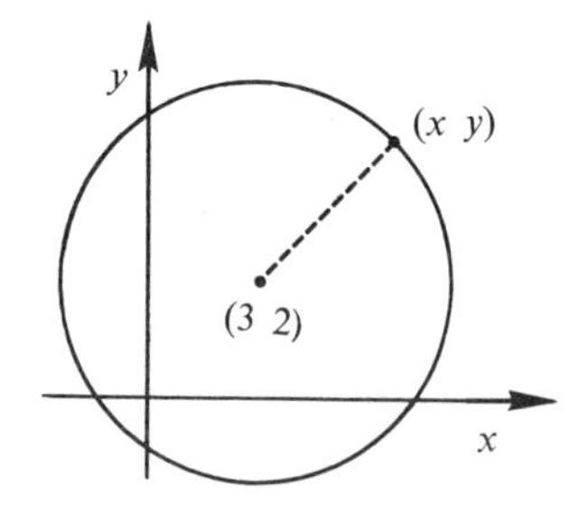

Figure 7.31

Several more such problems follow and then can come the query: What general problem does this suggest? The relation $(x - a)^2 + (y - b)^2 = r^2$ comes as a climax to the students experience. Even from this, expanded as $x^2 - 2ax + y^2 - 2by + a^2 + b^2 = r^2$, can come the question if this is the most *general kind* of equation of second degree, and if not, do others have geometric interpretations?

Of course, the teacher can readily propose different situations which lead to the equations for the parabola, the ellipse, and the hyperbola, and, indeed, the students might also invent other conditions to be imposed on the "point (x, y)." Simple examples as these may seem to have used only simple situations to illustrate student-centered approaches, but constant attention to the rich possibilities of such approaches, and the still richer results, places the student at the other end of the learning experience. Instead of being a receiver entirely of what others have done, he is an active producer of the *new* (to him).

Exercises

1. Work through the exercises above.
2. A straight line is at distance 12 units from the origin and the normal makes an angle of $45°$ with the x-axis. What is the equation of the line?
3. Given $x \cos \omega + y \sin \omega + C/B \sin \omega = 0$, see if you can compute $\cos \omega$ and $\sin \omega$ in terms of A, B, and C so the equation $Ax + By + C = 0$ can be transformed into a form still expressed only in terms of A, B, and C but which exhibits these functions of ω.
4. Consult a text or portion of a text on analytic geometry, select a topic, and try to plan a guide so that the content can develop by student contribution.

7.11. The Principle of Mathematical Induction

The principle of mathematical induction, as we know it, was first stated in 1545 by the Italian mathematician, Maurocylus.[6,7] For many years this topic was commonly found in college algebra texts, but recently it has been appearing in secondary school programs and now one or two ninth grade courses in algebra do work on mathematical induction.

An examination of texts yields fruitful suggestions to help students grow in the understanding and use of this principle. The climbing of a ladder provides a good analogy. It is reasonable to assume, the students readily agree, that one can climb an infinitely long ladder if it be shown only (a) that he can always get on the first rung and (b) that if he is on a certain rung he can always get to the next one. The *assumption* that conditions (a)

---

[6]G. Vacca, "Maurocylus, The First Discoverer of the Principle of Mathematical Induction," *Bulletin of the American Mathematical Society*, XVI (November, 1909), 71.

[7]Elizabeth Nypaver, *The Method of Mathematical Induction: Origin, Growth, and Place in Mathematics Education.* Unpublished Master's Thesis, Kent State University, Kent, Ohio, 1964. Chapter II.

and (b) are sufficient to insure the capability of climbing the "entire" ladder is analogous to the principle of mathematical induction as applied to statements about natural numbers.

Long before the study of mathematical induction, exercises should be done by the student to create a setting for the use of this principle. For example, some simple beginnings could follow this pattern:

1. Here are some sums of natural numbers.

   $1 = 1 =$ "sum" of the first number
   $1 + 2 = 3 =$ sum of the first two numbers
   $1 + 2 + 3 = 6 =$ sum of the first three numbers
   $1 + 2 + 3 + 4 = 10 =$ sum of the first four numbers

   Extend the table to include several more sums. Can you find a way to compute the sum for any number of numbers? Suppose n is the number of numbers to be added.

2. Does your formula hold for the sum of twelve natural numbers? Twenty? Does your formula enable you to find the sum of the first one hundred numbers *quickly*? What is this sum?

3. How could you *prove* that the formula $n(n + 1)/2$ is valid for *all* natural numbers?

Indeed, depending on his background, the student may give one of several valid arguments, but it has also been found effective to introduce the analogy of the ladder at this point. "Getting on the first rung" is analogous to testing the formula for 1. Now assume that you are on rung k; can you get to rung $k + 1$? If you can, we *assume* you can reach all the rungs. This is analogous to assuming that the formula is valid for k [i.e., analogous to assuming that $k(k + 1)/2$ is valid], and then showing that from this assumption it follows that $(k + 1)[(k + 1) + 1]/2$ is the formula for the sum of $k + 1$ numbers. Some exercises follow:

1. Is the formula $n(n + 1)/2$ valid for 1?

2. Now assume it valid for k; that is, assume that $k(k + 1)/2$ is valid, or that $1 + 2 + 3 + \cdots + k = k(k + 1)/2$. Does $1 + 2 + 3 + \cdots + k + (k + 1) = (k + 1)[(k + 1) + 1]/2$? Of course $1 + 2 + 3 + \cdots + k$ already equals $k(k + 1)/2$ because we *assumed* we were on the kth rung! Hence can you show directly that $[k(k + 1)/2] + k + 1 = (k + 1)[(k + 1) + 1]/2$? Try it.

This problem is one of the simplest. Students should have much experience with this kind of proof and with different kinds of problems. Some different kinds at the secondary level are the (a) ordinary number relations as found in problem lists in college algebra; (b) problems in divisibility as showing that $7^n - 1$ is divisible by 6, or that $a^n - b^n$ is divisible by $a - b$; (c) such problems as proving that $n + 1 = 1 + n$ for all natural numbers given that addition is associative; (d) perhaps the binomial theorem, and others. Reasonably frequent contact with mathematical induction over long periods of time gives a good working knowledge of the principle.

Skill in the use of the mathematical induction principle is, for most students, not an instant acquisition and, indeed, in this day of instant foods, instant responses of automated machinery, and instant appearances of items in dispensers, one should not forget that *learning* still requires patient nurturing of small skills to use in more complicated ones and the understanding of small processes to be synthesized later into methods to attack large complex problems.

In teaching the use of the principle of mathematical induction, teachers and students alike must be on guard against using a fallacious method which seems to appear in the work of some students in which conclusions are drawn which are unwarranted. In showing, for example, that $k(k + 1)/2 + (k + 1) = (k + 1)[(k + 1) + 1]/2$ some students *begin* with both members and change them into equivalent forms, as

$$\text{A}\left\{\begin{array}{ll} \dfrac{k(k+1)}{2} + (k+1) & \dfrac{(k+1)[(k+1)+1]}{2} \\[2ex] \dfrac{k(k+1) + 2(k+1)}{2} & \dfrac{(k+1)^2 + (k+1)}{2} \\[2ex] \dfrac{k^2 + k + 2k + 2}{2} & \dfrac{k^2 + 2k + 1 + k + 1}{2} \\[2ex] \dfrac{k^2 + 3k + 2}{2} & \dfrac{k^2 + 3k + 2}{2} \end{array}\right\}\text{B}$$

and, on observing that $(k^2 + 3k + 2)/2 = (k^2 + 3k + 2)/2$, will claim that the two original members are equal. This is faulty reasoning for false hypotheses can lead to "valid" conclusions, too, as, one learns when he studies implication. Some students may even place equal signs between the members in corresponding rows and hence, introduce the fallacy of "begging the question." The best which can be gleaned from this dangerous procedure is that, regarded as scratchwork, it can suggest a succession of steps to transform

$$\frac{k(k+1)}{2} + (k+1) \text{ into } \frac{(k+1)[(k+1)+1]}{2}$$

if the steps in Column B are reversible. Using the columns in this way, we can write, in a direct and correct manner, the following:

$$\begin{aligned}\frac{k(k+1)}{2} + (k+1) &= \frac{k(k+1) + 2(k+1)}{2} \\ &= \frac{k^2 + k + 2k + 2}{2} \\ &= \frac{k^2 + 3k + 2}{2} \\ &= \frac{(k^2 + 2k + 1) + (k+1)}{2} \\ &= \frac{(k+1)^2 + (k+1)}{2} \\ &= \frac{(k+1)[(k+1)+1]}{2}\end{aligned}$$

It is much better, in an induction proof, to state simply at the proper time, "Now we must see if we can show in a direct manner that $k(k+1)/2 + (k+1)$ can have the form $(k+1)[(k+1)+1]/2$ or that $k(k+1)/2 + (k+1)$ has also the name $(k+1)[(k+1)+1]/2$," and then proceed in a direct line of reasoning as follows:

$$\begin{aligned}\frac{k(k+1)}{2} + (k+1) &= \frac{k(k+1) + 2(k+1)}{2} \\ &= \frac{(k+1)(k+2)}{2} \\ &= \frac{(k+1)[(k+1)+1]}{2}\end{aligned}$$

Another emphasis which should be made by the teacher is that although the method being discussed here is known as the principle of mathematical *induction*, the process is entirely one of *deduction.* The principle appears either as a postulate or as a theorem and conclusions established by its use are *logical consequences* of a deductive pattern. As it has been pointed out in an earlier chapter, *scientific induction*, which has an important place in teaching to give rise to hypotheses suggested by the study of examples, is far different from proving a theorem as a consequence of the postulate of mathematical induction. Scientific induction *is* induction, mathematical induction uses deduction.

At the close of each "induction proof" the student should write a concluding statement with the reason "by the principle of mathematical induction"; this shows that the entire conclusion is dependent upon the mathematical induction principle either as a postulate or as a previously proved theorem.

One should not close this brief discussion of the induction principle without mentioning that this statement appears in several different ways. Some of these are:

(a) If a statement about natural numbers is true for 1 and if on assuming it true for k, it can be shown to be true for $k + 1$, then it is true for all natural numbers.

(b) Let P be a statement about natural numbers and let S be the set of natural numbers, $S \subseteq N$, for which P is true (valid). Now if

(i) $1 \in A$
(ii) $k \in S \rightarrow k + 1 \in S$

then $S = N$ where N is the set of natural numbers.

Some writers define and make use of *inductive sets*, or, *successor sets*. If $1 \in Q$ and a natural number $k \in Q \rightarrow (k + 1) \in Q$ then Q is a *successor set*. Then it is argued that the set of natural numbers is the "smallest" successor set or that it is the intersection of all successor sets. Hence if S be a set of natural numbers and if $1 \in S$ and the natural number $k \in S \rightarrow (k + 1) \in S$ then $N \subseteq S$ where N is the set of natural numbers. Now if, in a proof, S be taken as a subset of N and S can be shown to be inductive, then $S \subseteq N$, whence from $N \subseteq S$ and $S \subseteq N$ simultaneously we have $S = N$. In actual practice one simply tries to show that S is inductive; the mechanics of attempting to prove a statement by the use of methematical induction is basically the same now matter how the principle is stated. Readers interested in the "inductive set" approach might consult Fine's *An Introduction to Modern Mathematics*[8] or the National Council of Teachers of Mathematics *Insight into Modern Mathematics*.[9]

Closely associated with the principle of mathematical induction is the "least integer principle," which is the statement that in every nonempty set of natural numbers there is a *least*. This is often called the Well-Ordering Axiom. Indeed, the mathematical induction and least integer principles are logically equivalent in that beginning postulates for the natural numbers plus the induction postulate can lead to the least integer principle as a theorem and beginning postulates for the natural numbers plus the least integer principle as a postulate yields as a logical consequence the mathematical induction principle. A few teachers have become interested in employing the least integer principle instead of the postulate of mathematical induction in upper level high school classes. The least integer principle is

---

[8]Nathan J. Fine, *Introduction to Modern Mathematics* (Chicago: Rand McNally and Company, 1962) Chapter 4.

[9]National Council of Teachers of Mathematics, *Insight into Modern Mathematics*, Twenty-Third Yearbook (Washington, D.C.: National Council of Teachers of Mathematics, 1957), 51-54.

another tool by which to prove theorems about the natural numbers, and it is a step toward having fewer or less-encompassing postulates.

To prove the relation about natural numbers that

$$1 + 2 + 3 + 4 + 5 + \cdots + n = \frac{n(n+1)}{2}$$

by the least integer principle students might be guided in a manner similar to this:

Given: The postulate that "In every set nonempty S of natural numbers there is a least."

To Examine the Statement:

$$1 + 2 + 3 + 4 + \cdots + n = \frac{n(n+1)}{2}$$

Examination:

1. Either the statement is valid for all the natural numbers or ________.
2. In attempting to use the method of *reductio ad absurdum,* let us assume that there is a set S for which ________ $(S \neq \phi.)$
3. Now by the use of the "*least* integer" postulate the set S has ________, call this natural number k.
4. From the definition of k is the statement to be tested true for k?
5. Would the statement be valid for $k - 1$? Why?
6. Hence what does the statement become for $(k - 1)$?
7. Is it possible to prove that if the statement is valid for $k - 1$, it is valid for k? Try it. Show work.
8. If the answer to Problem 7 is "yes" what situation are we in now?
9. Hence complete the argument.

For some students there are advantages in this kind of proof. First the well-ordering postulate is simply stated—"in every nonempty set of natural numbers there is a least"—and it does not seem to have the complexity in wording possessed by the mathematical induction principle. Second, it gives additional practice in the use of *reductio ad absurdum* in nongeometric situations and the student can experience its application in a greater variety of settings. Many students in more advanced classes have difficulty in understanding "what the contradiction is" when much is left for the reader, and more experience in the use of this method decreases this danger. Third, the student develops also techniques in showing that a statement valid for $k - 1$ may lead to a statement valid for k, which is the very skill he needs in proofs by mathematical induction. Cautions to be taken in the arrangement of work and train of reasoning are the same as those discussed previously.

Exercises

1. In the proofs by the mathematical induction principle one always checks the statement whose validity is to be examined for 1. In proof by the least integer principle one should likewise make sure that the set S does not contain 1. Why must this be done? How can it be done?

2. Prove by the principle of mathematical induction that for all $n \in N$,

   (a) $1^2 + 2^2 + 3^2 + 4^2 + \cdots + n^2 = \dfrac{n(n+1)(2n+1)}{6}$

   (b) $1 + 5 + 9 + \cdots + (4n - 3) = n(2n - 1)$

   (c) $a + ar + ar^2 + \cdots + ar^{n-1} = \dfrac{ar^n - a}{r - 1}$, where $a, r \in$ Reals, $r \neq 1$.

   (d) $7^n - 1$ is divisible by 6.

   (e) $a^n - b^n$ is divisible by $a - b$.

   (f) $n^3 - n$ is divisible by 3.

3. Prove by the use of the principle of mathematical induction that for all $n \in N$

$$(a + b)^n = a^n + na^{n-1}b + \frac{n(n-1)}{2}a^{n-2}b^2 + \cdots + b^n$$

4. Prove by the use of the principle of mathematical induction that for all $n \in N$,

$$D_x ax^n = nax^{n-1}$$

   where D_x instructs one to form the *derivative* of ax^x.

5. Prove by the principle of mathematical induction that if a line segment of unit length be given, then a line segment of length $\sqrt{n}$ can be constructed with a straightedge and compass for any natural number n.

6. Prove several of those in Exercise 2 or other statements in this list by the use of the well-ordering principle of natural numbers.

7. When a student is asked to show, for example, that

$$1 + 2 + 3 + \cdots + n = \frac{n(n+1)}{2}, *$$

   or, that

$$\sum_{i=1}^{n} i = \frac{n(n+1)}{2}$$

he is really being asked to see if the conjecture indicated by $*$ can take on the stature of a theorem. Indeed, we deprive the student of an exciting experience by giving him the conjecture; he should, as often as possible, make up his own, then try to establish it as a theorem by deductive processes. In each of these cases, see if you can provide a conjecture on a formula for

(a) the sum of the odd natural numbers

(b) the sum of $1 + 8 + 13 + \cdots$ as we keep increasing by six

(c) $\frac{1}{1.2} + \frac{1}{2.3} + \frac{1}{3.4} + \cdots + \frac{1}{n(n+1)}$

8. Prove by the principle of mathematical induction that for all $n \in N$, $n(n+1)/2$ is a natural number also. One may make use of properties of addition and multiplication of natural numbers.

9. Do the same as in Exercise 8 for $n(n+1)(n+2)(n+3)/24$.

10. The principle of mathematical induction may be used to *create* statements also which helps to review skills developed previously. As an example let us begin a development of the formula for

$$\sum_{i=1}^{n} i^2 = 1^2 + 2^2 + 3^2 + \cdots + n^2$$

(a) We know that the $\sum_{i=1}^{n} 1 = n$

(b) We know that $\sum_{i=1}^{n} i = 1 + 2 + 3 + \cdots + n = \frac{n^2 + n}{2}$ which is an expression of the second degree.

(c) *Perhaps* $\sum_{i=1}^{n} i^2 = 1^2 + 2^2 + \cdots + n^2$ will be a *third* degree polynomial.

(d) Let us *assume* that $1^2 + 2^2 + \cdots + n^2 = An^3 + Bn^2 + Cn + D$ where A, B, C, D are to be determined.

(e) Let us *impose* on our assumption that it holds for 1, that is

$$1^2 = A(1)^3 + B(1)^2 + C(1) + D$$

(f) Also let us *demand* that when

$$1^2 + 2^2 + \cdots + k^2 = Ak^3 + Bk^2 + Ck + D$$

then the quantity

$$1^2 + 2^2 + \cdots + k^2 + (k+1)^2 = A(k+1)^3 + B(k+1)^2 + C(k+1) + D$$

must hold also.

(g) Hence we have imposed that

$$(k+1)^2 = A(k+1)^3 + B(k+1)^2 + C(k+1) + D$$
$$- 1^2 - 2^2 - 3^2 - \cdots - k^2$$

or that

$$(k+1)^2 = A(k+1)^3 + B(k+1)^2 + C(k+1) + D$$
$$- Ak^3 - Bk^2 - Ck - D.$$

See if you can determine from the last equation in (g) values for the coefficients A, B, C, and D which result from the demands made above. If these can be determined the expression in (d) will give the desired formula.

7.12. Developing Approaches

There has been a constant attempt in this writing to emphasize that student-centered approaches bring forth in the classroom a creative, productive, and enjoyable interest in mathematics. The contriving of stories, real or artificial, and the manufacture or recognition of problem situations help the student to move with ease into the invention and understanding of mathematics new to him. All of this helps to make mathematics more "reasonable" and tends to keep anxiety, frustration, and fear at a minimum. At all levels, attempts should be made to invent situations from which more mathematics emerges or which leads the student gradually into new fields. These situations may well come from mathematics itself, especially for more advanced students.

The story or contrived situation has been found helpful in logic. Let us suppose that the student is acquainted with conjunction, disjuction, negation, and DeMorgan's Laws, and the teacher wishes to introduce the connective known as *implication.* The teacher can write the truth table on the chalkboard and simply say, "This table defines implication." Indeed, it would be correct and in keeping with the way it is usually done for the implication connective is often defined by table or simply by the equivalence

p	q	$p \to q$
T	T	T
T	F	F
F	T	T
F	F	T

$$p \to q \leftrightarrow {}^{\sim}p \vee q$$

Although the students still make an arbitrary definition, it is interesting to pursue the following approach:

1. Let use suppose that I come into the room and say, "If *P*am comes, then I *q*uit ($p \to q$)!" how could you challenge or test me to see if I really meant what I said?

Students will say, "Have Pam come and see if you quit."

2. Does "having Pam here and I do not quit ($p \wedge {}^{\sim}q$)" seem equivalent to "$p \rightarrow q$" or to "${}^{\sim}(p \rightarrow q)$" or neither? Which?
3. From the above discussion, how could we reasonably *define* $(p \rightarrow q)$?
4. Can you use logical algebra on ${}^{\sim}(p \rightarrow q) \leftrightarrow p \wedge {}^{\sim}q$ to deduce an expression for $p \rightarrow q$?

Note that rather than *defining* $p \rightarrow q \leftrightarrow {}^{\sim}p \vee q$ the story suggests defining

$$ {}^{\sim}(p \rightarrow q) \leftrightarrow p \wedge {}^{\sim}q $$

from which the usual definition and table come as a theorem. The cases in the table which seem strange to consider as part of the definition now appear as consequences to a definition which seems quite acceptable to the student. We must emphasize that no matter how implication is begun, it is necessary to formulate an arbitrary definition.

The ordered-pair-of-natural-numbers approach to the integers can be begun by a situation also.

Johnny has a job at a store and Mr. Jones, the proprietor, explained that he would be away for several days so Johnny should take over. Mr. Jones left hurriedly and did not explain to Johnnie how to record the day's business. At first Johnnie wrote "took in $5" and "paid out $3" but then it occurred to him just to write (5, 3) to mean that he took in $5 and paid out $3.

1. How would Johnny write "took in $10 and paid out $4?"
2. What is the actual gain described in Problem 1? Suggest several other pairs which would indicate the same gain.
3. What would be a good name or characterization for two ordered pairs which represent the same gain?
4. Is the pair (5, 3) used above an *ordered* pair? Why? Is this a convenient notation?

From the suggestions above can come the idea of "equivalence" of ordered pairs of natural numbers. Indeed the students can now be led to formulate a definition:

Equivalence of (a, b) and (c, d) where $a, b, c, d \in N$

Let it be said that $(a, b) \sim (c, d)$ if and only if $a + d = b + c$. Even now we can investigate to see if the equivalence of ordered pairs of natural numbers is an equivalence relation. Soon can follow considerations like this:

5. Monday's business was summarized as (7, 2) and Tuesday's business as (3, 1). What ordered pair summarizes the first two days' business?

After such beginnings the class can suggest and agree on a definition for *addition* of ordered pairs of natural numbers. Perhaps they will say "Let it be said that $(a, b) + (c, d) \sim (a + c, b + d)$. Where $a, b, c, d \in N$." Many questions can arise now. Is the addition as defined commutative? Associative? Is the sum unique? Is there an additive identity element? Does each ordered pair of natural numbers possess a pair which serves as an additive inverse? And all of this came from a method devised by Johnny to keep records until Mr. Jones got back.

In three days in succession the business summary read (7, 5) (7, 5) (7, 5). Now Johnny knew that each of these meant a gain of 2, hence the total gain was *six*. Next Johnny decided to use (4, 1) as *three* and he wondered if he could manipulate the symbols 7, 5, 4, 1 so that $(7, 5) \times (4, 1)$ would yield an ordered pair which represents the number *six*. The student is asked to devise a method to "multiply" these ordered pairs and a definition is born! This begins much investigation on the part of the student and teacher alike and a logical structure is begun on which, as the reader knows, is based the multiplication of integers. In one class a student created the definition that $(a, b)\cdot(c, d)$ by given by the ordered pair $(ac - ad, bc - bd)$ and, indeed, the definition seemed to work in all cases in which it was tested (by the use of repeated addition). This was known as "Susan's definition." The teacher wisely kept silent. Not until the next day was the definition brought into question by a student because, he pointed out, "Subtraction of natural numbers is not always possible." Clashes of minds, revision of definitions, testing, and sometimes rejecting, conjectures are most effective ways for the student to learn and to have realized in him the goals of mathematics instruction. With approaches as described above the student begins in a simple setting in which there is no feeling of uneasiness or desperation on his part because of possible pressures on him to think abstractly immediately. Indeed, the student helps to explore and contributes to the growth and development of he subject itself.

Roberts suggests the mistake made by a boy in solving an equation as a means of introducing and developing a sequence of rationals to approximate (or to define, depending on the purpose and level of the student) the square root of 2.[10] Relate the experience of this boy in the form of a *story*:

Johnny had the task of solving the equation $x^2 = 2$ and, being uncertain about what to do, he wrote $x^2 - 1 = 1$, and then $(x - 1)(x + 1) = 1$, and finally $x - 1 = 1/(x + 1)$, and

$$x = 1 + \frac{1}{x + 1} \tag{1}$$

[10]J. B. Roberts, *The Real Number System in an Algebraic Setting* (San Francisco: W. H. Freeman and Company, 1962), p. 133 ff.

Now since "x" had the "other name" $1 + 1/(x + 1)$, he placed in the right-hand member of (1) this solution or value of x and obtained

$$x = 1 + \cfrac{1}{1 + \cfrac{1}{x+1} + 1} = 1 + \cfrac{1}{2 + \cfrac{1}{x+1}} \tag{2}$$

But there was still an "x" in the right-hand member so he did this again. The right-hand member of (2) became

$$x = 1 + \cfrac{1}{2 + \cfrac{1}{1 + \cfrac{1}{x+1} + 1}} = 1 + \cfrac{1}{2 + \cfrac{1}{2 + \cfrac{1}{x+1}}}$$

He tried again and obtained

$$x = 1 + \cfrac{1}{2 + \cfrac{1}{2 + \cfrac{1}{2 + \cfrac{1}{x+1}}}}$$

The reader might "finish the story" but Johnny developed an expression from which can be made successive approximations to $\sqrt{2}$ in a manner suggested by Fig. 7.32 and they are $1, 1\frac{1}{2}, 1\frac{2}{5}, 1\frac{5}{12}, 1\frac{12}{29}, \ldots$ A class in algebra can use this as another method to approximate the square root of 2; a more advanced class can use the above sequence to witness and to emphasize how a nonrational real number can be defined in terms of an intrinsically convergent sequence of rationals.

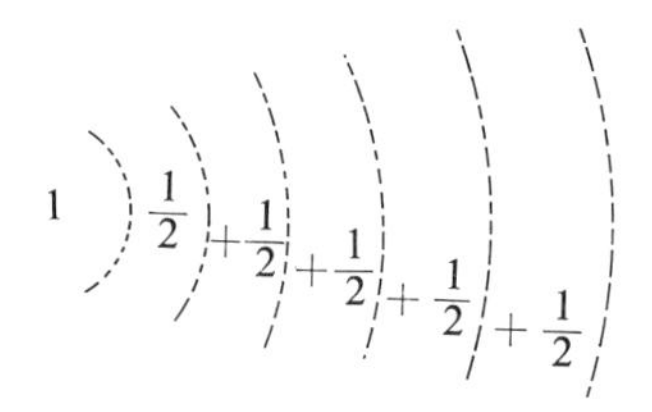

Figure 7.32

Exercises

1. See if you can detect a pattern by which you can construct more terms of the sequence $1, 1\frac{1}{2}, 1\frac{2}{5}, 1\frac{5}{12}, 1\frac{12}{29}, \ldots$ You might like to try to examine this sequence for convergence.

2. In a manner similar to that above determine a sequence which will yield successive approximations to the square root of $\sqrt{5}$.

3. Perhaps the reader recalls that a rational number may be defined on the basis of an equivalence class of ordered pairs of integers where $(a, b) \sim (c, d)$ if and only if $ad = bc$ and where a, b, c, d are integers. See if you can develop a situation or a story which will guide the student to the definition of the *sum* of two rationals: $(a, b) + (c, d) \sim (ab + bc, bd)$.
4. Develop a story or situation which will lead the student to constructing the graph of inequations of the type $3x + 4y < 12$.
5. The problem of linear programming is to maximize or minimize a linear expression which varies under given restraints which are also linear. Make up a story or a situation from which a simple problem might emerge.

We urge that the teacher consider the possible use of such approaches as those illustrated; it should not be thought that the use of "situations" from which mathematics can be developed contribute in any way to a lack of scholarliness in the study of mathematics, for, indeed, it is through physical or contrived situations that mathematics has developed. From the practical questions created by the flooding of the Nile and matters of mensuration and problems imposed by commerce to the present "I wonder if..." relative to the most profound and puzzling questions in mathematics, the field grows by the consideration and investigation of *situations*. Why deprive the student from seeing mathematics grow through need and challenge by keeping him at all times the passive acceptor of what appears in a book? It is the urgent task and compelling challenge for the teacher to color his classroom methodology so that the student becomes engaged and absorbed in attacking problem situations by relooking at what he has learned or by inventing new mathematics or by becoming identified, somehow, with one who *is* investigating a problem. Such methodology does not carry with it, as has been emphasized before, any neglect of necessary practice and drill in the proper manner, but it does generate genuine interest and fruitful creativity in which the student finds abiding joy in making a contribution and/or experiencing the growth of the subject before his very eyes. Again, we quote from the great teacher, Polya, who writes

> ... mathematics 'in statu nascendi,' in the process of being invented, has never before been presented in quite this manner to the student, or to the teacher himself, or the general public.[11]

7.13. Some Sample Study Guides

A Study of Quadratic Expressions

1. *Review.* Find the solutions of these equations by means of the quadratic formula.

[11]G. Polya, *How To Solve It* (Princeton: Princeton University Press, 1945).

(a) $x^2 + 7 + 12 = 0$ (d) $2x^2 + 7x - 30 = 0$ (g) $x^2 - x - 7 = 0$
(b) $x^2 - x - 6 = 0$ (e) $2x^2 - 7x - 30 = 0$ (h) $x^2 - 3x + 1 = 0$
(c) $x^2 + x - 12 = 0$ (f) $6x^2 + 20x + 21 = 0$ (i) $x^2 - x - 2 = 0$

2. *Text* (selected problems)
3. *Review.* Construct the partial graph of each of these quadratic expressions.

   (a) $y = x^2 - 4x$ $(-1 \leq x \leq 5)$
   (b) $y = x^2 - 4x + 6$ $(-1 \leq x \leq 5)$
   (c) $y = x^2 + 6x + 7$ $(-8 \leq x \leq 2)$
   (d) $y = x^2 + 2x$ $(-4 \leq x \leq 4)$
   (e) $y = x^2 - 3x + 4$ $(-1 \leq x \leq 4)$

   Carefully identify the value of x at which the curve has a minimum value of y.
4. Study the results of Problem 3 above. Is there a method by which you can *predict* the value of x for which $y = x^2 + bx + c$ has a point of minimum? Can you *prove* your conjecture so it becomes a *theorem*?

A Study of Quadratic Expressions

1. *Review.* Use the quadratic formula to find solutions for these equations.

   (a) $3x^2 + 5x + 1 = 0$ (b) $3x^3 - 7x^2 - 204 = 0$
   (c) $3x^3 + 10x^2 - 8x = 0$
2. *Review.* Use the conjecture (theorem) made the last time to determine values of x for which each of these quadratic expressions has a minimum value. Note particularly problems (c), (d), (e); how shall we treat these?

   (a) $y = x^2 + 4x + 5$ (d) $y = 3x^2 - 12x + 10$
   (b) $y = x^2 - 8x + 16$ (e) $y = 3x^2 + 18x + 4$
   (c) $y = 2x^2 + 8x + 16$ (f) $y = x^2 - 4$
3. *Text* (selected problems)
4. One number is 20 more than another, as x and $x + 20$. What pair of such numbers has a product which is a *minimum* or which has the smallest value? Some pairs are (2, 22), (5, 25) whose products are 44 and 125 respectively. [Ans. $(-10, +10)$]
5. Construct a partial graph of the function $y = 20x - x^2$. Does y possess a relative maximum or a relative minimum in 0, 20. Which? Can you formulate a conjecture to determine the value of x for which expressions

as $y = c + bx - x^2$ has a *relative maximum*? See if you can test this conjecture by deduction.

A third sheet in this series, not written here in detail might well include suggestions and questions and problems as these:

1. Review problems on relative maxima and relative minima.
2. *Word* problems seeking the relative maxima or minima of quadratic expressions.
3. Maxima and minima problems from the text.
4. Would it be possible to formulate a theorem on quadratic expressions $y = ax^2 + bx + c$ to include statements about both "relative maximum and minimum possibilities?"
5. List steps which you suggest to determine the value of x for which the expression $y = ax^2 + bx + c$ has a relative maximum or minimum.

Bibliograhpy

Baker, B. L., "Method of Differences in Determination of Formulas," *School Science and Mathematics*, LXVII (March 1967), pp. 217-20.

Bell, Clifford, "What Every Teacher Should Know About the Uses of Mathematics," *The Mathematics Teacher*, LVI (May, 1963), pp. 302-6.

Edwards, R., "Curve Straightening and Formula Determination," *School Science and Mathematics*, LXVII (February, 1967), pp. 149-61.

Forbes, J. E., "Most Difficult Step in the Teaching of School Mathematics: from Rational Numbers to Real Numbers with Meaning," *School Science and Mathematics*, LXVII (December, 1967), pp. 799-813.

Ranucci, E. R., "Anyone for Tennis?" *School Science and Mathematics*, LXVII (December, 1967), pp. 761-5.

Van Engen, H., and R. W. Cleveland, "Math Models for Physical Situations," *The Mathematics Teacher*, LX (October, 1967), pp. 578-81.

8

The Teaching of Calculus

> I want all the students, not merely . . . (the) . . . privileged better ones to have a chance to know what it is all about.
>
> Murnaghan[1]

8.1. Introduction

It is not the purpose of this chapter to discuss at length the place of calculus in high school. Surely the student deserves as much introduction to this powerful tool as he does to new ideas in chemistry, biology, physics, sociology, history, computer science, and to all fields which help to make him a more informed citizen. Obviously he cannot wait until he gets to the university for all subjects; it is possible to channel his eagerness to learn into acquiring correct ideas of the calculus at the high school level, with the fine points to be examined when the student is more mature in the subject. Indeed some college texts even now require a mathematical maturity which many students do not have and cannot develop at the rate required to gain the desired depth and breadth in the calculus.

One argument against the presentation of any calculus at all in the secondary school is that many teachers only show the student a few me-

[1]F. D. Murnaghan, *Differential and Integral Calculus* (Brooklyn: Remsen Press, 1947), p. iv.

chanical processes devoid of understanding. Whether or not this is corrected in university teaching may be deduced by observing students' use of sample problems, study outlines, and guides to working problems. Some claim, furthermore, that the study of a small amount of calculus in the secondary school gives the college student false confidence and that, as a result, he often flounders in university work. The writers suggest that this danger is always present when a student pursues a subject in college which he has begun in high school and that this claim merely emphasizes that little or no learning in mathematics should reduce to mechanical processes devoid of understanding.

Of course, the liberal arts point of view is important in the high school because not all of these students continue in the college or university and fewer still pursue the study of mathematics. That the student should be denied ideas of calculus because he cannot comprehend in a rigorous way certain theoretical beginnings has serious consequences. Marshall H. Stone says,

> The educated man, whom we envisage as the end-product of our elaborate educational process, should not be left some 200 years behind the times in mathematics, merely because he is not to be a specialist in science or mathematics—least of all in an era when mathematics is growing so vigorously and penetrating so profoundly into so many different domains of thought.[2]

The writers have developed study guides to help the student experience beginnings of the calculus based on concepts gained in advanced algebra. Use is made of the theorem $S_\infty = a/1 - r$ which seems to satisfy the inquiring mind of the student until, with teacher help, one wishes to move into more basic considerations. Some properties and uses of the derivative and integral are developed and the student is led to the conjecture that an integral can be evaluated quickly by antidifferentiation. Indeed, this same approach has been used in university classes and sets the stage for more mature investigation. Students can be gaining practice in manipulation while examining further the foundations of the subject. When honestly done, with fundamental assumptions made explicit, such an approach seems to pay dividends. As Professor E. A. Maxwell points out,

> It is always hard to know what degree of rigour can be sustained, and there are necessarily a number of "gaps" left to be filled in by intention at school level. It is important that, where possible, the existence of a gap should be made clear; but it is equally

[2]Organization for European Economic Co-operation, Office for Scientific and Technical Personnel, *New Thinking in School Mathematics* (Washington, D. C., O. E. E. C. Mission, Publications Office, 1961), p. 16.

important not to confuse the beginner by needlessly introducing difficulties which are basically beyond him.[3]

The purpose of this chapter is to present some approaches which have been found effective in the classroom. They have been designed to encourage student contribution and to help the student become a part-creator of the calculus. Such approaches have helped students to understand the calculus and this is the stated aim of all texts. So pressing is this problem that, as long ago as 1907, Saxelby, in writing about his approach to differentiation, suggested that "it too often happens that a student . . . acquires a merely *fatal facility* (italics ours) in differentation, regarding it as a mechanical juggling with symbols, but having no conception of its relation to experience."[4]

Present writers frequently imply that there is need for improvement in the teaching of the calculus and make claim that their texts will meet this need. We agree with Murnaghan, who writes in his introduction, "I want *all* the students, not merely . . . (the) . . . privileged *better ones* to have a chance to know what it is all about."[5] We suggest student-centered approaches to be effective in helping the students learn "what it is all about." Approaches which encourage student contribution, answer Dale's description of teaching when he says, "Teaching is a sharing process, a two-direction process—not a one-way affair. It is intercommunication. It includes the collision, the creative interaction of minds . . . learning blossoms in a mood of mutuality."[6] We hope to show some of this kind of methodology at work.

8.2. Speed of a Body Where Motion Is Accelerated

There are several different ways to introduce the idea of *rate of change*. In the high school classroom, and sometimes for university freshmen, an apparatus has been set up as that shown in Fig. 8.1 where the weights W are equal and weight w is small so that constant force caused by w produces an accelerated motion. A seconds pendulum and a meter stick make up the rest of this simple equipment. The students sense immediately that the body (W, w) falls faster and faster. "Is it possible that we could determine the speed at the end of (let us say) the third second after (W, w) is released?," asks the teacher. It is interesting to hear students' suggestions on attacking

[3]*Ibid*. p. 87.

[4]F. M. Saxelby, *A Course in Practical Mathematics* (London: Longmans, Green, and Company, 1907), p. v.

[5]F. D. Murnaghan, *op. cit*.

[6]Edgar Dale, "What Does It Mean to Teach?" *The News Letter* (published by the Bureau of Education Research and Service, The Ohio State University, Columbus, Ohio), XXX (March, 1965).

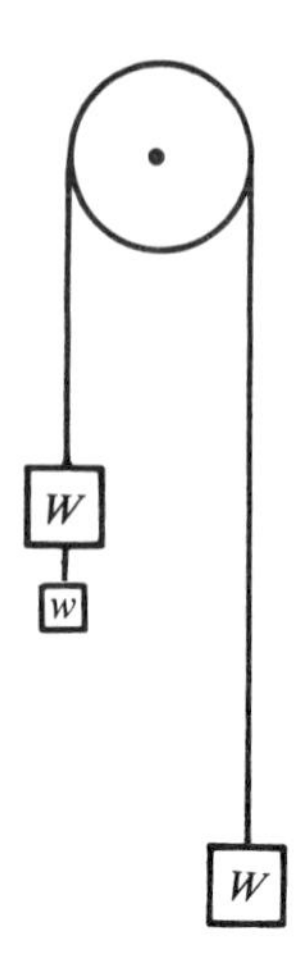

Figure 8.1

this problem. These may be devising mechanical and electronic equipment to furnish measurements by which a tentative conclusion can be reached using arithmetic computation, or perhaps attempting to find an algebraic relation which can be assigned as representing in an ideal way the motion of the body.

For the purpose at the moment, the teacher is more interested in the approaches which involve measurement. These are like this: "If we could measure the distance the body falls between $t = 3$ sec and $t = 4$ sec then this average speed could be considered as an approximation to the speed at $t = 3$; the distance between the times $t = 3$ sec and $t = 3\text{-}1/2$ sec divided by $1/2$ sec would give a still better approximation!" Of course, there is mechanical difficulty in making measurements of distance for smaller time intervals because of the difficulty in making marks rapidly enough and in calling "time" for fractional seconds, so the student finally wishes he had a formula for the distance S in terms of time t! Truly here is a situation approach which emphasizes a need for more powerful tools from which some important mathematics can emerge. It does not take long to give up the above approach and seek to devise a formula which is assumed to describe the motion. We have discussed this in Chapter 1, Section 7.5. One will conclude that $S = kt^2$ seems to describe the distance relative to time t.

In practice the students soon recall that the accepted $s = 16t^2$ for the distance of a freely falling body affected only by gravity is a "square relation" too and our attention is soon given to the speed at the end of the third second in this case. Now we have a *formula* by which to compute distances and hence the desired average speeds. A formula developed for the W, W, w-pulley system serves just as well. Suggestions by students direct our consideration to something similar to the average speed between $t = 3$ sec and $t = 4$ sec, between $t = 3$ sec and $t = 3\frac{1}{2}$ sec, between $t = 3$ sec and $t = 3\frac{1}{4}$ sec, and so on. One obtains a sequence of average speed. Some study-guide helps appear as follows:

1. If the distance s traversed in time t be taken as $s = 16t^2$, what is the distance fallen at the end of the third second, or, for $t = 3$?
2. What is the distance traversed by the time $t = 4$?
3. Compute the average speed between $t = 3$ and $t = 4$.
4. How could one obtain a better approximation to the speed at $t = 3$? Compute as you suggest.
5. Compute a still better approximation.
6. Can you get a still better approximation? Compute.
7. Will it even be possible to compute the speed at $t = 3$ sec? Explain or elaborate on this.
8. From the results above can you suggest a number which might be assigned to define the speed at $t = 3$ sec?

The essential part of the above approach is to have the time intervals selected with some regularity. For example, intervals of 1 second, $\frac{1}{2}$ second, $\frac{1}{4}$ second, $\frac{1}{8}$ second, $\frac{1}{16}$ second, and so on, provide a sequence of average speed which leads to an easily recognized number which might be assigned as the speed at $t = 3$. Intervals from $t = 3$ sec to $t = 4$ sec, then $t = 3.0$ sec to $t = 3.1$ sec, $t = 3.0$ sec to $t = 3.01$ sec, and so on, might be used also. Indeed, students should explore several such plans. Not every student will use initially such regularly chosen time intervals, but the ones who do can point out an advantage which is readily seen. "Teaching is . . . intercommunication . . . the creative interaction of minds."[7] Guides for student study may now continue in a manner like this:

1. Make a series of approximations to the speed of a falling body at the end of the *fourth* second ($t = 4$) by the use of the intervals $t = 4$ sec to $t = 6$ sec, $t = 4$ sec to $t = 5$ sec, $t = 4$ sec to $t = 4\text{-}1/2$ sec, $t = 4$ sec to $t = 4\text{-}1/4$ sec, $t = 4$ sec to $t = 4\text{-}1/8$ sec. What number does this set of approximations or average speeds seem to define for the speed at $t = 4$?
2. Use another set of time intervals, chosen with some regularity, to arrive at a number to define speed at $t = 4$ sec.

There are several variations at this level and they may be pursued as the class seems to demand. If more practice and elaboration are needed one

[7] *Ibid.*

can come to numbers to define speeds at the end of "different" seconds and for different "laws of motion" as $s = 16t^2 + 3t + 6$. Soon, however, there comes a time for an elementary kind of *reflection* and *abstraction* from these concrete examples. In Problem 1 just preceding, forexample, the series of resulting approximations is

160 ft/sec, 144 ft/sec, 136 ft/sec, 132 ft/sec, 130 ft/sec

and the student might be invited to think about his results in a way like this:

1. Does there come a time at which you can predict what the next approximation will be? Predict the sixth, seventh, and eighth approximations to speed at $t = 4$.
2. What might such a succession of numbers be called, as 160, 144, 136, 132, 130, 129, 128-1/2, 128-1/4, ...? What name might be given the "128" which this set of numbers seems to define?

From a discussion such as suggested here arise words as "sequence," "approximating numbers," "approximating sequence," "series," "succcssion," "pattern," along with "approach," "narrow down," "limit," and others. These words are idea-filled and the student has first an idea which he is trying to describe and name; he is creating reflectively and actively from something he has experienced rather than receiving passively a formal definition which he may not understand. Exercises for the student might continue as follows:

3. What does it mean to say that the sequence 160, 144, 136, 132, 130, 129, 128-1/2, 128-1/4, ... "approaches" 128, or that the "limit" of the given sequence is 128?

Students give various answers: "every number in the sequence gets 'closer' to 128," the "differences between 128 and successive numbers are smaller and smaller," the "differences between successive pairs of numbers become smaller and smaller." Note that two usual definitions of convergence of a sequence are hidden in these suggestions. Explorations can be made here, too. What term is 1/4 away from the limit? (the eighth one) What term is 1/8 aways from the limit? (the ninth one) What term is $(1/2)^k$ away from the limit? [the $(k + 6)$th one]. This, with teacher help, can lead to a definition of a sequence approaching a limit, for, in addition to the above questions, the teacher (or guide sheet) may ask, "If I name any 'closeness' to what you suggest the limit is—as $(1/2)^{10}$ away from 128—can you always point out to me a term which is just $(1/2)^{10}$ away and such that the next term will be closer? Is *this* what we mean by the given sequence approaching 128?" The teacher will want to explain that mathematicians find this idea very important and that they express it this way. We have the ϵ, N-definition for $\{a_n\} \rightarrow A$, or, the students may to look into textbooks to see how other writers and workers express this idea. We are now ready to begin a student-arranged list of terms, perhaps as follows:

D1. Sequence

D2. Limit of a Sequence

D3. Average Speed

D4. Speed

But we are not yet through with this idea. There remains more for student exploration and hence strengthening of skills and concepts in new settings. Through problems and discussion the teacher might keep the student cautious about deciding what the limit is and finally come to this question:

4. Is there any way by which we can study sequences to make sure that the number we suggest for the limit *is* the limit?

Students have suggested a procedure like the following, and the writer has found its use effective in elementary calculus. The sequence 160, 144, 136, 132, 130, 129, 128-1/2, 128-1/4, . . . can be written as 160, 160 − 16, 160 − 16 − 8, 160 − 16 − 8 − 4, 160 − 16 − 8 − 4 − 2, . . . and hence as 160, 160 − 16, 160 − (16 + 8), 160 − (16 + 8 + 4), 160 − (16 + 8 + 4 + 2), . . . 160 − (16 + 8 + 4 + 2 + . . .). But 16 + 8 + 4 + 2 + 1 + . . . is a geometric progression and $S_\infty = a/1 - r$ becomes

$$\frac{16}{1 - 1/2} = \frac{16}{1/2} = 32$$

Hence 160 − (16 + 8 + 4 + 2 + . . .) becomes 160 − 32 = 128. Therefore, the limit of the sequence is 128.

5. If the above method is used what assumption is being made or what assumed data is one using at the outset?

Of course, the above method makes use of the fact that it has been proved previously that

$$\lim_{n\to\infty} \frac{a - ar^n}{1 - r} = \frac{a}{1 - r}.$$

$$|r| < 1$$

In the use of this method we are circumventing "limit-of-sequence" theory and the student should be made aware of the fact that he is borrowing powerful tools from another field.

Other students may suggest that to make sure 128 is the limit one must show that the ϵ, N-definition or its equivalent is satisfied. In considerations of the calculus for part of a year in high school the "geometric progression" approach seems very workable, but one would not wish to deny experience with the ϵ, N-definition and this is good to encourage.

As we grow in constructing sequences to define speed, we must yet help students to ask this question:

6. Given $S = 9t^2 + 3t + 10$ is there a way to construct a sequence which will define the speed at *any* time t, such as $t = a$?

Students will suggest, let the time intervals be $t = a$ to $t = a + 4$, $t = a$ to $t = a + 2$, $t = a$ to $t = a + 1$, $t = a$ to $t = a + 1/2$, and so on. It is interesting to study the resulting sequence of average speeds. Finally,

7. Given $s = 9t^2 + 3t + 10$ can we find a number by which we agree to define the speed at $t = 2$ *without* the use of intervals $t = 2$ to $t = 4$, $t = 2$ to $t = 3$, and so on, but rather by the use $t = 2$ to $t = 2$ to $t = 2 + h$?
8. See if you can use a similar method to define speed at $t = a$ given that $s = 9t^2 + 3t + 10$.
9. If s be given as $pt^2 + qt + r$ can you develop a number which can define the speed at $t = a$?
10. In the methods used above are there any assumptions which we are making which should be investigated further. For example, what postulate or theorem do we have in our development thus far which justifies our saying

$$\lim_{h \to 0} \{18a + 9h + 3\} = 18a + 3?$$

Discussion on such questions as those in Problem 10 above help the to examine his work from a logical point of view (he attempts to discern what he has assumed from what he has proved) and this is one of the goals of mathematical study. Our logical organization may now look like this:

Pr. Real Number System (taken as primitive)

D1. Sequence

D2. Limit of a Sequence

D3. Average Speed

D4. Speed

P1. $S_\infty = \dfrac{a}{1 - r}$ from $S_n = \dfrac{a - ar^n}{1 - r}$ (Pl meaning Postulate One)

P2. $\lim_{h \to 0} \{kh\} = 0$ (k a constant)

T1. $S = pt^2 + qt + r \to \text{speed} = 2pa + q \quad (\text{at } t = a)$
(Tl meaning Theorem One)

Of course one must always be scrutinizing his logical structure to see if any of the postulates can be deduced from previous postulates or theorems, or to see if some simpler postulates can be formulated so this can be done. Indeed, the writer is not proposing that beginning logical organization must look like the above; the point is that students should help to develop such an organization, and that it may be different with different groups.

Exercises

1. Work out several of the problems which construct sequences of average speeds to arrive at a number by which to define the speed at a certain time.

2. Begin again with $S = 16t^2$ and set up a sequence of average speeds to attempt to find a number to define the speed at $t = 4$ by using the intervals $t = 2$ sec to $t = 4$ sec, $t = 3$ sec to $t = 4$ sec, $t = 3\text{-}1/2$ sec to $t = 4$ sec, $t = 3\text{-}3/4$ sec to $t = 4$ sec and so on.

3. Can you think of any other suggestion a student might make to attack the above problem?

4. Given $S = pt^2 + qt + r$ use time intervals $(a, a + 2)$, $(a, a + 1)$, $(a, a + 1/2)$, $(a, a + 1/4)$, $(a, a + 1/8)$, and so on, to arrive at a number by which to define speed at $t = a$.

5. Given $s = pt^2 + qt + r$, use the time interval $(a, a + h)$ and discuss what happens to the average speed as $h \to 0$ to arrive at an expression for speed at $t = a$.

6. What kind of expression for s might you next introduce in a class? Suppose it is $S = 4t^3 + t$ and you wish the speed at $t = 2$. Can one use the method which makes use of $S_\infty = \dfrac{a}{1-r}$? Try it.

7. If your students develop the expression for speed at $t = a$ from s (distance) $= mt^3 + nt^2 + pt + q$ and it turns out to be speed $= 3ma + 2na + p$, does this result have the status of a *theorem* or of a *conjecture*. Which? Why?

8. If your students decide from the study of examples that if s (distance) $= mt^n + pt^{n-1}$ then the formula for speed must be $mnt^{n-1} + (n - 1)pt^{n-1}$; what is the status of this expression?

9. If your students have had advanced algebra should they be able to *prove* that if $s = mt^n$ then speed at $t = a$ can be defined as mna^{n-1}? Why? Are there any postulates necessary?

10. If your students *prove* that $s = mt^n \to$ speed at $t = a$ is nma^{n-1} and that $s = qt^r \to$ speed at $t = a$ is rqa^{n-1} then does it simply follow that if $s = mt^n + qt^r$ that the speed at $t = a$ is $mna^{n-1} + qrat^{r-1}$? Why? If not, could this be proved with what the student now has? If so, how?

11. If one never *deduces* the relation in Exercises 8 above, then it must be

how listed in the logical organization? Might it one day be a conjecture and possibly the next day become a theorem?

12. At the beginning of this section we used a pulley and a system with mass $2W + w$ accelerated by a force of w pounds; this introduced a varying speed and the problem was to find the speed by the end of the second second. Can you think of one or more other physical devices which could have been utilized to introduce this problem? Your suggestion should be practical and usable in the classroom.

13. Prove that if h takes on values 1, 1/3, 1/9, 1/27, ... then $\lim_{h \to 0} \{h\} = 0$. Show that for any "specified closeness" (expressed as powers of 1/3) you can find a term which is precisely that close and that the terms beyond are still closer.

14. Prove by the conventional N, ϵ-method that

$$\lim_{n \to \infty} (1/3)^n = 0$$

It should not be thought that the student does no more problems than those suggested in this section. These show one possible flow of development. Exercises are necessary for two reasons: to maintain skills useful in later work and to develop new skills which may help a new concept to emerge. All this requires teacher planning in actual classroom practice. The emphasis in this discussion is the use of concrete beginnings and the student-centered approach in which the class as a whole brings its thought to bear on the problem at hand. Studies in learning concur that

> ... initial learnings come from experience (physical and mental experiments), constructive methods, not from definitions. It is the dynamic aspects of events that aid learning. Whatever is to be learned must have its roots in some challenging problem-presenting situation.[8]

We know also that there is much merit in the students' attempting to organize his own findings into logical organization. This leads to understanding, which

> ... refers to something that is in possession of an individual. The individual who understands is aware of a satisfying feeling, a psychological closure, which results from having fitted everything in its proper place ... (he is) in possession of the cause and effect relationship—the logical implication and sequences of thought that unite two or more statements by means of the bonds of logic ... understanding is an organizational process.[9]

---

[8]National Council of Teachers of Mathematics, *The Learning of Mathematics; Its Theory and Practice*, Twenty-First Yearbook (Washington: The National Council of Teachers of Mathematics, 1953), p. 70.

[9]*Ibid.*, pp. 74-6.

Some problems which have appeared on one of the first study guides prepared by the writers are:

1. An auto started from rest and during the second between $t = 1$ sec and $t = 2$ sec it went 30 feet. During the first half-second between $t = 1$ and $t = 2$ it went 10 feet and during the first fourth-second between $t = 1$ and $t = 2$ it went 4 feet.
   (a) What was the average speed between $t = 1$ and $t = 2$ sec? Between $t = 1$ and $t = 1\text{-}1/2$ sec? Between $t = 1$ and $t = 1\text{-}1/4$ sec?
   (b) Which of these is the "best approximation" to the speed at the end of the first second ($t = 1$)?
   (c) How could one obtain a better approximation to the speed at the end of the first second ($t = 1$)?
   (d) Are there any difficulties which need to be overcome in making very good approximations to the speed of the auto at the end of the first ($t = 1$) second? If so, what?
2. A wire is bent in the form of a circle with a diameter of 14 inches. A bead on the wire is released from the top of the circle and it slides to the bottom in 3 seconds.
   (a) What is the average speed in inches per second?
   (b) Would the speed be a "changing" speed or would it be a "constant" speed?
   (c) How could one attempt to find the speed at the end of the third second?
3. Explain in words how one finds *average* speed.

From a guide sheet which suggests the use of time intervals becoming smaller in geometrical progression, we have

1. Apply the method which you have just developed to estimate the speed of a falling body at the end of the third second. Show all the work neatly arranged.
2. Apply the method developed above to estimate the speed of a falling body at the end of the fourth second. Arrange work neatly.
3. A body is traversing distance in such a manner that $s = t^2 + 3t$. What is the speed (which you would suggest) at the end of the third second?
4. A body moves such that distance s and the time t are related by this expression: $s = 3t^2 - 2t$. What speed would you suggest describes the behavior of the body at the end of the second second?
5. Reconsider Problem 1 but this time see if it makes any difference if you use another method to make the time interval smaller. Show the work in good form.
6. Use another method to make the time interval smaller in Problem 1.
7. Try to use a method which approaches $t = 3$ from "beneath" it, e. g.,

use the interval from $t = 2$ to $t = 3$, the interval from $t = 2\text{-}1/2$ to $t = 3$, the interval from $t = 2\text{-}3/4$ to $t = 3$, etc.

8. What do you conclude from your work in Problems 1, 5, 6, 7 about the effect of making the interval smaller? Make a complete statement on this.

Later the "h-method" is introduced and guides contain exercises as

1. By the method just developed estimate the speed of a falling body at the end of the third second. Show work. Use a figure if you wish.
2. Find the speed of a falling body at the end of the fourth second. Show work.
3. Find the speed at the end of the second second by the method developed in this section if

   (a) $s = t^2 + 3t + 2$

   (b) $s = t^3 - 2t^2$

   (c) $s = 2t^2 + 4t$

4. Find the speed at the end of the third second if

$$s = \frac{1}{t}.$$

5. Find the speed at the end of the fourth second if

$$s = \frac{t - 4}{t}.$$

6. By the method developed above estimate the speed at the end of the times indicated if the distance functions s of the time t are given as follows:

   (a) $s = 3t^2 + 6^t - 5$ at $t = 2$ sec

   (b) $s = \dfrac{1}{t^2}$ at $t = 3$ sec

   (c) $s = \dfrac{t^2 + 4}{t}$ at $t = 2$ sec

   (d) $s = \dfrac{t + 3}{t - 4}$ for $t = 5$ sec

   (e) $s = \dfrac{t^2 + 4}{t + 1}$ for $t = 3$ sec

7. On re-examing the "h-method" for finding speed at the end of $t = a$. What things or facts are there about which we should be more *sure* in order to place our work on a firm foundation? What assumptions do we seem to be making?
8. We have made an estimate of the speed at the end of the second by four

methods. List these and add a descriptive phrase about each. What features do these have in common? Again, what assumptions do we seem to be making?

Some hypotheses which students have suggested near the beginning of our study are illustrated by this list.

H1: Each of the sequences of average speeds (in problems we are working) approaches a limit ("limit" undefined).

H2: This limit of average speeds is the speed we seek.

H3: One can reduce the length of the time intervals in any regular manner whatsoever; one will get different sequences, but the limit of each will be the same.

H4: One can approach $t = a$ from above or from below; he will get different sequences, but the limit of each will be the same.

Later this same group formulated

D1: Speed at any time $t = a$ is defined to be the limit of a sequence of average speeds, with each time interval beginning at $t = a$, as the time interval approaches zero.

8.3. Slope at a Point on a Curve

As teachers we know that this is an important concept in the calculus and it is one of the topics usually studied first. Of course, the lecturer can define the slope at $x = a$ at the very outset simply (Fig. 8.2) by saying that

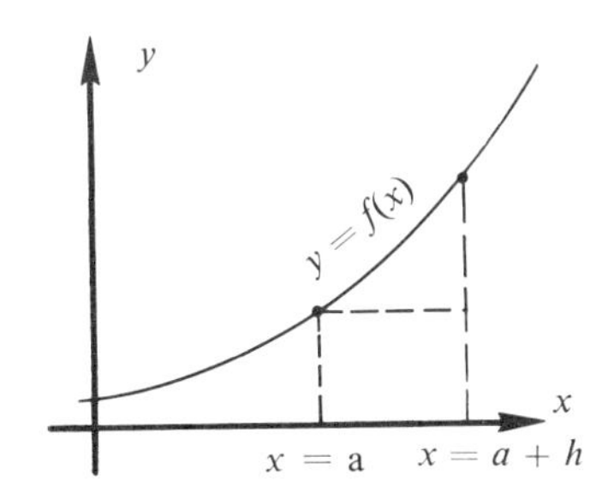

Figure 8.2

it is the

$$\lim_{h \to 0} \left\{ \frac{f(a + h) - f(a)}{h} \right\}$$

or

$$\lim_{(x_2 - x_1) \to 0} \left\{ \frac{f(x_2) - f(x_1)}{x_2 - x_1} \right\}$$

and the student can memorize it. Another approach seeks to invite the student to consider this problem on the basis of the experience he has had with slopes of straight lines. Indeed, a portion of a study may review slopes of lines and in a very nonconspicuous way mention the subject of slopes of curves at given points.

1. *Review* (Fig. 8.3). Compute the slopes of these line segments $\overline{PQ}$

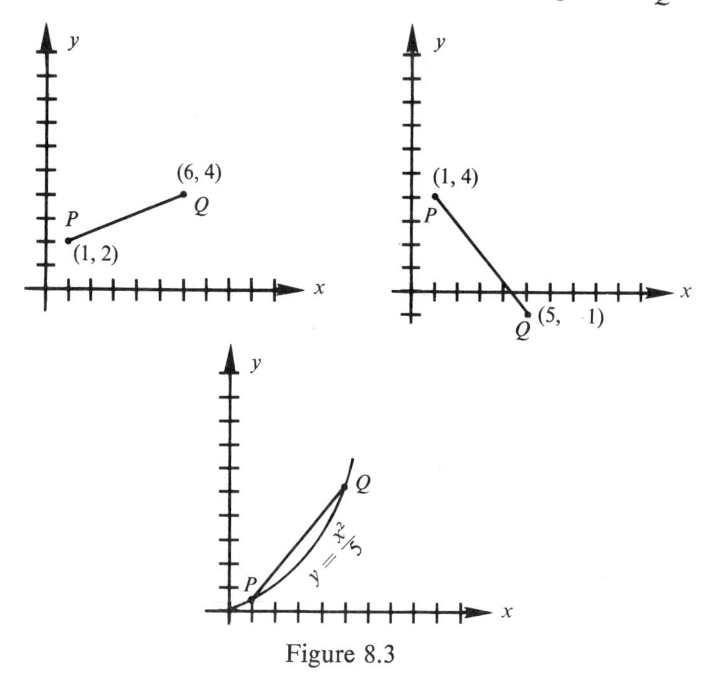

Figure 8.3

2. Slope is defined as a property of straight lines (Fig. 8.4). Yet curves have steepness too. What do you suggest as an approach to "steepness" of a

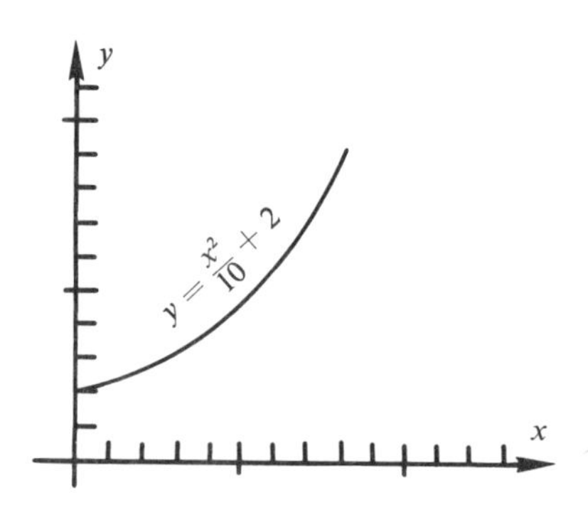

Figure 8.4

curve? Use your approch to say something about the "steepness? at $x = 0$? at $x = 2$? at $x = 5$? at $x = 8$? You might wish to use a ruler to help make estimates.

One does not know what the class response will be, but the teacher must be ready to pursue any suggestion. A student may use the "concrete-method" approach and use his ruler to draw a line which he considers as having the same "slope" as the curve and then compute the slope by dividing some height by a corresponding length, as in Fig. 8.5. In this case it is interesting

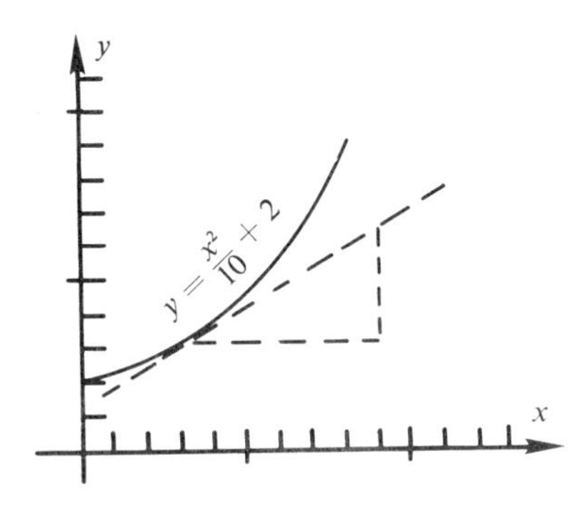

Figure 8.5

to ask him to demonstrate his method before the class. Immediately arises the question of how accurate his eye is in drawing the tangent to the curve. Different measures of steepness presented for the same point enable the teacher to emphasize that we have a challenging problem which may require an approach new to us. Another student may suggest computing the slope of the line segment determined by two points on the curve which are "close together." The teacher can ask for a better approximation. We might well give this problem consideration on the next study guide:

1. Yesterday we learned that the data secured through observation and used in determining the slope of a curve is not a very reliable way to arrive at what we think the "slope of a curve" might be. Can you suggest any approach which might give a more certain result?

Some students may suggest the sequence-of-slopes-of-secants approach, and, indeed, some might have suggested this at the outset. If not, the teacher may ask, "Could we devise a method similar to that used to determine *speed* at $t = a$ when we know the formula for distance s in terms of time t?" With or without teacher help immediately as the class requires, the students may consider problems such as

1. Consider the suggestion of computing slopes of secants set up in some

regular manner to arrive at a number which may be assigned as the "slope" at $x = 5$ (Fig. 8.6).

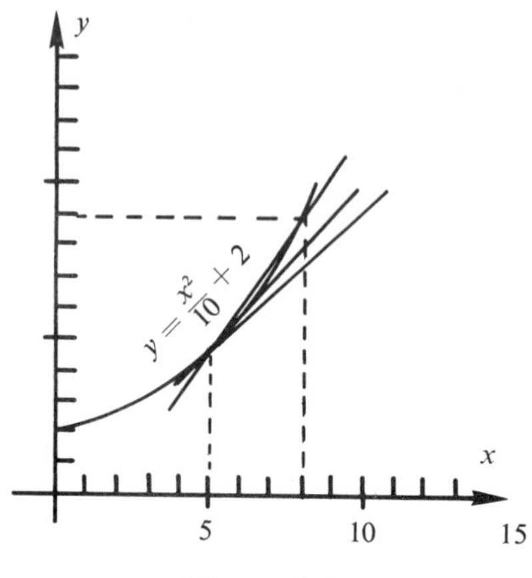

Figure 8.6

One may use the same pattern of development as that used for speed, only now it proceeds more rapidly. Students may find the slope at $x = a$ and even invent an algorithm for finding slope at $x = a$ for curves expressed as polynomials in x. Indeed, at most any time, one can seek values of x for which the slopes of curves is zero, or equal to 1, or equal to any number. This is an invitation to *explore*.

In our attempt at logical development, we meet some of the very words which originated in the study of *speed*: sequence, limit, "approximating sequence," and others. A definition of slope of a curve which students often suggest is

> The slope of a curve at any point $x = a$ is defined to be the limit of the sequence of slopes of secants, all going through $x = a$, as the x-interval approaches zero.

Indeed, we now have several definitions and we may have many conjectures yet to be tested. One conjecture may be that if $y = 3x^2 + bx + 10$ then the *algorithm* which yields the slope expression at $x = a$ is $6a + b$ (obtained by what we know as the differentiation algorithm).

The reader should note that we started with a concrete object, a curve, which we wished to study relative to an as yet undefined but intuitively accepted *slope*. Some students used rulers and measurement, some used arithmetic (coordinate geometry) immediately, and finally the use of a sequence of approximating slopes. The student witnesses and helps the mathematics to grow and he gains practice in old skills in new settings while he and the mathematics are developing. Here we see *sources* and *causes* of mathematics. Indeed Polya emphasizes the value of this when he quotes Leibnitz as saying, "Nothing is more important than to see the *sources* (italics ours) of invention which are, in my opinion, more interesting than the inventions themselves."[10] Yes, the role of the teacher is to have students experience, in part, how mathematics *came to be* and to help them attack problems in their own way and then to glean from these attempts many of the concepts and skills which will further the students' program in mathematics. As early as 1911, J. W. Young gave some advice on teaching geometry and algebra, but it applies to calculus as well:

> The trouble in brief is that authors of practically all our current textbooks lay all emphasis on the formal logical side, to the almost complete exclusion of the psychological, which latter is without doubt far more important at the beginning of a first course in algebra or geometry . . . it is necessary to arouse his (the student's) interest and then let him think about the subject in his own way.[11]

Or, as Young continues,

> Let the teacher once fully realize that his science . . . is alive and growing . . . and he will bring to his daily teaching a new enthusiasm which will greatly enhance the pleasure of his labors and prove an inspiration to his pupils.[12]

Indeed, following suggestions of students is a good way to keep the subject alive as one continues in a given mathematical discipline.

Exercises

1. Given the curve whose equation is $y = x^2 + x$, set up a sequence of slopes of secants emanating from (1, 2) from $x = 1$ to $x = 10$, then from $x = 1$ to $x = 4$ then $x = 1$ to $x = 2$, $x = 1$ to $x = 1\text{-}1/3$, so that the length of each interval is being reduced by one-third. Suggest a limit and prove that it is.

2. In defining the slope of the curve in Exercise 1 above, set up the sequence of slopes of secants going from $x = 1$ *back* to $x = -11$, then to $x = -3$, then to $x = -1/3$, then to $x = 5/9$ and so on. Find the limit of this sequence. Does one get the same limit as in Exercise 1? If so, what conjecture might be make?

---

[10]G. Polya, *How to Solve It* (Princeton: Princeton University Press, 1946), p. 112.

[11]J. W. Young, *Fundamental Concepts of Algebra and Geometry* (New York: Macmillan Company, 1911), p. 5.

[12]*Ibid.*, p. 7.

3. Suggest another way to set up a numerical sequence for Exercise 1. What conjecture comes from this?

4. State the δ, ε-definition for

$$\lim_{h\to 0}\left\{\frac{f(a+h)-f(a)}{a+h-a}\right\} = A$$

5. State in terms of δ, ϵ the meaning that

$$\lim_{h\to 0}\left\{\frac{(a+h)^2-a^2}{h}\right\} = 2a$$

For this case, what choice of δ may be taken for a given ε?

8.4. Rate of Change

At some time during the discussion of the "slope" of a curve one can ask questions such as these: "Were there any parts of the procedure to arrive at a number to define *speed* like those to arrive at a number to define the *slope* of a curve?" How is the slope of a curve or line like speed? These with other questions, if necessary, help the student to formulate that both speed and slope are *rates of change*. Indeed, in class it is suggested that the symbol $R_x y$ be used for slope and $R_t s$ be used for speed and that $R_v u$ be used to mean the rate of change of u with respect to v. Now slope and speed have properties in common and they are not at all unrelated for they are representatives of certain kinds of rates of change and, as one student put it, "our faith that a certain number is the answer to a speed or a slope or any rate of change problem is just as firm as our faith that a certain sequence approaches a certain limit." These are student's words and not textbook assertions! Now a hitherto unexplored field opens: what is the rate of change of *volume* with respect to the *radius* of a sphere, say, at $r = 3$? at $r = 4$? Given the desired volume of a rectangular box along with prices per unit area for materials to make the sides, top, and bottom, and with a relation between the length and the width of the box, what is the rate of change of the cost of materials with respect to the width x? Also, simply given $y = x^3 + 2x^2 + 3x + 4$ what is $R_x y$ at $x = a$? Some typical study-guide questions can be:

⋮

4. Given $y = x^3 + 2x^2$, compute $R_x y$ at any point x by the "$a + h$" method.
5. Given $y = x^2 + 7x + 12$ compute $R_x y$ at any point x by the "$x + h$" method.

⋮

9. Given $y = 1/x$, compute $R_x y$ at any point x.

10. Is it possible to formulate a method by which $R_x y$ at any point x can be written *at sight*? If such a "rule" is discovered it may well be given the name of the student who first stated it. This procedure lends interest in the class and it hints at how mathematics grows.
11. Hence, what conjecture do you make on $R_x y$ if $y = ax^n$? See if you can *deduce* your conjecture from the definition of $R_x y$ by the "$x + h$" method. If this can be done, we have a theorem to add to our logical outline.

The use of inductive quizzes as discussed in Chapter 3 can help to build ideas in a very effective manner. Perhaps this sample will help to illustrate (Fig. 8.7).

(a)

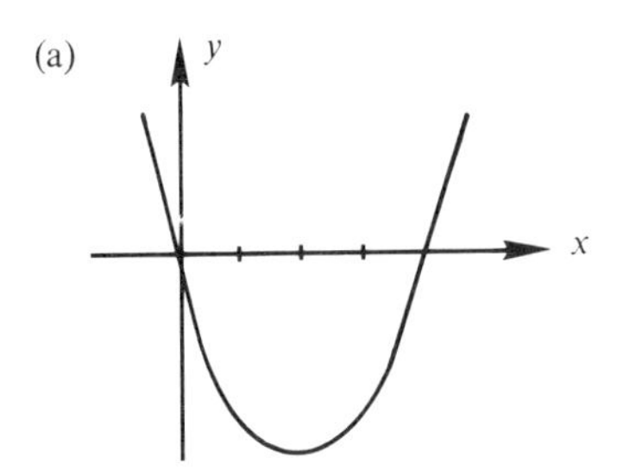

or (b)

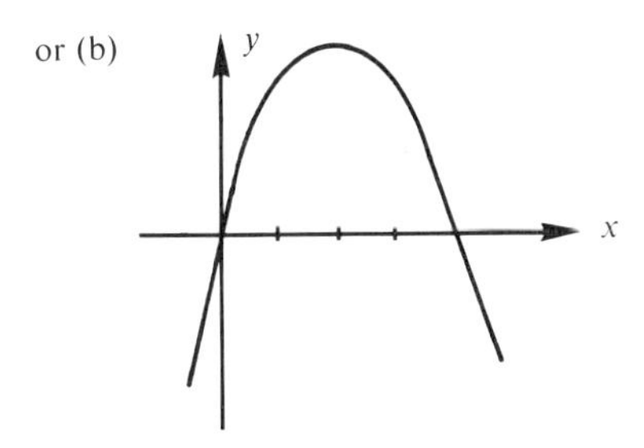

Figure 8.7

1. Given $y = 3x^2 - 12x$, what is $R_x y$?
2. What is the value of $R_x y$ at $x = 5$?
3. For what value of x, if any, is $R_x y = 0$?
4. What is $R_x y$ at $x = 3$?
5. What is $R_x y$ at $x = 1$?
6. Does (a) or (b) in Fig. 8.7 more nearly represent the situation learned from Problems 3, 4, and 5?
7. Hence does the function have a point of relative minimum or relative maximum at $x = 2$. Which?
8. Find the value of x for which $y = x^2 - 4x$ has zero slope and determine if it is a point of relative minimum or relative maximum. Show argument.

The teacher can introduce and develop many ideas in this way, each of which can be studied further by the student in materials prepared for the next class.

If a student does not raise the issue then the teacher must ask this question at some suitable moment, perhaps while one is computing at sight $R_x y$ for a polynomial in x:

> Our *theorem* that if $y = ax^n$ then $R_x y = nax^{n-1}$ permits us to write $R_x y$ only for a *monomial* in x. Can we *prove* that it is permissible to compute $R_x y$ for a polynomial by *adding* the rates of change for each separate term? That is, what logical permission do we have to write
>
> $$R_x(x^2 + 4x^3) = R_x x^2 + R_x 4x^3?$$
>
> Is the R_x-process distributive over addition? What statement or possible theorem do we need to investigate?

On the next study guide might well appear

1. Examine by deduction the conjecture that

$$R_x [f(x) + g(x)] = R_x f(x) + R_x g(x)$$

at $x = a$.

The term "rate of change" keeps the student mindful at all times that the differential calculus is a study of *rates*. Not until the course has progressed well need the term "derivative" be introduced unless a student asks about it. Sometimes just, "Let's use '$R_x y$' for awhile" is enough. The use of the term "rate of change" makes the consideration of rate-of-change problems quite natural.

1. Given $V = x^3 - 2x^2 + 24x + 10$, what is $R_x V$? Hence, how fast is V changing with respect to x when $x = 3$? when $x = 2$?
2. From Problem 5 we note that V is changing 28 times as fast as x when $x = 2.0$. Now if there is a slight change in x from $x = 2.0$ to $x = 2.1$ suggest what *approximate* change there is in V.

A very helpful figure to demonstrate the relation between the rate of change of $y = f(x)$ at $x = x_b$ and the approximate and actual changes in y as x_b becomes x_f is shown in Fig. 8.8. Too, one notes from the figure the error introduced in using $R_x y$ at $x = x_b$ to approximate the change in y; it is easily seen that the product $(x_f - x_b) \cdot R_x y$ does not equal the actual change in y.

An attempt at developing exercises and approaches to encourage student contribution should be made on as many topics as possible. The "product rule" is no exception. Exercises as these might appear on a study guide:

1. Given $y = (5x + 3)(4x + 7)$, compute $D_x y$.
2. Given $y = (3x + 4)(2x + 5)$, compute $D_x y$.
3. Given $y = (7x + 2)(4x + 3)$, compute $D_x y$.
4. Does it appear that there is any way to arrive at the expression for $D_x y$ without multiplying first? Test with $y = (ax + b)(cx + d)$.

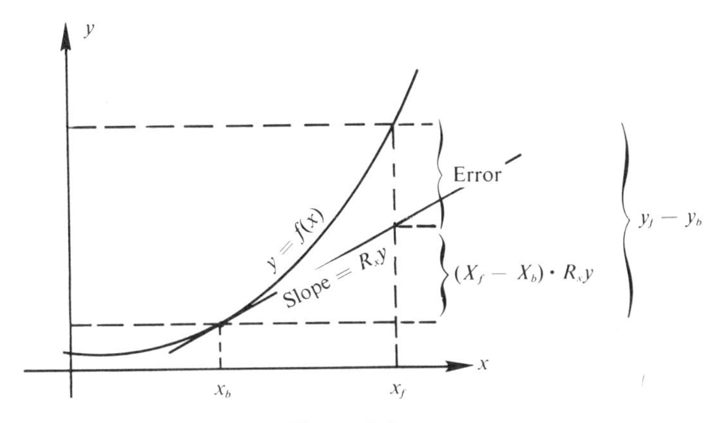

Figure 8.8

A conjecture and a subsequent theorem may develop and one never knows what it will be. Next, of course, if we have as our objective the knowledge of the product rule (and if it were not mentioned yet) we should introduce problems as

1. Given $y = (3x^2 + 4)(4x + 7)$, compute $D_x y$.

Several days may be required for the "product rule" to appear, but much growth and exploration and *invention of conjectures* with subsequent examination for special cases are going on as the class, for the moment, gives its main concentration to other topics. Soon comes the "product rule conjecture" and the students pursue its proof with purpose and interest. They should become very proficient at this proof. Now an interesting exercise can follow:

Using a technique similar to the proof for the derivative of a product, start with the definition for the derivative of a *quotient*,

$$D_x \frac{f(x)}{g(x)} \equiv \lim_{h \to 0} \left(\frac{f(a+h)}{g(a+h)} - \frac{f(a)}{g(a)} \right) \Big/ h \qquad \text{at } x = a$$

and see if you can develop an expression for it.

Some students may be able to complete this development with little help.

Another "rate of change" topic which can be introduced so the student makes a usable conjecture rather quickly and whose logical scaffolding can be erected as the course progresses is that of $D_x \sin x$. After some discussion on how one might attack this problem, exploration may be suggested in this manner:

1. Let us construct a careful graph of $y = \sin x\ [0, 4\pi]$ (Fig. 8.9).

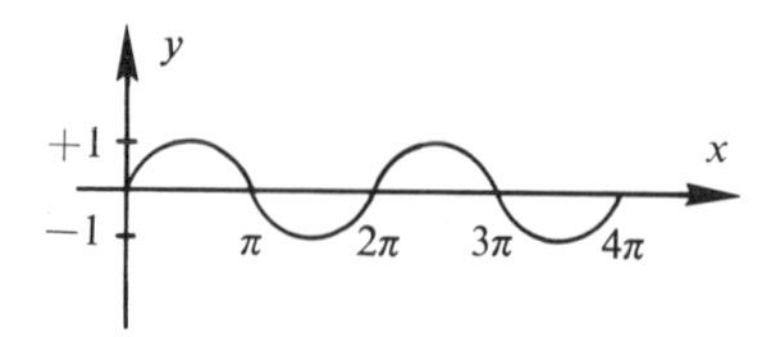

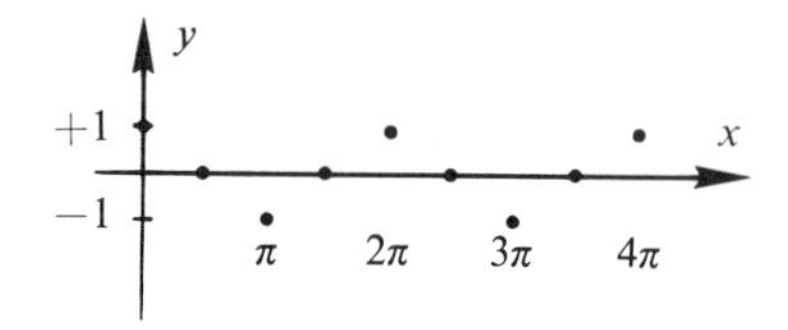

Figure 8.9

2. See if it is possible to estimate the *slope* of the $y = \sin x$ curve at several points and then construct a graph of the *slopes* of $y = \sin x$ relative to x. Is the graph like that of any function you recognize? What?
3. Can you make a conjecture on an expression for $D_x \sin x$?
4. Try to confirm your conjecture by deduction from the definition.

It sometimes happens that a student suggests computing the derivative of $\{(x, \sin x)\}$ from the definition at the very outset but the difficulty which one encounters as the development proceeds makes quite rewarding the immediately obtainable conjecture sought in the above exercises. The conjecture is that $D_x \sin x = \cos x$. Now an adventurous field opens! Our study guide might now have exercises as:

1. Using the *conjecture* that $D_x \sin x = \cos x$, compute

$$D_x y \text{ if } y = x^2 + x \sin x.$$

2. Can you make some conjecture on $D_x \cos x$? Show your work.

If one has studied *chain rule* then $D_x \cos x = D_x(1 - \sin^2 x)^{1/2}$ and no *conjecture* on $D_x \cos x = -\sin x$ is needed. Now while the detailed proof of $D_x \sin x = \cos x$ continues, students are also developing laws for $D_x \tan x$, $D_x \cot x$, and others, largely on their own strength. The *proof* that $D_x \sin x = \cos x$ comes largely from calculus texts, but even the supporting parts of this proof can be developed by exercises and "inductive quizzes."

Someday some exercises as these may appear on a study guide:

If Johnny can run twice as fast as George and George can run three times as fast as Harry, then Johnny can run how many times as fast as Harry?

If y changes $3x^2$ times as fast as x, but x changes z^3 times as fast as z, then what is the rate of change of y with respect to z?

If y changes dy/dz times as fast as z, but z changes dz/dx times as fast as x then y, we *conjecture* would change how many times as fast as x? Hence it appears that $dy/dx =$ ______

and now

if $y = 3x^2 + 2x$, but $x = 4z^2$ then $D_z y =$ ______

and the students have intuitively arrived at a conjecture which they can use to differentiate forms in which the use of the chain rule is helpful. Indeed, constantly there should be exercises asking students to compute the derivatives by several methods, as:

Compute $D_x(x^2 + 3x)^2$ by first expanding the binomial and then differentiating

and

Compute $D_x(x^2 + 3x)^2$ by renaming $x^2 + 3x$ as u where u is a function of x and then using the "chain rule"

or

Given that $y = x^2 + 4x$ but $x = 3t^2 + 6t + 10$, compute $D_t y$ by first expressing y as a function of t

and

Given that $y = x^2 + 4x$ but $x = 3t^2 + 6t + 10$, compute $D_t y$ by the use of the "chain rule,"

to demonstrate that the chain rule or the method of differentiating a composite function is a convenience. Logical substantiation of this conjecture may be carried on at the level to which the teacher and students aspire. In texts and in articles in journals there is discussion of the topic on several levels.[13]

Exercises

1. Make up a story or situation which will lead to locating points on a plane by means of *polar coordinates.*

2. Devise several questions on an "inductive quiz" which will provide properties of simple equations in polar coordinates as $r = 10$, $\phi = \pi/3$, $5 = r \cos \phi$, $r = \phi$, and so on.

3. Is $D_\theta r$ a measure of slope of a curve whose equation is given in polar coordinates as $D_x y$ when the equation is given in rectangular coordinates? Suggest several problems to help your students find out.

---

[13] Lyman S. Holden, "Providing Motivation for the Chain Rule," *The Mathematics Teacher,* LX (December, 1967), 850-4.

4. Plan a study guide to help your students discover the use of the *second derivative* in helping to tell the nature of curves at various points when the equation is given in rectangular coordinates.

5. Continue with Exercise 4 to help students formulate a method to use the second derivative in testing a point of zero slope to see if it is possibly a point of maximum or minimum value of the function.

Before leaving the subject of maximum and minimum points of functions it should be suggested that, in teaching, the student must also have experience with points of zero slope which turn out not to be points of relative maximum and minimum. Indeed, the student should be made aware that if $R_x y$ exists at each point in an interval (a, b) and if $y = f(x)$ has a point of relative maximum or relative minimum in (a, b) then the abscissa of such a point will (simply) be found among those values of x for which $R_x y = 0$. Each value of x for which $R_x y = 0$ must be studied to learn if it is the abscissa for a point of relative maximimum or relative minimum or neither. This situation is similar to the solving of equations in algebra where we say that "if $f(x)$ has zeros they will be found among the zeros for $g(x)$ where $g(x)$ has been produced from $f(x)$ by the usual algorithms of equation solving."

Another interesting topic relative to rates of change is one which can be developed like this:

(a) Note secant from $x = 0$ to $x = 3$ in Fig. 8.10. Does it appear that there is a line tangent to the curve which has the same slope as the secant?
(b) At what value of x does this tangent line appear to be?

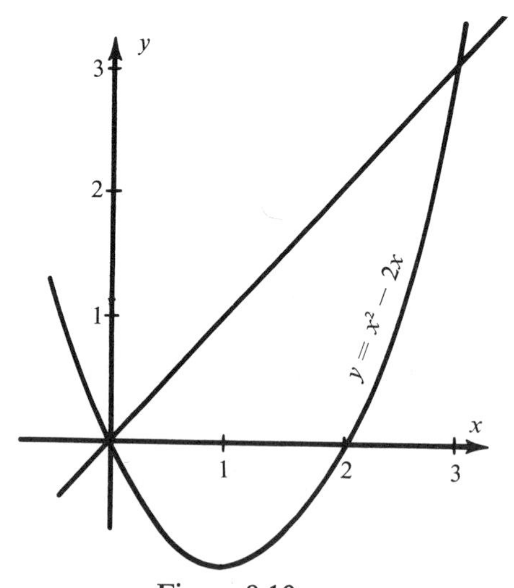

Figure 8.10

(c) Can your estimate be checked by computation?

Several problems can be done. Later the considerations of "Is there always such a tangent and such a point?" and "What conditions on the function will insure the existence of such a point?" have created a strong conjecture —perhaps named after the student who formulates it—about an interesting point $x = \xi$ between $x = a$ and $x = b$ such that $f'(\xi) = f(b) - f(a)/b - a$. What shall we call this conjecture? Some suggest "the average slope-conjecture," or "the secant-tangent conjecture." Whatever name is suggested is usually easily transformed to the "mean-value conjecture for derivatives." Devising or consulting references for a proof can follow; indeed, texts become sources for reference, continued study, "seeing a different point of view," and exercises, while the course itself proceeds with emphasis on initial student exploration and investigation and formulation of significant problems. The mean-value theorem for derivatives is useful in replacing a form as $f(b) - f(a)$ by one involving the *derivative* of $f(x)$—and this should be emphasized by means of exercises so that the student can do this easily and readily. Students should "feel at home" in considerations like this: Given $y = x^3$ write a relation which could be used to replace $8^3 - 2^3$? (Ans. $3\xi^2 \cdot 6$ where $2 < \xi < 8$). This skill is used in various places in the calculus.

Exercises

1. The mean-value theorem is important theoretically. One is less interested in *computing* the $x = \zeta$ for which $\dfrac{f(b) - f(a)}{b - a} = f'(\zeta)$ for any given situation than he is in knowing that such an $x = \zeta$ exists. Consult texts on the calculus for the proof of this theorem. Could you devise exercises in proof which would help the student later to prove this theorem by using results of the exercises?

2. Note how the mean-value theorem is used in developing a formula for computing arc length when the equation of the curve is expressed parametrically.

3. Construct an "inductive quiz" or a study guide to lead the student to computing the equation of the line tangent to a curve at a given point; the line normal to a given curve at a given point.

4. Outline an experimental approach to the derivative of $y = \log_{10}x$. One *can* start as we did in arriving at a conjecture on $D_x \sin x$. Perhaps the properly oriented student can make the conjecture that the slope D_xy, varies inversely as x. Investigate.

5. A piece of cardboard is 20 inches long and 16 inches wide. A box is to be made by cutting squares from the corners and folding up the parts left for the sides and ends (Fig. 8.11). What height x must be used so the box will

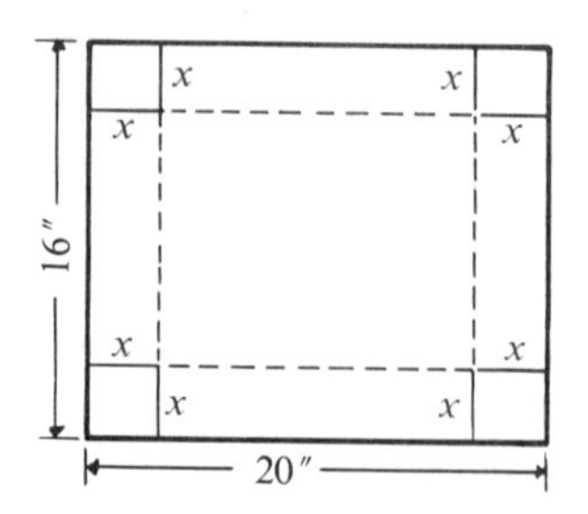

Figure 8.11

hold the most. How can you approach this problem in the classroom? What suggestions might you get? The method of calculus "comes to the rescue." Could this problem be done without the calculus? How?

6. Suggest several pedagogical advantages in having six or seven *models* of boxes with different heights in Exercise 5. What would be the advantage of "guess-and-check" methods in Exercise 5?

7. Devise a study guide or a succession of exercises or situations on such guides from which will emerge the idea of *antidifferentiation* and several uses of this process.

8.5. Curvature of a Curve at a Given Point

Nearly every topic in the calculus can have an experiential or concrete beginning. The subject of curvature invites such consideration. At some convenient time an inquiry like this may appear on a study guide or exercise sheet:

Fig. 8.12 illustrates several curves. In which case does the curve at P

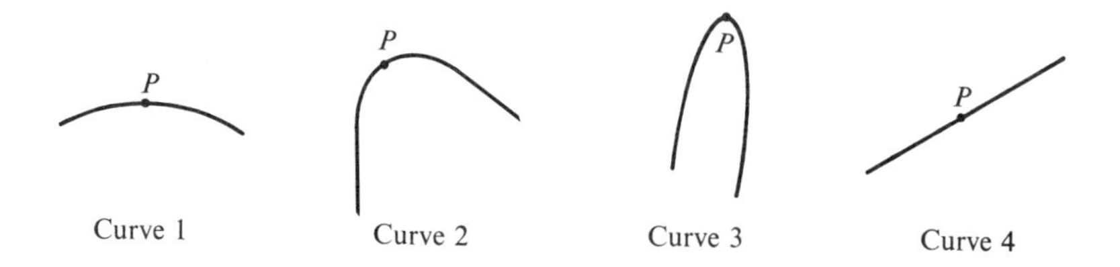

Figure 8.12

seem to be the "most curved"? We expect the student to say "Curve 3" and "this is all" on the topic for the first day. On the succeeding day's study guide this may appear:

Yesterday it was decided that at the point P, Curve 3 was the "most curved." What suggestion can you make to devise a *measure* of "curvedness"?

Almost invariably someone suggests that "if the tangent or *slope* of the curve changes much as x changes, then the curve is "quite curved." Hence "the second derivative is perhaps a good measure," is the generally accepted agreement. How can this definition of curvedness (or curvature) be put to test? Discussion yields that the curvature of a straight line is zero and that the curvature of a circle is constant. Students investigate their proposed definition guided by exercises as these:

Using d^2y/dx^2 as a measure of "curvature" see if you obtain 0 for the curvature at any point x on the line $y = mx + b$. Test your definition on the circle $x^2 + y^2 = r^2$.

For $y = mx + b$, $d^2y/dx^2 = 0$, but from $x^2 + y^2 = r^2$ one obtains $d^2y/dx^2 = -r^2/y^3$ which is still a function of y. Hence the proposed definition fails. The student perceives that mathematics does not always come out just right and that there is struggle, failure, groping, the wrong conjecture, none of which is evident in the carefully organized, synthetic, and expository treatment in the usual text. Such investigation strengthens him mathematically. Above all there is review in attempting to formulate physical properties in the language of mathematics as well as review of implicit differentiation. As other work continues we have this question, "Can you suggest another possible way to express the curvedness of a curve at a point P?" "Let us try the rate of change of the slope (tangent) with respect to arc length" is sometimes a second suggestion.

$$\frac{d}{ds}\left(\frac{dy}{dx}\right) = \frac{d}{dx}\left(\frac{dy}{dx}\right) \cdot \frac{dx}{ds}$$

$$\frac{d^2y}{dx^2} \cdot \frac{1}{\sqrt{1 + (dy/dx)^2}}$$

This expression taken as "curvature" fails the test also as students learn that for the circle $x^2 + y^2 = r^2$ we have $-r/y^2$ which is still a function of y and not constant at all.

We seek more suggestions. If no further suggestions are forthcoming, the rate of change of ϕ with respect to arc length s may be suggested as a possibility (see Fig. 8.13). This proposed definition seems to satisfy the requirements that the curvature of a straight line be zero and that for a circle be a constant. With the practice in the use of the chain rule in other problems and investigations, the student now quite easily develops $D_s\phi$ where $\phi = \text{arc tan}\,(dy/dx)$. This yields the definition of curvature found in texts.

Students sometime suggest $D_s\beta$ as a measure of curvature at P, but are

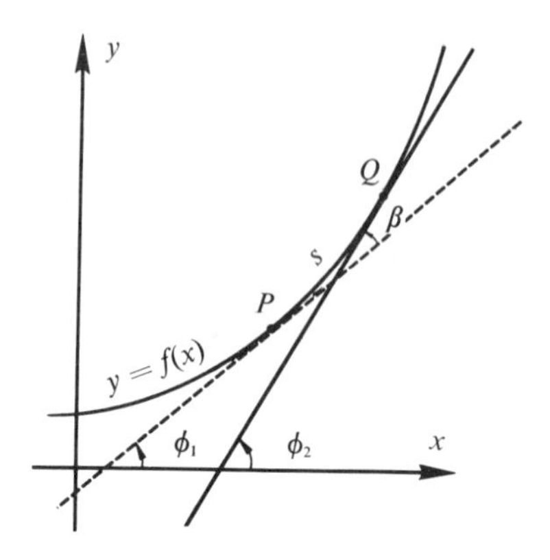

Figure 8.13

slightly discouraged when other difficulties arise. Of course, the students could be directed to "curvature" in the text at the very outset, but there is stimulation in mathematics *growing* (even with failures sometimes) and in creating definitions and conjectures which seem to fit. Also, now, as we have emphasized before, the student can read the works of others with deeper understanding for he, too, has developed some mathematics. It is not wrong to start with concrete situations as we did with a simple question about "curvedness" for it begins the student at the very bottom and, as he thinks of the problem in his own way, fundamental ideas take root and the foundation becomes firm. Here the student investigates a definition of his own making. Polya says, "The mathematical experience of a student is incomplete if he has never had an opportunity to solve a problem invented by himself."[14] A methodology which encourages mathematics to be created from simple experiential beginnings and then urges the investigation of conjectures, with the possible subsequent construction of a logical system appropriate to the level of the student, is far more effective than placing the student in a textbook sea filled with the bewilderment caused by overrigorous definitions and theorems at the very beginning. This approach to teaching is defended by Saxelby when he says, "This intuitional direct vision method is intended, not to take the place of, but to prepare the way for a more rigorous analytical study of the subject."[15] Courant also seeks to help the student when he says we must give "due credit to intuition as the source of mathematical truth."[16] As early as 1894, Fletcher Durell supported the

[14]G. Polya, *op. cit.*, p. 66.

[15]Saxelby, *op. cit.*, pp. v, vi.

[16]Richard Courant, *Differential and Integral Calculus,* translated by E. J. McShane (New York: Interscience Publishers, Inc., 1937), Vol. I, p. vi.

methodology of starting with something *concrete* when he explained that the student "in each new advance is to begin with the concrete object, something which he can see and handle and perhaps make, and go on to abstractions only for the sake of realized advantages."[17] In emphasizing this point Durell continues by explaining how Gauss and Cauchy conceived the idea of representing a complex quantity by a pair of numbers and by a point in a plane and then argues his case further:

> If the greatest mathematical minds feel such an aid necessary in their work, why should not the corresponding aid be offered to the farmer's boy, who is engaged, year after year, in struggling with numbers of whose relations to sensible objects he can have no clear conception?[18]

Exercises

1. Can you suggest another approach to the "curvature" of a curve?
2. One student suggested this (Fig. 8.14):
   Suppose you create a measure for

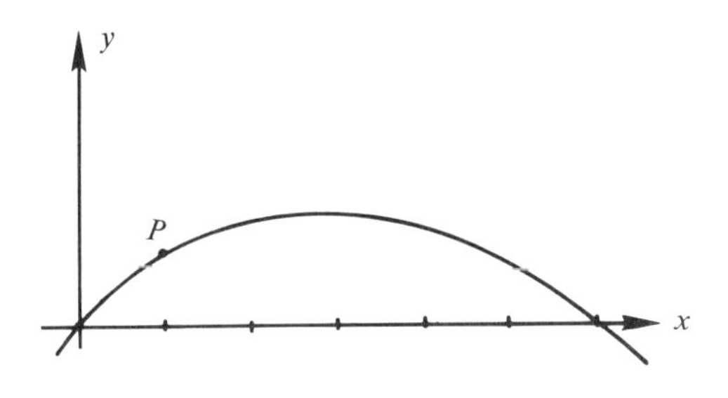

Figure 8.14

   curvedness in this way: take values of x at one unit on each side of the abscissa of the point in question. Compute $D_x y$ at each point. Define the absolute value of the difference of slopes as the measure. Investigate this proposed definition.
3. Another student-suggested approach to the "curvedness" of a curve at the point $P(x, y)$ is to let R and S be points on the curve with abscissae $x - 1$ and $x + 1$, respectively, and then define the perpendicular distance from

---

[17] Fletcher Durell, "Applications of the New Education to the Differential and Integral Calculus," *American Mathematical Monthly*, I (January, 1894), 15.
[18] *Ibid.*, p. 41.

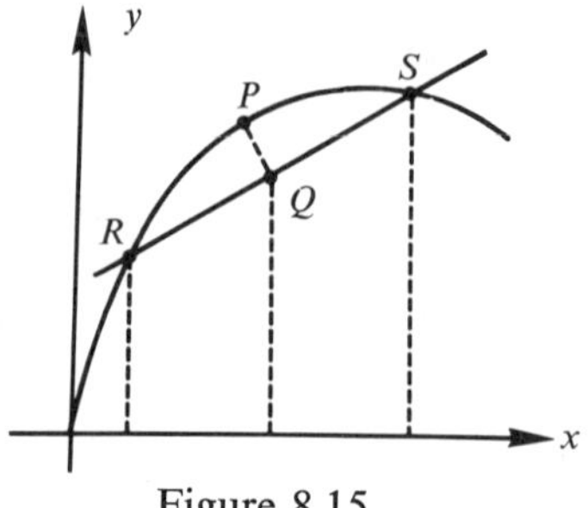

Figure 8.15

P to the line through R and S to be a measure of the curvedness at P (Fig. 8.15). Does this definition seem acceptable?

8.6. The Area Problem

This interesting problem can be started at almost any time after the student has been introduced to constructing the partial graphs of functions as $y = x^2$, $y = x^2 + 3$, $y = 2x^2 + 7$, and the like. On some particular study guide might appear this simple-looking question: Note the partial graph of $y = x^2$ in Fig. 8.16. Suggest some methods by which you could determine an approximation to what one might call the "area" under the curve, above the x-axis and bounded by $x = 0$ and $x = 3$.

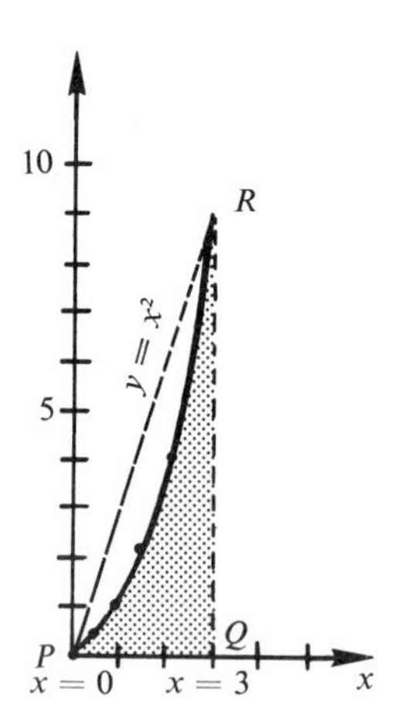

Figure 8.16

Several suggestions will come: cutting the figure from cardboard and weighing against unit squares cut from the same kind of cardboard, drawing the curve on graph paper and counting the unit squares "under the curve," and using graph paper with smaller squares with appropriate calculations to change the number of very small squares to the originally agreed upon unit squares. Some students may suggest the use of the area of the right triangle *PQR* is an approximation and immediately can come the query from the teacher or others, "Can we make an approximation to area which is better than that given by the use of the right triangle?" Along with current work in the course, the student can pursue various suggestions and, indeed, different students can carry on investigations along different lines. Weights of the heavy cardboard parabolic areas enclosed by $y = x^2$, the x-axis, and lines $x = 1$, $x = 2$, $x = 3$, $x = 4$, $x = 5$, and so on, may be studied to see if an empirical relation between the weights and x can be found. Or an empirical relation may be sought between the number of squares counted versus x. Every approach gives some facet of understanding which deepens meaning for the student or more clearly defines or emphasizes the profoundness of the problem.

Using as approximations to the intuitively accepted "area" under the curve first the area of a right triangle, then the sum of the areas of a right triangle and a trapezoid, and then the sum of the area of one right triangle and three trapezoids, and so on, yield interesting results fully within the capacity of every student to understand and which gives every student an opportunity to make a contribution (see Fig. 8.17). Hence, in a study guide

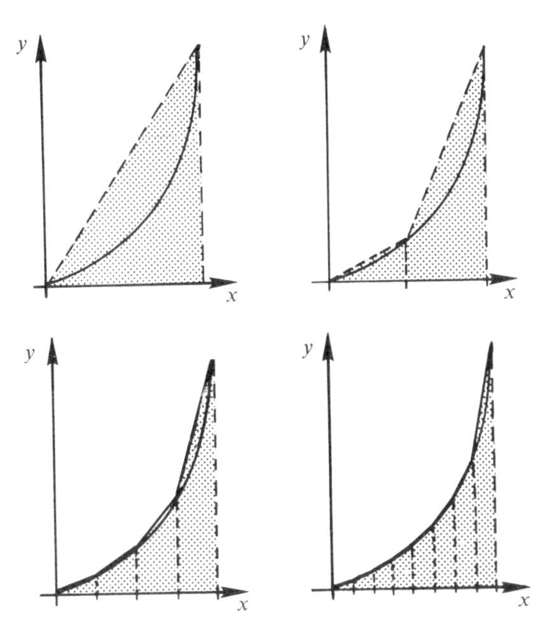

Figure 8.17

can appear exercises as these (the unit lengths on both scales are not the same):

Compute the approximation to the "area" by the use of the right triangle shown in Fig. 8.18.

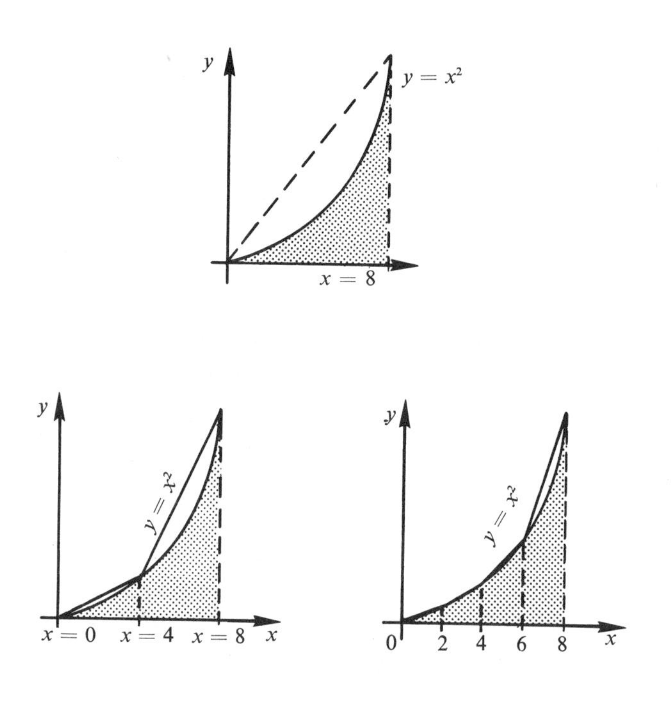

Figure 8.18

Now use the sum of the area of the right triangle and the trapezoid.

Use now one right triangle and three trapezoids constructed on the partition [0, 2, 4, 6, 8].

Note the opportunities for arithmetical review in the computation of areas and in the use of fractions. All of this helps in improving necessary skills in new and purposeful settings. On summarizing results of the work of the students, one has this set of approximations: 256, 192, 176. The question of whether or not we have enough data to make a prediction is a very pressing one. It is decided to make two or more approximations and the students expect the direction for a next assignment to compute the area using one right triangle and seven trapezoids. Tediousness, yes, but understanding and appreciation of the power of mathematics is more deep when it is realized what amount of labor it replaces and how the development of its outward face enables us to do quickly what formerly was nearly impossible. The perseverance and patience and imagination of those who have labored tediously have furnished the power for progress in mathematics, as one has said, "We must never forget that most of the footprints in the sands of time have been made by *workshoes*." Hence, our students, either individually or with others, should make several more computations. Now the question appears:

Study the sequence of approximations to "area": 256, 192, 176, 172, 171, . . . , and see if you can predict the next term. Let us attempt to check the "prediction" by combined class effort. Is there any way by which we can arrive at the limit of this sequence?

The previous experience with the use of $S_\infty = a/(1 - r)$ helps the student to adapt very easily this formula to computing the limit of the sequence {256, 192, 176, 172, 171,· · ·} He rewrites it as {256, 256 − 64, 256 − (64 + 16), 256 − (64 + 16 + 4),· · ·256 − (64 + 16 + 4 + 1 + · · ·) and finally has

$$64 + 16 + 4 + 1 + \cdots = S_\infty = \frac{64}{1 - \frac{1}{4}} = \frac{256}{3} = 85\tfrac{1}{3}$$

whence the limit of the given sequence is $256 - 85\frac{1}{3}$ or $170\frac{2}{3}$. Is "$170\frac{2}{3}$ square units" the area under the curve? Will we ever know? We *define* the area under the curve to have a measure of $170\frac{2}{3}$ square units. Students should understand well why areas under curves must be defined arbitrarily. Students may receive more practice by the use of variations on this problem as to find a number which one can assign as the measure of area under $y = x^2 + 5$, or under $y = x^2 + 2x$, and others. Rather soon, however, comes the challenge (and it is hoped that it is student-suggested) of finding a number to assign as the measure of the area under $y = x^2$ from $x = 0$ to $x = a$. If this can be done, the class has proved a *theorem* on the number to assign as area under $y = x^2$ from $x = 0$ to $x = a$. We know it is $a^3/3$. Our theorem, however, is based on the accepted validity of $S_\infty = a/(1 - r)$ for the "infinite sum" of a geometric series.

At this point there are many opportunities for investigation: we can study many different kinds of polynomials to arrive at a number to assign to "area" under the curves which they represent bounded by $y = 0$, $x = 0$, and $x = a$. Exercises for the student which might now appear are as follows:

Note the numbers for measures of area arrived at in the cases you have investigated. Is there a way by which one might tell the number *quickly* to be assigned, say, to the area under $y = x^2 + 3x$ between $y = 0$, $x = 0$, $x = a$? to the area under $y = x^3 + 4x^2$? to the area under $y = x^2 + 6x + 4$? What notation do you suggest to indicate that we have computed an "area-number" just as $D_x y$ represents or tells that we have computed a "slope-number."

Some students suggest the notation A_x. Hence $A_x(x^2 + 3x)$ from $x = 0$ to $x = a$ is given by $a^3/3 + 3a^2/2$. Some use the notation $A_{x=0}^{x=a}(x^2 + 5) = a^3/3 + 5a$. Notation invented by the student is most meaningful and it can easily be altered to meet the preference of others. The student should note the property that

$$A_{x=0}^{x=a}(x^3) + A_{x=0}^{x=a}(x^2) = A_{x=0}^{x=a}(x^3 + x^2)$$

also that

$$A_{x=0}^{x=a}(kx^3) = kA_{x=0}^{x=a}(x^3)$$

and so on. Now some logical organization can follow:

D. $A_{x=0}^{x=a} f(x)$

T. $A_{x=0}^{x=a} (k) = ka$

T. $A_{x=0}^{x=a} (x) = \dfrac{a^2}{2}$

T. $A_{x=0}^{x=a} (x^2) = \dfrac{a^3}{3}$

T. $A_{x=0}^{x=a} (x^3) = \dfrac{a^4}{4}$

T. $A_{x=0}^{x=a} (kx) = \dfrac{ka^2}{2} = kA_{x=0}^{x=a} (x)$

⋮

T. $A_{x=0}^{x=a} (x + x^2 + k) = \dfrac{a^2}{2} + \dfrac{a^3}{3} + ka$

⋮

Conjecture: $A_{x=0}^{x=a} (x^n) = \dfrac{a^{n+1}}{n + 1}$

Conjecture: $A_{x=0}^{x=a} (kx^m + rx^n) = \dfrac{ka^{m+1}}{m + 1} + \dfrac{ra^{n+1}}{n + 1}$

Of course, these are only suggestions, but they are typical of what students can do. This is knowledge organized by the learner.

Exercises

1. Consider the sequence 256, 192, 176, 172, 171 . . . See if you can set up the nth term and prove that $170\frac{2}{3}$ satisfies the ε, N-definition of limit.

2. Suggest another way to partition the interval [0, 8] under $y = x^2$ and see if you can get a sequence whose limit one may use to define area. Perhaps not all the subintervals need have the same length.

3. The $A_{x=0}^{x=a}$ number is really the limit of a sequence of ________ (sums of products). In comparable language, what is the derivative?

4. Suggest how one might approach each of these problems in an elementary way:

   (a) $A_{x=1}^{x=4}\left(\dfrac{1}{x^2}\right)$

   (b) $A_{x=0}^{x=\pi}(\sin\theta)$

5. The "approach to area" made by the reader might have been through the use of rectangles; first the sum of one rectangle, then two rectangles, then three rectangles, then four rectangles, and so on (as in Fig. 8.19). Use the

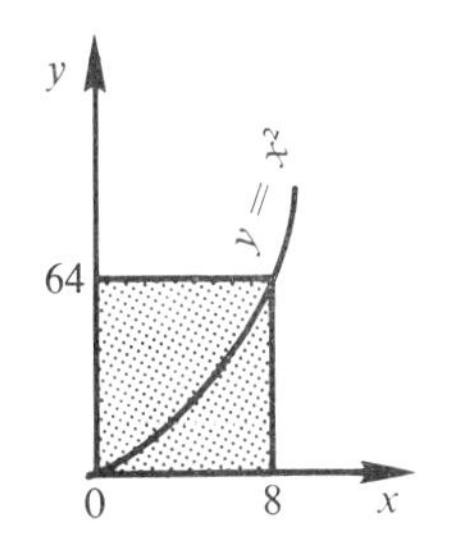

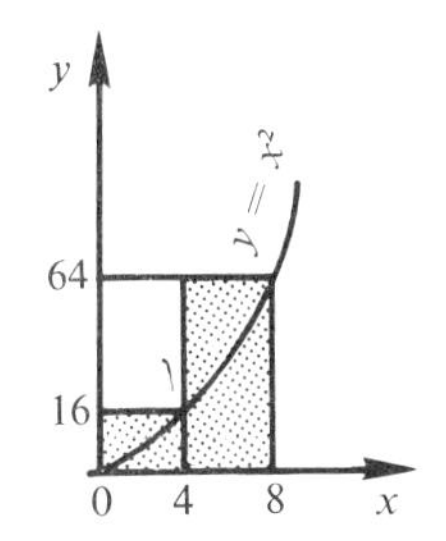

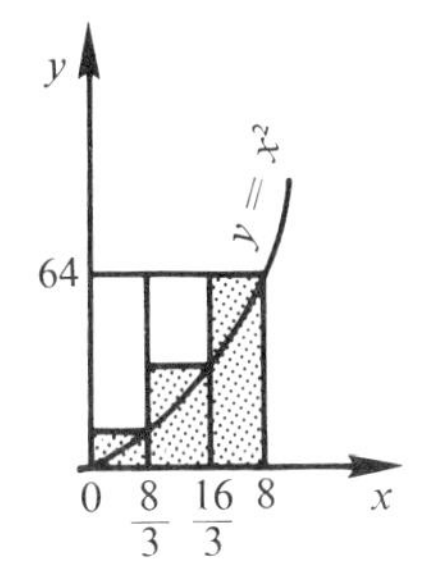

Figure 8.19

right-hand end points of the equal subdivisions on the x-axis to determine the heights of the rectangles and set up a numerical sequence of sums which, it is hoped, will lead to a number which may be assigned as the area under $y = x^2$ from $x = 0$ to $x = 8$. Can you determine easily the limit of such a sequence? It begins as: 512, 320, $\cdots$

6. Reexamine the method of Exercise 5, but use partitionings

   [0, 8], [0, 4, 8], [0, 2, 4, 6, 8] [0, 1, 2, 3, 4, 5, 6, 7, 8], $\cdots$

   and see if the sequence is more easily handled.

7. What advantage, if any, does the method of approximation by a sequence of sums of areas of trapezoids, as used in the foregoing material, seem to have over the "sum of areas of rectangles method?"

8. Reexamine the "trapezoid method" of the text by setting up the term of the approximating sequence which arises from using n equal subdivisions in the x-axis from $x = 0$ to $x = 8$ to arrive at a number to assign as the measure of the area under $y = x^2$. Can you determine the limit of this sum as $n \to \infty$?

9. Reexamine the "rectangle method" of Exercise 6, but this time use n equal subdivisions in $[0, 8]$. See if you can determine easily the limit of the sequence from a study of this sum of n rectangles.

10. Do Exercise 9 again, but let the left-hand endpoint of each subdivision determine the height of each (now *inscribed*) rectangle. There will now be $(n - 1)$ rectangles and a sum of $(n - 1)$ numbers different from zero.

8.7. The Problem of Work Done on a Spring

The use of concrete materials is very helpful here. Weights of various magnitude lifted through different distance require varying numbers of foot-pounds of work. This must be done, for many students even in our modern day have not studied physics nor do they recall much about "work" from courses in general science. Now the spring is introduced and it has been found helpful to create a homemade one from wire large enough so the entire class can see it. A "law" of the spring can be made up as F(force) $= 32s$ (distance) which means that the force required to stretch the spring s feet from its position of rest is given by $32s$.

Exercises on study-guide sheets may appear as follows, perhaps while the students are studying areas under curves.

The force-stretch formula for a certain spring is $F = 32s$ where F is in pounds and s is in feet. How many pounds force will be required to hold the spring stretched to 3 feet from its position of rest? 2 feet from its position of rest?

As one pulls the spring from a stretch of 2 feet to a stretch of 3 feet, does the force remain constant? If not, in what way does it change?

This setting is sowing seeds for some interesting investigation to come very soon and these "seeds" are simple enough at the very beginning to help students gain fundamental concepts easily. The use of weights, springs, yardsticks, help us to see what "work" is. The student will easily sense the difficulty we have in attempting to compute the work done on a spring. The use of models and apparatus is quite in keeping with Comenius' insistence that learning is helped by *non verba, sed res.*[19]

---

[19]National Council of Teachers of Mathematics, *The Teaching of Mathematics in the Secondary School*, Eighth Yearbook (Washington, D. C.: The National Council of Teachers of Mathematics, 1933), p. 226.

The next day's guide sheet may contain, among other problems, this exercise:

Yesterday we saw that to produce a stretch of 3 feet in the spring required 96 pounds force. At 0 feet stretch there were 0 pounds of force, but by the time the stretch was 3 feet the force required was 96 pounds. We surely have a "force times distance" or work
How might we arrive at an answer to the "work done" in stretching the spring from 0 feet stretch to 3 feet stretch?

The experienced teacher knows that there may be several suggestions or there may be none. If there are none one may ask, "What would be the approximate work done if we used 96 pounds force as the force *throughout* the entire interval from 0 to 3 feet stretch?" (Ans. $96 \cdot 3 = 288$ foot-pounds.) Would this approximation be too small? Could we make a better approximation? Could we make an approximation by using the sum of "two works": the work using the force at $s = 1\frac{1}{2}$ feet multiplied by $1\frac{1}{2}$ plus the force at $s = 3$ feet multiplied by $1\frac{1}{2}$ feet?" (See Fig. 8.20) The answer would be $48 \cdot 1\frac{1}{2} + 96 \cdot 1\frac{1}{2} = 72 + 144 = 216$ foot-pounds. A third approximation would be that obtained as the sum of *four* "little works" com-

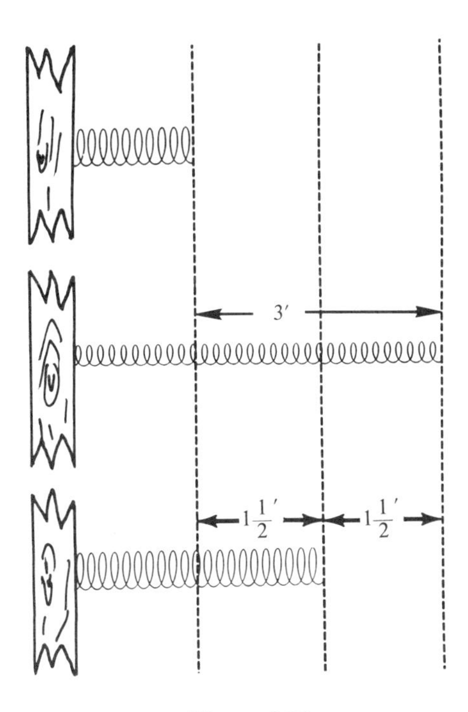

Figure 8.20

puted as

$$(32 \cdot \tfrac{3}{4}) \cdot \tfrac{3}{4} + (32 \cdot 1\tfrac{1}{2})\tfrac{3}{4} + (32 \cdot 2\tfrac{1}{4})\tfrac{3}{4} + (32 \cdot 3)\tfrac{3}{4}$$
$$= 18 + 36 + 54 + 72 = 180 \text{ foot-pounds of work.}$$

We now have the beginning of a *sequence* of approximations to the work done in stretching the spring from a stretch of 0 to one of 3 feet: 288, 216, 180, . . . and the reader is invited to confirm that the approximate work done is 162 foot punds if there are *eight* subdivisions and, therefore, eight "little works" to be added. This is tedious, but the ground work is being laid for a most important concept, that of the definite integral. Again the sequence 288, 217, 180, 162, . . . can be treated as

$$288,\ 288 - 72,\ 288 - (72 + 36),$$
$$288 - (72 + 36 + 18), \cdots$$
$$288 - (72 + 36 + 18 + \cdots) \cdots$$

and, with the help of $S_\infty = a/(1 - r)$, the limit can be determined to be $288 - 144 = 144$ foot-pounds of work.

The interested and resourceful teacher can now suggest many investigations similar to those used in the study of *area* under a curve. Could we use the forces at the *left* end points to determine work? Will different methods of subdivision alter the answer? Is $W_{x=0}^{x=a}(32x)$ equal to $32\ W_{x=0}^{x=a}(x)$?—and many like questions. Indeed, an entire "logical structure" can be built up similar to the one which the student helped to construct in arriving at measure-numbers for *areas under curves*. Of course, one can make up more complicated force-stretch formulae and, indeed, the question may arise if *all* springs have the simple $F = ks$ linear formula to describe their behavior.

Exercises

1. Construct an approximating sequence for work done on a spring if $F = 32s$ from $s = 0$ to $s = a$. Use first one, then two, then four, then eight (and so on) equal subdivisions. Use the right-hand endpoint of each interval to determine the force assumed to be effective throughout the interval.

2. Do the same as in Exercise 1, but use the left-hand endpoint of each interval to determine the force assumed to be operative throughout the interval.

3. Construct the approximation to the work done in stretching the spring whose formula is $F = 32s$ from $s = 0$ to $s = a$ by the use of n equal subdivisions. Use the right-hand endpoint of each interval to determine the force assumed to be operating throughout the interval.

4. In Exercise 3 as n is considered to be 1, then 2, then 3, then 4, . . . there is defined a *sequence* of approximations. Determine the *limit* of this sequence.

5. Can you devise an elastic body for purposes of demonstration whose force-stretch formula is not linear?

8.8. The Definite Integral

A study-guide sheet may direct the student to compare methods for computing area-measure-numbers and for determining work-measure-numbers by the use of review exercises:

Review. Determine by numerical methods (setting up a sequence of approximations and so on) the number assigned as the measure of area under the curve $y = 3x^2$ from $x = 0$ to $x = 8$.
Review. Determine by numerical methods the number assigned as the work done to stretch a spring whose force-stretch formula is $F = 3s^2$ from $s = 0$ to $s = 8$.

What elements in common are observed in the two problems above? Can both the above problems be worked by antidifferentiation also? Students sense that these two problems, although about different situations are *abstractly the same.* In each case, the number sought is in principle the limit of a sequence of sums of products (with certain finer conditions which can be added). Here we have a *big idea* which the students have helped to construct; an abstraction of the idea of work and area but for which as yet there is no name. Two apparently separately conceived ideas have merged dramatically into one great principle! This is mathematics really growing and there is nurture here of the inward-looking face of mathematics. What shall we name this idea? Some students suggest the "sum-idea" or the "limit-of-a-sequence-of-sums-of-products-idea." Some have invented the notation S for this idea; $S_{x=0}^{x=3}\, x^2$ becomes $3^3/3 - 0^3/3 = 9$; $S_{x=b}^{x=a}\, x^3 = b^4/4 - a^4/4$.
The same symbolism can now be used for both area and work problems.

The students are now ready to "reorganize" their logical organizations. What previously had been lists of agreements and theorems on "area under a curve" and "work done on a spring" now merge into *one* all inclusive list on the properties of the S-process.

Yes, there comes a time when the teacher may suggest that scholars in mathematics have come to call this "sum-process" by the name "integral" and that whereas we write that from the result of a computation $S_a^b x^2 = b^3/3 - a^3/3$ the accepted notation is

$$\int_a^b x^2 dx = \frac{b^3}{3} - \frac{a^3}{3}$$

where the dx may be used to help us to remember that the subdivisions were made along the x-axis. (Notations by some writers do not include the "dx.")

Finally the students help to *define*

$$\int_a^b f(x)dx \text{ as } \lim_{n\to\infty}\left\{\sum_{i=1}^{n} f(x_i)\Delta_i x\right\}$$

where

$$\sum_{i=1}^{n} \Delta_i x = b - a.$$

Of course, this definition could have been given at the outset and in a very formal manner, but the gradual growth of the student nourished and nurtured by experiences in pursuing his own practical and common sense suggestions deepen his understanding and make the above definition of the definite integral seem quite natural.

Now come many questions on the properties of integrals. There is much opportunity here for student investigation and the continuing erection of the logical structure, especially if the student has been developing a background in work on limits.

It is urgently desirable, however, that the student be aware that although $\int_a^b f(x)$ is *defined* as $\lim_{n\to\infty}\left\{\sum_{i=1}^{n} f(x_i)\Delta_i x\right\}$ where $\sum_{i=1}^{n} \Delta_i x = b - a$ and where be described as agreed upon in the course, yet he has already *proved* for certain cases that he can find the value of the integral quickly by a process which involves antidifferentiation. The student has not yet *proved* that

$$\lim_{n\to\infty}\left\{\sum_{i=1}^{n} f(x_i)\Delta_i x\right\} = D^{-1}f(x)\Big|_a^b$$

for any expression $f(x)$. Being remined of this often and in various ways will help the student to keep the "logical structure" in good order. What is assumed, what is defined, and what is deduced should always be items of careful distinction. Also such emphasis helps to maintain an air of expectancy in the classroom that perhaps some of the conjectures we may soon be able to prove.

Too, at this stage in the development of the calculus it is rewarding to have in the study guides exercises like these:

$\int_0^4 (x^2 + 4x)\,dx =$ _______. Is the answer a function of x or is it simply a number? $\int_0^t (x^2 + 4x)\,dx =$ _______. Is the answer a function or is it simply number? If it is a function, it is a function of what variable?

And later, exercises like these: $D_t \int_0^t (x^2 + 4x)\,dx =$ _______

$D_t \int_0^t (x^3 + 4x^2 + 6x + 10)\, dx =$ ______

The time is becoming ripe for a student-formulated conjecture which will lead to the Fundamental Theorem of the Calculus. Also on study guides, problems can appear which will review properties of integrals and at the same time provide a readiness for the proof of the conjecture which becomes the Fundamental Theorem:

$\int_a^b f(x)\, dx$ has what relation to $\int_b^a f(x)\, dx$? ______

$\int_a^b f(x)\, dx + \int_b^c f(x)\, dx =$ ______

$\int_a^c f(x)\, dx - \int_a^b f(x)\, dx =$ ______ $(a < b < c)$

$\int_2^{r+s} f(x)\, dx - \int_2^r f(x)\, dx =$ ______ $(2 < r;\ s > 0)$

$\int_a^{t_1+h} f(x)\, dx - \int_a^{t_1} f(x)\, dx =$ ______ $(a < t_1;\ h > 0)$

In teaching it is so helpful to have students acquainted with "little techniques" which are used in proofs and in problem solving. Exercises as the preceding assist effectively in preparing the student so the newness of the proof along with the strangeness of a hitherto unknown technique do not overpower him.

Exercises

1. Consult several different calculus texts and note their definitions of the (definite) integral $\int_a^b f(x)dx$. You may detect some differences.
2. Some texts prefer to use the term "integral" to denote the "limit" of a sequence of sums of products and never do use the term "indefinite integral" which is really synonomous with the term "antiderivative." Consult several texts on this point. Which do *you* prefer—"antiderivative" or "indefinite integral?"
3. What changes can be made on the definition of $\int_a^b f(x)dx$ given in the discussion above to make it conform with some of the definitions found in the texts which you have examined? Of course, in your teaching, you may wish to help alter student-invented definitions to conform with those of the text being used.
4. Prepare a series of exercises which will lead the student to formulate a conjecture which will later become the Mean-Value Theorem for Integrals.

8.9. Some Fundamental Theorems

All the theorems which contribute to the structure of calculus are "fundamental" but the one usually regarded as the *Fundamental Theorem* is

$$D_t \int_a^t f(x)\, dx = f(t)$$

or some equivalent statement. The proof of this theorem depends on the Theorem of Mean-Value for Integrals and the proof for the latter theorem depends on Rolle's Theorem. It has been found practical in teaching to accept the Mean-Value Theorem as a conjecture and use it to prove the Fundamental Theorem. As the course continues one can try to substantiate the mean-value conjecture. Of course the student can be directed in his thinking by appropriately devised study-guide exercises or he can be referred to texts to see what other investigators have done, as the teacher chooses. The point is that much groundwork can be laid even for proofs of theorems by exploratory exercises. A "little" exercise like this starts the students to the mean-value conjecture:

1. The integral $\int_0^5 x^2 dx$ defines the measure of area under the curve $y = x^2$ and bounded by $x = 1$, $x = 5$, and $y = 0$. What would be the height of a rectangle built on the segment from $x = 1$ to $x = 5$ which would give the same area?

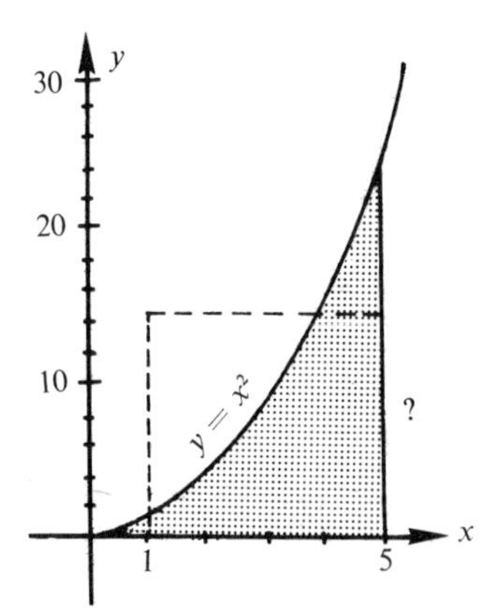

2. Do the same for $\int_1^5 (x^2 + 4x)\, dx$.

And, perhaps on the next day, this thinking can be carried a little further as students work currently on other topics,

In Number 8 on yesterday's study guide, at what value of x does the height

$10\frac{1}{3}$ occur? In Number 9 at what value of x does the required height occur? Do the values of x occur between $x = 1$ and $x = 5$?

Finally one may ask this question, "From the results of our work on areas under curves and finding abscissae for heights of equivalent rectangles with the base given by the segment between the endpoints of the integral, what conjecture would you be willing to make on

$$\int_a^b f(x)\,dx$$

relative to the rectangle constructed on the segment determined by $[a, b]$?"

Of course, the student may need help in wording this conjecture and the accepted version may be the result of class discussion along with teacher guidance, but mathematics is *growing* and being *created.* (At this writing in the authors' class in calculus, this is known as the "Mark Coenen mean-value conjecture," named for the student). Finally must come this consideration,

Can you think of examples of functions for which the mean-value conjecture does not hold? Illustrate; explain why the conjecture would fail.

What conditions, therefore, does it seem we must place on $f(x)$ in $[a, b]$ so that the mean-value conjecture holds. Now let us try to reword our conjecture.

We leave it to the reader to review the proof of the Fundamental Theorem of the calculus and other fundamental theorems from texts on the calculus. We wish to emphasize again that a student can be "conditioned" for proofs by well-planned exercises and experiences. All of this makes the study of mathematics more enjoyable and excitingly fruitful and hence the teaching and learning alike more effective, more dramatic, and filled with unexpected adventure. Indeed the organization of the chapter suggests how the ideas of the "A-process," the "W-process," the "S-process," the integral, the antiderivative, can be developed simultaneously with conjectures made in an atmosphere in which there arises a dominant feeling of something about to happen—the support of the Fundamental Theorem. *Students* express their feelings about this great theorem in this manner:

> "We start with sequences and go from there to rate of change, and then to the derivative and the antiderivative. On the other side of the picture we go from sequences up through areas to the integral. Here we go through the mean-value theorem to the fundamental theorem. Here the integral and the antiderivative come together.

[20]Kenneth B. Cummins, "A Student-Experience-Discovery Approach to the Teaching of Calculus," unpublished Ph. D. dissertation. The Ohio State University (1958), pp. 137-8.

The fundamental theorem, therefore, is the link which holds the two chains together."

"The fundamental theorem is the missing link which connects the integral calculus to the differential calculus."

". . . The derivative and the integral . . . the concepts are different . . . The real connection occurs by the fundamental theorem."

"From the integral we develop a mean-value theorem and from derivatives we go to the antiderivative and the two branches meet in one culminating theorem."

Exercises

1. Plan a series of exercises which will lead to a conjecture which will become *Rolles' Theorem.*

2. Begin with the *definition* of the derivative,

$$D_t \int_a^t f(x)dx = \lim_{h \to 0} \left\{ \frac{\int_a^{t_o th} f(x)dx - \int_a^{t_o} f(x)dx}{h} \right\}$$

at $t = t_o$

and prove by the support of theorems on integrals, the mean-value conjecture, theorems on limits, that

$$D_t \int_a^t f(x)dx = f(t)$$

3. Prepare a study sheet to help student write some of this proof by themselves.

4. From the Fundamental Theorem

$$D_t \int_a^t f(x)dx = f(t)$$

prove the (useful) corollary that

$$\int_a^b f(x)dx = F(b) - F(a)$$

where $F(x)$ is an antiderivative or "primitive" of $f(x)$.

5. Consult several texts on the calculus and compare their statements of the Fundamental Theorem. Do you have a preference? If so, what? Why?

6. The first formulation of the Fundamental Theorem is usually credited to Isaac Barrow. Consult histories in mathematics to learn his place in the development of mathematics.

7. The proof of $\int_a^b f(x)dx = F(b) - F(a)$ depends on the Fundamental Theorem as given above, the proof of the latter is based on the mean-value theorem for integrals, and this, in turn, has as one of its "pillars" Rolle's Theorem. Consult texts on the proof of Rolles' Theorem and trace its foundations back through several stages. What seem to be the ultimate theorems and definitions in the calculus?

8.10. Student-Centered Approaches in Other Topics

Every topic in the calculus can be introduced by some physical or numerical setting from which generalizations and abstractions can emerge. The teacher, as an "engineer of experiences," must plan carefully for some of these but the dividend is very high. Student-contribution and the student-thinking-in-his-own-way approach, along with timely teacher guidance, create a point of view and a productivity in mathematics which the hurried, matter-of-fact acceptance, and often frustration-producing following of the text seldom can accomplish.

"Moments of area," for example, after preliminary discussion of moments through exercises, of course, can invite this question: In Fig. 8.21 the area changes with x. How might we find the "moment of area" around the y-axis?

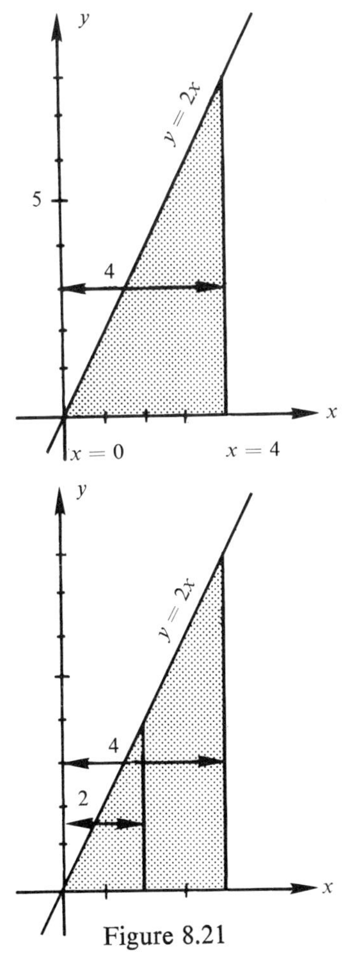

Figure 8.21

Class discussion brings out that if one uses as the moment arm the distance from 0 to 4 then the moment will be too large. Yet $4 \cdot 4 \cdot 8 \cdot \frac{1}{2} = 64 = M_1$ is an *approximation* to the moment (in square units, units). A better approximation comes by using the partition [0, 2, 4] and using as the moment arm the distance from $x = 0$ to the right most end of the respective interval. Hence $M_2 = 2 \cdot 2 \cdot 4 \cdot \frac{1}{2} + 4 \cdot 2\,(4 + 8) \cdot \frac{1}{2} = 8 + 48 = 56$ is a better approximation. A still better one is given by using the partition [0, $1\frac{1}{3}$, $2\frac{2}{3}$, 4] and it is not long until one has the suggestion that the moment of area might be *defined* to be

$$M = \lim_{n\to\infty} \left\{\sum_{i=1}^{n} x_i f(x_i)\Delta_i x\right\},$$

Indeed the above definition comes quite naturally from the previous experience of the student in setting up sequences. It is the practice of the writers to encourage students always to set up such (and other) problems as

$$M = \lim_{n\to\infty} \left\{\sum_{i=1}^{n} x_i f(x_i)\Delta_i x\right\} = \int_0^4 x f(x)\, dx$$

where

$$\sum_{i=1}^{n} \Delta_i x = 4 - 0$$

which helps to keep the "logic of the calculus" in good order. On guide sheets students are still invited to see if they can arrive at the answer by finding the limit of a numerical sequence. This gives a sense of accomplishment in addition to a deep feeling of assurance that the integral is really working!

The subject of volumes is interesting, also, and the student may be invited to try his hand at finding the volume of the solid made by rotating the shaded portion about the x-axis by setting up a numerical sequence (Fig. 8.22). A first approximation is $4^2 \cdot \pi \cdot 2 = 32\pi$ but the volume of the cylinder is much too large; a second is the *sum* of volumes of *two* cylinders: $1^2 \cdot 1 \cdot \pi + 4^2 \cdot 1 \cdot \pi = 17\pi$. If the partitioning on the x-axis is (0, $\frac{1}{2}$, 1, $1\frac{1}{2}$, 2) and if the right-hand endpoint of the interval is used to determine the radius, one gets $[(\frac{1}{2})^4 \cdot \frac{1}{2} + 1^4 \cdot \frac{1}{2} + (1 \cdot \frac{1}{2})^4 \cdot \frac{1}{2} + 2^4 \cdot \frac{1}{2}]\,\pi$ or $11\frac{1}{16}\pi$.

The numerical sequence may not be readily treated but the student can use the concept involved to suggest to him to write

$$V \equiv \lim_{n\to\infty} \left\{\sum_{i=1}^{n} (x_i)^4 \pi\, \Delta_i x\right\}$$

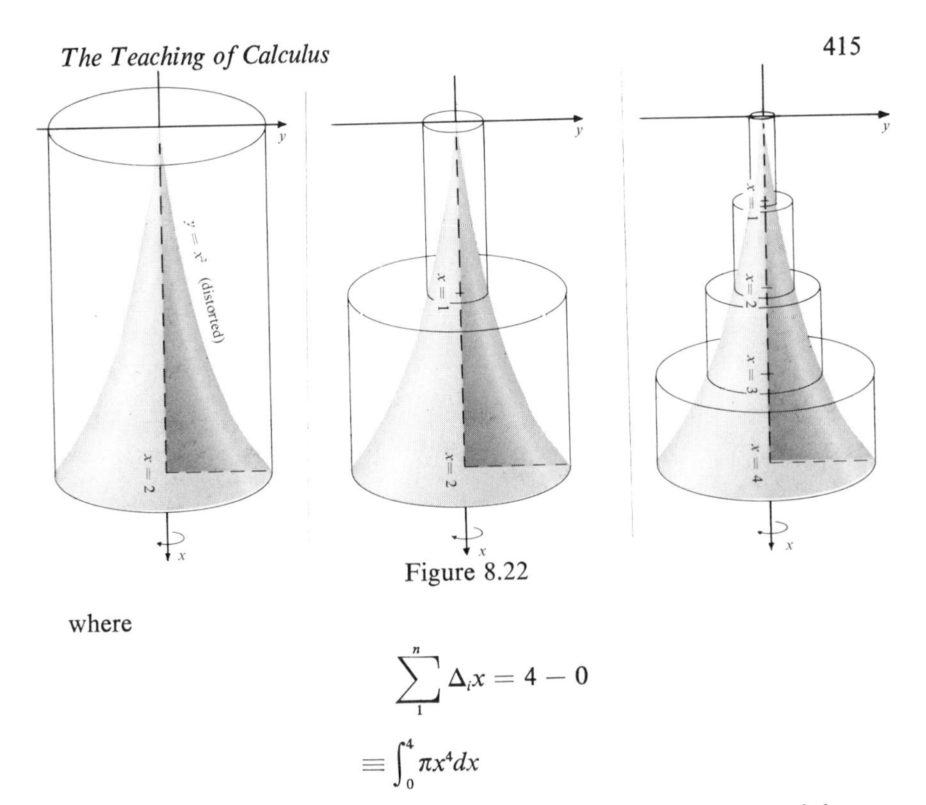

Figure 8.22

where

$$\sum_{1}^{n} \Delta_i x = 4 - 0$$

$$\equiv \int_0^4 \pi x^4 dx$$

The teacher and student-suggestion may encourage students to regard the x_i at the left-hand endpoints of the intervals. Indeed the student should learn that no matter where the x_i are taken, the limit of the sequence which it helps to define is always the same and defines the same integral.

Arriving at a number to assign to measures of volumes by setting up sequences of approximations can lead to the same encouraging depth of understanding as the treatment of area described in the foregoing pages. Such attempts, of course, lead to the double integral. A study-guide exercise may appear as follows:

In Fig. 8.23 is a portion of a surface whose equation is $z = 16 - 2x - y$. What do you suggest as a first approximation to the volume of the solid under the surface bounded by the planes $x = 4$ and $y = 4$ and in the first octant?

Students often construct a rectangular solid as shown and name $4 \cdot 4 \cdot 4 = 64$ cubic units as a first approximation.

If the region in the xy-plane bounded by $x = 4$ and $y = 4$ be partitioned into four subregions as shown, what is the approximation to the volume? Let us use the "nearest-to-the-reader-right-hand-most point" as the one to determine the height of the rectangular solid.

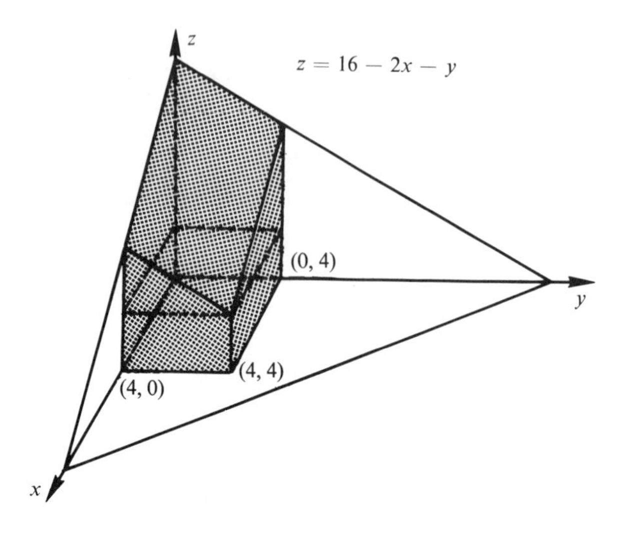

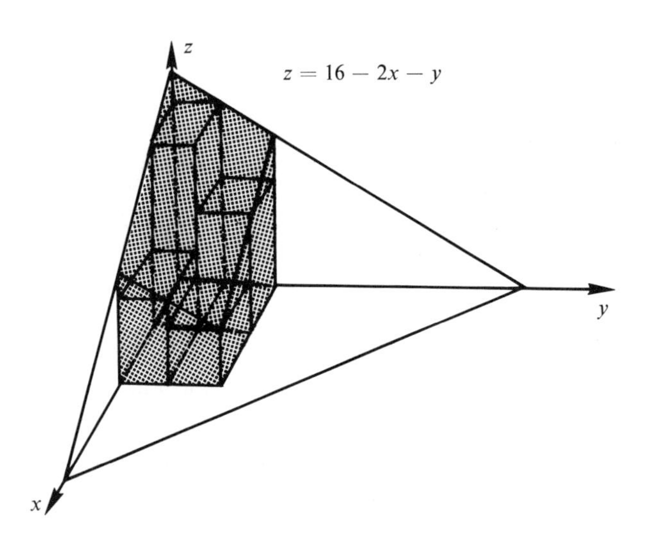

Figure 8.23

Hence a second approximation arises by using the points whose coordinates are (2, 2), (2, 4), (4, 2), (4, 4) to yield respective heights 10, 8, 6, 4 which, along with the area of 4 square units for each subregion, determine the volumes of the four rectangular solids. These measures of volumes are then added:

$$10 \cdot 4 + 8 \cdot 4 + 6 \cdot 4 + 4 \cdot 4 = 112 \text{ cubic units.}$$

A third approximation to the volume in question arises by partitioning the region $x = 0$ to $x = 4$, $y = 0$ to $y = 4$ in the xy-plane into sixteen equal subregions. One then has the tedious task of computing the volumes of sixteen rectangular solids. In this problem their sum is 136 cubic units. The performance of these meticulous and perhaps tiring calculations to compute closer approximations are parceled out to class members, but all this is worth the cost when a usable sequence is manufactured. Of course there is the challenge of setting up n^2 subregions and finally arriving at a limit of the sequence by studying the expression for the sum of the volumes of n^2 rectangular solids, and much mathematics is reviewed. Mathematics "comes to the rescue" when it is asked "Is there some kind of 'antidifferentiation process' we could perform on $16 - 2x - y$ to yield our arithmetically conjectured limit?" A discovery or a demonstration shows it can be done and we have begun another exciting and powerful chapter in the calculus, a chapter which develops and substantiates what we have started on the double integral. It seems that many such topics begun in a concrete and simple way far in advance of the time at which the teacher wishes to consider them more systematically not only gives the student a clear picture of what the question is but also helps him to anticipate what the theoretical discussion may try to support and prove. Too, he becomes a partner in attempting to place on a firm logical foundation what he himself has helped to create.

Exercises

1. Set up the nth term (n equal subdivisions) of the sequence whose limit defines the moment of area around the y-axis for the shaded portion in Fig. 8.24. Use it to evaluate the limit of the sequence.
2. Now set up the moment in Exercise 1 in terms of x_i, $\Delta_i x$, et cetera, and then the integral it defines, and compute the integral antidifferentiation.
3. Set up the nth term (n equal subdivisions) for the sequence whose limit defines the volume made by rotating the shaded portion about the x-axis. By using this term compute the limit of the sequence.
4. Compute the volume in Exercise 3 by integration.
5. In the volume problem just discussed in the text confirm that the approximation made by adding the volumes of the sixteen rectangular solids each of whose bases has an area of one square unit and whose heights are the lengths of the lines erected at (1, 1), (1, 2), (1, 3), (1, 4), (2, 1), (2, 2), (2, 3), (2, 4), (3, 1), (3, 2), (3, 3), (3, 4), (4, 1), (4, 2), (4, 3), (4, 4) is 136 cubic units.
6. Compute another approximation if you wish (64 subdivisions) and from

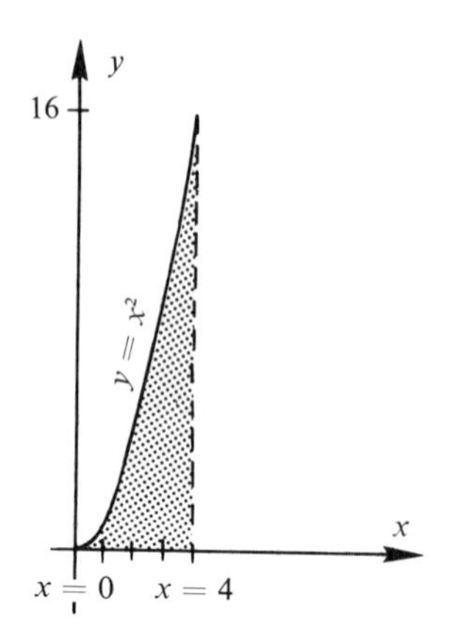

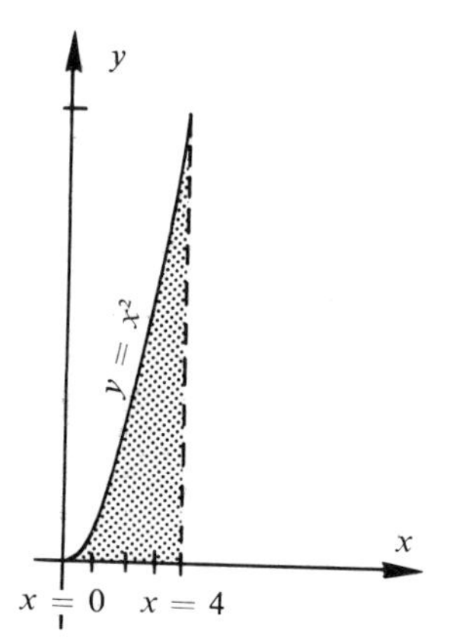

Figure 8.24

the pattern of the sequence which emerges, use $S_\infty = a/1 - r$ to suggest a number to express the measure of volume.

7. Pursue the suggestion made in the text to set up the sum of the volumes of n^2 rectangular solids with congruent bases under the plane $z = 16 - 2x - y$ and bounded by the planes $x = 0$, $x = 4$, $y = 0$, $y = 4$. Consider the *limit* of this expression as $n \to \infty$.

8. Compute the volume under the plane $z = 16 - 2x - y$ in the first octant and bounded by the planes $x = 4$ and $y = 4$ by the use of integration.

9. Try a similar approach to determine the volume in the first octant under the surface whose equation is $z = 36 - x^2 - y^2$ which is contained by the planes $x = 4$ and $y = 4$.

10. Select another topic or kind of problem in the calculus and arrange a series of exercises to invite the student to explore the idea in his own way. Furnish guiding exercises.

Student approaches to the length of a segment of a curve are interesting, too. Responses to a "bottom-of-sheet" question as "What suggestions can you make on arriving at a number to assign as the length of the segment of $y = x^2$ from $x = 0$ to $x = 4$?" may furnish well considered plans which are surprising to the teacher (Fig. 8.25). Some suggest constructing a sequence

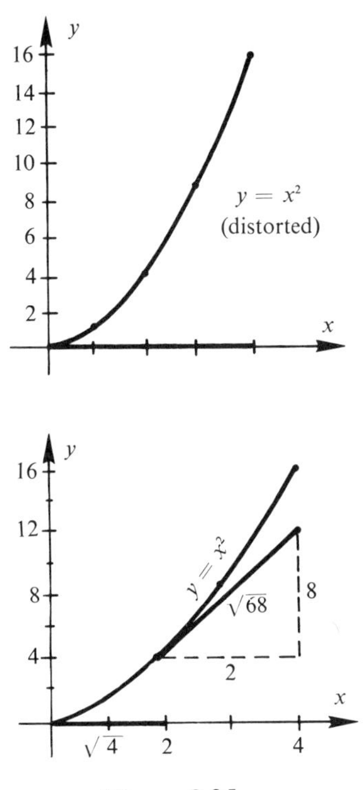

Figure 8.25

of sums of chords from P to Q and attempting to find the limit. One has suggested a method using the principle illustrated here when there are four subintervals: at $x = 0$, the slope of the curve is 0; at $x = 1$ the slope is 2. Hence at $x = 1$ the "rise over the run" is 2 and the length of the second segment tangent to the curve is $\sqrt{5}$; the length of the third segment (Fig. 8.26), found in similar way, is $\sqrt{17}$, et cetera. Hence by the use of this method the approximate length of the curve-segment when there is just one subinterval is 4, when there are two equal subintervals it is $\sqrt{4} + \sqrt{68}$, and for four equal subintervals we have $\sqrt{1} + \sqrt{5} + \sqrt{17} + \sqrt{37}$. Some of the approaches may seem not to be very fruitful but the student *is* creating, he is thinking in his own way, and he is experiencing the usual successes and failures of any one who carries on investigations. Moreover, certain facets of many approaches can be utilized and/or altered to introduce

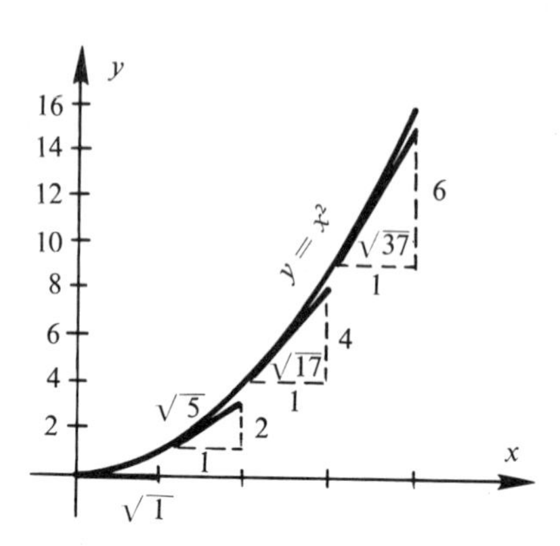

Figure 8.26

in a more natural and more familiar manner some of the more conventional ways of attacking the problem. In student-centered approaches there is little to lose but much to gain.

The power series and the experimental study of convergence can be introduced very early and in a very casual and nonassuming manner simply by asking the student to devise another name for $1/1 - x$ and study by trial for what values of x one may use $1 + x + x^2 + x^3 + x^4 + \cdots$ to approximate $1/1 - x$. The idea of interval of convergence is born and the student may accumulate much data before the formal study is begun. Indeed some of the formal study can be motivated by seeking answers and logical support to observations made on such series. Such an approach helps to provide a mathematics readiness for deeper study into which the student, all too often, is thrust immediately, sometimes with disastrous results. How interesting it is to learn later that the same series $1/1 - x = 1 + x + x^2 + x^3 + \ldots$ can be set up with the aid of *derivatives*!—the Maclaurin expansion.

Someday this question can be introduced: "Is it possible to express $x^2 + 6x + 8$ as ____ $(x - 3)^2 +$ ____ $(x - 3) +$ ____, where numbers in the blanks are to be determined?" By experimenting a bit it is found that $x^2 + 6x + 8$ can be written as $(x - 3)^2 + 12(x - 3) + 35$, but for what purpose is one interested in such new names? Such problems and discussion help the student to perceive a very good reason for pursuing the study of Taylor expansion. All of this is done through student exploration before the Taylor expansion is formally considered. Students can be invited to attempt to formulate an *algorithm* to transform the name for the expression $x^2 + 6x + 8$ into the new name in powers of $(x - 3)$ to learn later that this can be done also by the use of derivatives.

The subject of partial fractions can have an equally exciting beginning. Even in advanced algebra the student can be challenged to reverse the process of addition of two fractions. After adding $3/x + 2$ and $4/x + 6$ to get $7x + 26/x^2 + 8x + 12$ one can suggest, "I wonder if there is a way to begin with the sum and find out what fractions we added?" The student thus begins on an interesting journey into mathematics, the results of which will be very useful later. Much can be done with simple experimental beginnings to develop concepts which set the stage for subsequent consideration of more complicated matters. Topics begun in an almost imperceptible manner and without realization of the consequences can blossom slowly and clearly into most significant fundamental parts of the mathematical structure.

Exercises

1. Perhaps you can suggest another approach to the length of a segment of a curve. Try to set up a sequence of sums of lengths of line segments whose limit, it seems, if it exists, might well define the length of the curved line segment. Perhaps you can state this in the language of integrals.
2. Divide 1 by $1 + x$ to obtain an infinite series for $1/1 + x$. For what values of x can the new expression be used to yield approximations to $1/1 + x$?
3. Construct the Maclaurin series for $1/1 + x$ by the use of derivatives.
4. Work out another name for $x^2 + 7x - 12$ expressed in powers of $(x - 5)$; also in powers of $(x - 2)$; also in powers of $(x + 2)$.
5. See if you can develop an algorithm to perform the tasks requested in Exercise 4.

8.11. A Challenge to the Teacher of Calculus

The teacher who has just completed courses in analysis and in advanced calculus and who has a sense of the overwhelming beauty of the logical structure of the calculus when it is assembled in an imposing synthetic manner, may note omissions of some topics in the approaches in the preceding pages. Indeed one aim of student-centered approaches is to encourage the student to creat a logical structure *of his own*. The teacher may raise questions which will include as much material as he wishes and as the student can handle at his stage of development within the time available. Experience has shown that although the student can study and regurgitate proofs of the highest order, yet as Bruner writes,

> What is most important for teaching basic concepts is that the child be helped to pass progressively from concrete thinking to the utilization of more conceptually modes of thought[21]

[21]Jerome S. Bruner, *The Process of Education* (New York: Vintage Books, a Division of Random House, 1960), p. 38.

and that

> ... it is worth the effort to provide the growing child with problems that tempt him into the next stage of development.[22]

How well these thoughts, as well as researches on this method of teaching, urge approaches which may begin with the simple limits of slopes of secants but end with scholarly N, ϵ-definitions of convergence of a sequence. How promising, and in accord with thoughts of those who have much concern for basic understanding, is the approach which begins with concrete numerical approximations to the area under a curve, which may well lead to sequences of "upper sums," sequences of "lower sums," to the possible existence of a common limit, and to the precise definition of the Riemann integral along with the ϵ, δ definition of the "limit of a function." It is the decision of the teacher to determine just how far he and his class can go and it is the ingenuity and interest of the teacher on how he can help his students experience, generalize, summarize, and organize findings of their common effort into a synthesis at the level to which both teacher and students aspire.

The approaches discussed above have been found to be effective in classroom practice. Topics not mentioned may be treated similarly. Almost any item can be introduced in a very concrete way at any time. "... any subject," says Bruner, "can be taught effectively in some intellectually honest form to any child at any stage of development"[23] and, for example, the seeds leading to definition and computation of volumes by the use of the triple integral can be shown at most any time with introductory numerical work. Such introductions encourage the student to use his "characteristic way of viewing the world and explaining it to himself"[24] and such exploratory experiences started far in advance of the "textbook time schedule," as it has been repeatedly emphasized, provide a natural setting and backdrop for increasingly learned and mature discussion of the subject and finally the triple integral is born. Such an approach may use the treatment in the text as a crowning point of climax. Contrast this with the directive to "study the next several pages and work the problems." Indeed, some of the very beautiful facets of calculus which the reader may feel have not been treated above are the very topics for which the student is made ready through the approaches described. Our aim has been to urge the the use of a kind of classroom methodology designed to produce a general understanding of the structure of the calculus. On this scaffolding the individual teacher may continue to plan experiences so that the student, with his own contribution, with study of the writings of others, and with classroom student-teacher

---

[22] *Ibid.*, p. 39.
[23] *Ibid.*, p. 33.
[24] *Loc. cit.*

discussion will see how these topics are all related and hence will be able to reflect upon his results with confidence and pleasure. This most powerful and useful tool which serves in an outward-looking role can be made as much as the teacher and student desire into a beautifully strong and logically coherent structure by giving attention to the inward-looking face of mathematics.

8.12. Some Sample Study Guides

Several study sheets selected from those prepared in connection with a recent course in the calculus appear as follows:

Review and More on Areas

1. *Review.* Compute

   (a) $D_x^{-1}(x\sqrt{x^2+6})$ (b) $D_x^{-1}(2x^2\sqrt{x^3+3})$ (c) $D_x^{-1}\dfrac{2x}{\sqrt{x^2+4}}$

2. *Review.* Use $S_\infty = a/1 - r$ to find the limits of these sequences:

   (a) 4, 8, 10, 11, $11\frac{1}{2}$, $\cdots$ (b) 12, 8, 6, 5, $\cdots$

3. *Review.* In a previous exercise you arrived at a sequence whose limit, we agree, will define the measure of the area under the curve $y = x^2 + 3x$ from $x = 0$ to $x = a$. Use the method of Problem 2 just above to arrive at the limit.

4. *Review.* In Problem 3 above, were we to be finding the number to assign to the area under $y = x^2 + 3x$ from $x = 0$ to $x = b$ what would that number be?

5. Now if $a < b$ then the area under $y = x^2 + 3x$ from $x = a$ to $x = b$ would be assigned what number?

6. Hence if $a = 4$ and $b = 6$ what is the number assigned as the measure of the area?

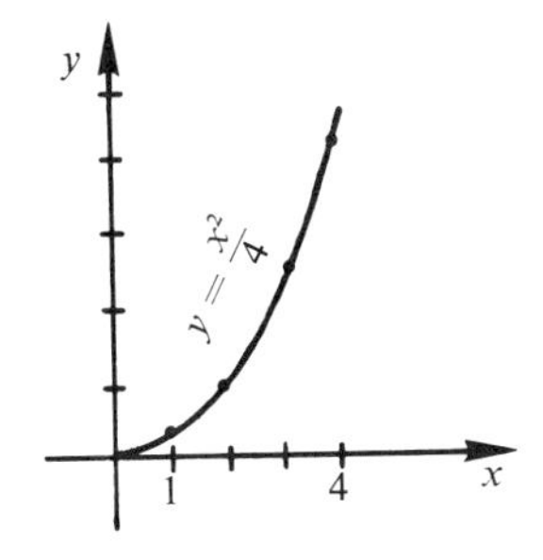

7. Suggest another method to arrive at a sequence whose limit we would expect to define a number to assign as the measure of area. Let the curve be the one whose equation is $y = x^2/4$ and let the bounds be $x = 0$ and $x = 4$. Show the work which you do (refer to figure on page 423).

8. Use the five-step rule to show that if $y = f(x) \cdot g(x)$, then $D_x y = f(x) \cdot g'(x) + f'(x) \cdot g(x)$.

9. A rectangular field is to have an area of 400 square rods. The "lengthwise" fences cost \$10 per rod and the "widthwise" fences cost \$6 per rod. What dimensions will make the cost of fencing the least?

10. *Quick Review.*

    (a) $6^0 =$ ____ (b) $36^{-\frac{1}{2}} =$ ____ (c) $36^{-\frac{3}{2}} =$ ____

    (d) $(25x^0)^{-\frac{1}{2}}$ ____ $=$ (e) $(a + b)^{\frac{1}{2}}$ ____ (first four terms)

Review and More on Areas

1. *Review.* Compute.

   (a) $D_x^{-1}(2\sqrt{4x+3})$ (b) $D_x^{-1}(3\sqrt{4x+3})$ (c) $D_x^{-1}\dfrac{x+3}{\sqrt{x^2+6x+10}}$

2. *Review.* We have *defined* the area under a curve to be the *limit* of a ____ of ____ of ____ as ____.

3. *Review.* It turns out that it seems reasonable to define the area (measure) under $y = x^2$ between $x = 0$ and $x = a$ (and above the x-axis) as ____; between $x = a$ and $x = b$ $(a < b)$ as ____.

4. *Review.* It turns out that it seems reasonable to define the area under the curve whose equation is $y = x^2 + 3x$ between $x = a$ and $x = b$ $(a < b)$ to be ____. Indeed it seems also that we can obtain this result *quickly* by ____ the expression $x^2 + 3x$ and ____. Hence we might make the conjecture that the ____ is the same as ____.

5. Use the conjecture reviewed in Problem 4 to write an expression which we would expect to define the area measure number for areas under each of these curves (and above the x-axis) between $x = a$ and $x = b$.

   (a) $y = x^2 + 7x + 10$ (b) $y = 5x^2 + 3x + 12$
   (c) $y = x^3 + 4x^2 + 6x + 15$

6. *Text.* Use the conjecture of Problem 4.

7. Let us use rectangles *again* to study the area problem but this time let us set up the sum of not one, or two, or four, or eight rectangles, but the sum of n rectangles, immediately. Also, this time, let us use

rectangles which extend out over the curve $y = x^2$ as shown in the figure.

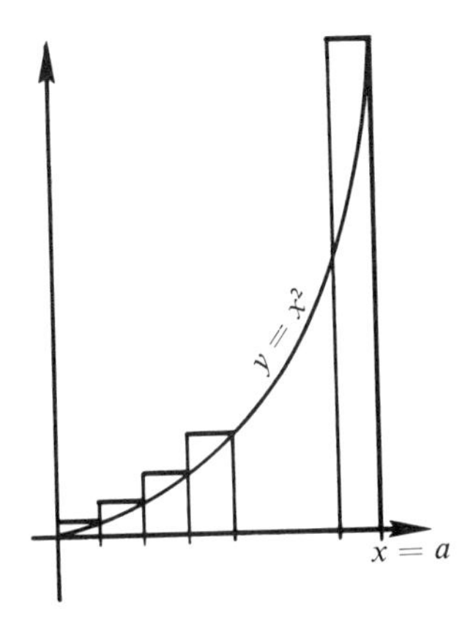

(a) Each subdivision has what length on the x-axis?

(b) Write the abscissae of each of the partitioning point. They

$$\frac{a}{n}, \frac{2a}{n}, \frac{3a}{n}, \_\_\_ \_\_\_ \_\_\_ \_\_\_ \_\_\_ \cdots \frac{na}{n} = a$$

(c) Compute the height of the first rectangle. (Ans. a^2/n^2) Now compute the height of the next, then the next et cetera.

(d) Write the *sum* of the area of these rectangles. Write as

$$S = \frac{a^3}{n^3} + \_\_\_ + \_\_\_ + \_\_\_ + \_\_\_ + \ldots + \_\_\_.$$

(e) See if you can write the expression for S in a simpler and more compact way. Recall that

$$1^2 + 2^2 + 3^2 + 4^2 + 5^2 \neq \cdots + i^2 = \frac{i(i + 1)(2i + 1)}{6}$$

(f) Now what happens to the value of your answer in (e) as n becomes larger and larger?

(g) Hence again what number is suggested to be used as the definition of the measure number for area under $y = x^2$ from $x = 0$ to $x = a$ (and above the x-axis)?

8. What is the meaning of the expression in (e) for $n = 1$? for $n = 2$? for $n = 3$? for $n = 4$?

9. Hence as n increases we really do have a _______ of numbers the *limit* of which we have defined as the _______.

10. The line PQ is assumed perpendicular to the line whose equation is

$2x + y = 4$. Compute the *length* of the segment PQ if P has coordinates (3, 4).

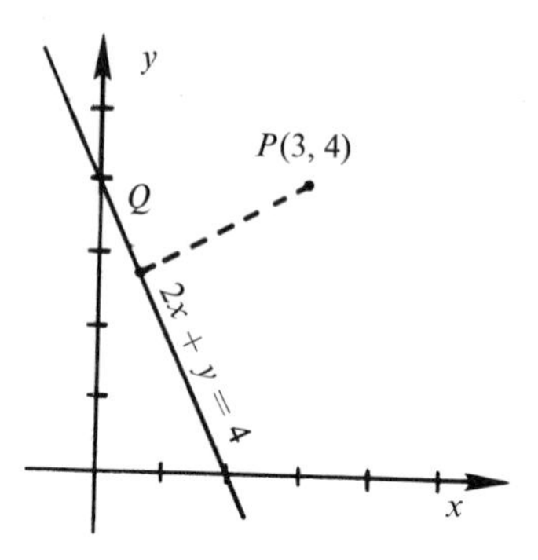

Review. Derivative of sin X

1. *Review.* A certain spring has a rather complicated force-stretch formula, namely $f = 12s^2 + 3s$. What work is done in stretching it from a stretch of 0 inches to one of 3 inches? Let f be expressed in pounds.
2. *Review.* Every problem which involves a "limit of sequence of sums of products" might show promise of being treated by means of the ________.
3. In the cistern in the drawing each thin slice of water is regarded as being lifted to the top. What is the *work done* for the slice which is 7 feet in radius and which is $\Delta_i h$ feet thick if water weighs 64 pounds per cubic foot? The slice is lifted through h_i feet. The *total work* can be defined as what number? Can this be expressed as an integral?

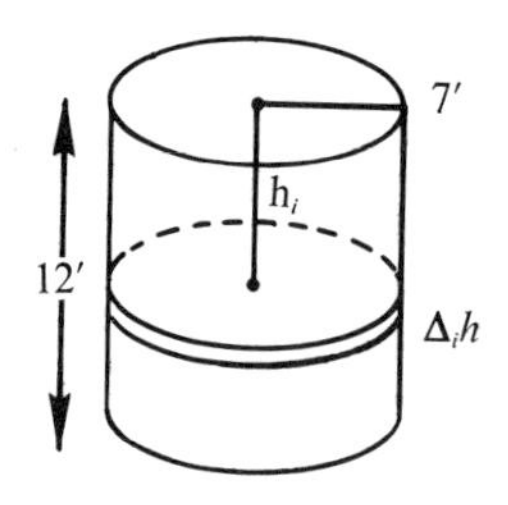

4. The conjecture from our work on the graph of $(x, \sin x)$ is that $D_x \sin x = \cos x$. Hence $D_x (2 \sin x) = 2 \cos x$, while $D_x \sin 2x = 2 \cos 2x$ (chain rule). Write the derivatives of these expressions. Use the chain rule where it is applicable.

(a) $\sin 3x$	(f) $x^2 + (\sin 5x)^3$
(b) $\sin 4x$	(g) $x^3 + (\sin 7x)^4$
(c) $(\sin x)^3$	(h) $3x^2 + (1 + \sin x)^2$
(d) $(\sin 4x)^2$	(i) $3x^2 + \sqrt{(\sin x)^3}$
(e) $\sin x + x$	(j) $\sqrt{1 + (\sin x)^2}$

5. Try to develop the derivative of $y = \sin x$ at $x = a$ by the five-step rule for computing derivatives. Compute Δy, $\Delta y/h$, $\lim_{h \to 0} \Delta y/h$.

6. *Assuming* that $D_x \sin x = \cos x$ and that $\cos x = \sqrt{1 - \sin^2 x}$ compute $D_x \cos x$.

7. Assuming that $D_x \cos x = -\sin x$ compute $D_x y$ for each of these expressions (use the chain rule):

   (a) $y = 3 \cos x$ (b) $y = 3(\cos x)^2$
   (c) $y = 4 + 3 (\cos x)^2$ (d) $y = \sqrt{1 - \cos^2 x}$

8. Use $D_x \sin x = \cos x$ and $D_x \cos x = -\sin x$ and the *quotient rule* to show that $D_x \tan x = \sec^2 x$.

9. Hence compute $D_x y$ for each to these expressions.

   (a) $y = \tan 3x$ (b) $y = \tan 4x$ (c) $y = (\tan x)^2$
   (d) $y = 3 \tan^2 x$ (e) $y = 3 \sin x + \tan x$ (f) $y = \tan x \sin x$

10. Using $D_x \tan x = \sec^2 x$ develop an expression for $D_x \cot x$.

Bibliography

Aeberly, J. J., "Calculus: a Trigonometric Procedure," *School, Science and Mathematics*, LVIII (January, 1958), pp. 44-52.

Allendoerfer, C. B., "Case Against Calculus," *The Mathematics Teacher*, LVI (November, 1963), pp. 482-5.

Blank, A. A., "Remarks on the Teaching of Calculus in the Secondary School," *The Mathematics Teacher*, LIII (November, 1960), pp. 537-9.

Buchanan, O. L., "Opinions of College Teachers of Mathematics Regarding Content of the Twelfth-Year Course in Mathematics," *The Mathematics Teacher*, LVIII (March, 1965), pp. 223-5.

Cummins, Kenneth. "A Student Experience-Discovery Approach to the Teaching of Calculus," *The Mathematics Teacher*, LIII (March, 1960) pp. 162-70.

Ferguson, W. E., "Calculus in the High School," *The Mathematics Teacher*, LIII (October, 1960), pp. 451-3.

Francis, R. L., "Placement Study in Analytic Geometry and Calculus," *Educational and Psychological Measurement*, XXVI (Winter, 1966), pp. 1041-6.

Graesser, R. F., "Note on the Fundamental Theorem of Integral Calculus," *School, Science and Mathematics*, LXIV (June, 1964), pp. 523-4.

Grossman, G., "Advanced Placement Mathematics," *High Points*, XLIII (October, 1961), pp. 22-3.

Hight, D. W., "Limit Concept in the SMSG Revised Sample Textbooks," *The Mathematics Teacher*, LVII (April, 1964), pp. 194-9.

Holden, L. S., "Providing Motivation for the Chain Rule," *The Mathematics Teacher*, LX (December, 1967), pp. 850-5.

Karst, O. J., "Limit," *The Mathematics Teacher*, LI (October, 1958), pp. 443-9.

Kucinski, R. A., "Introduction to Calculus for Junior High School Students," *The Mathematics Teacher*, LII (April, 1959), pp. 250-5.

Laper, L. M., "It's Time to Review Calculus," *School, Science and Mathematics*, LXIV (March, 1964), p. 202.

Pieters, R. S., and E. P. Vance, "Advanced Placement Program in Mathematics," *The Mathematics Teacher*, LIV (April, 1961) pp. 201-8.

Smith, L. T., "Could we Teach Limits?" *The Mathematics Teacher*, LIV (May, 1961), pp. 344-5.

Steinbrenner, A. H., "Teaching Aid For the Generalized Law of the Mean," *School Science and Mathematics*, LVIII (January, 1958), p. 9.

Stretton, W. C., "Straight-Line Tunnels Through the Earth." *The Mathematics Teacher*, LX (January, 1967), pp. 12-13.

Viertel, W. K., "Visual Aid for an Elementary Applied Maximum Problem in Calculus," *The Mathematics Teacher*, LXI (January, 1968), p. 29.

Williams, H. E., "Demonstration of Indeterminate Forms Using Finite Methods," *The Mathematics Teacher*, LVII (December, 1964), pp. 537-8.

INDEX

INDEX

A fortiori, 222
Aaboe, Asger, 202
Abacus, 99-101
Absolute geometry, 271
Ad hominem, 222
Addition,
  integers, 182-184
  physical aids, 99-105
  rationals, 132-135
Additive inverse, 157
Adler, Irving, 12, 85, 89, 92, 145, 149
Aeberly, J. J., 427
Affine geometry, 275
Affine transformation, 281
Ahmes papyrus, 201
Albaugh, A. H., 309
Algebra,
  points of view, 149-151
  study guide, 70
Allendoerfer, C. B., 427
Altshiller, Court N., 310
Analogy, 222
Analytic teaching approach, 18-26
Archimedes, 212
Area problem, 398-404, 423-426
Argumentation, 219
Arithmetic,
  curricula, 84-91
  defined, 82-83
Arrowsmith, William, 79, 221
Art, 10
Atkins, R. A., 12
Auer, A., 43
Avers, P. W., 310

Bacon, Marjorie, 145
Baker, B. L., 311, 367
Ball, W. W. Rouse, 209
Ballew, H., 198
Banks, M. S., 310
Barker, Stephen F., 226
Barnett, I. A., 198
Base five, 128-129
Bassham, Harrell, 91
Bassler, O. C., 310
Baumgartner, W. S., 12
Beaumont, Andrie, 145
Beberman, Max, 58
Bedford, Fred L., 312
Bell, Clifford, 12, 367
Bengtson, Ray, 43
Bhushan, V., 198
Bierberbach, Ludwig, 261
Biggs, John, 48
Binary system, 129
Binomials, 195-197
Biological sciences, 8
Birkhoff, George D., 287
Black, J. M., 310
Blank, A. A., 427
Bolyai, 270, 271
Bowen, J. J., 43
Bowie, H. E., 43
Bridges, F., 311
Brinkmann, E. H., 310
Brown, J. A., 45
Brown, O. R., 310
Brumfiel, Charles F., 300
Bruner, Jerome S., 421
Buchanan, O. L., 427
Buck, R. C., 315
Bulletin boards, 5
Burington, Richard S., 12
Butler, Charles H., 12, 43, 145, 198, 238, 310
Byrne, M. M., 310

Cain, R. W., 12
Cairns, Stewart Scott, 12
Calculus,
  challenge to teacher, 421-423
  student-centered approaches, 413-421
  study guide, 72
Callanan, C., 199
Cambridge Conference on School Mathematics, 88-89, 263, 299
Carrol, E. C., 310
Casmir, F. L., 79
Chaney, J. M., 310
Chemistry, 8
Chirko, Thomas, 145
Christofferson, Halbert C., 43, 310

Circulo probando, 222
Clark, John R., 17, 23, 32
Clarkson, D. M., 310
Cleveland, R. W., 367
Clifford, Edward L., 13
Clock arithmetic, 130-132
Cochran, B. S., 145
Coffin, Joseph G., 261
College Geometry Project of the University of Minnesota, 303
Coltharp, F. L., 310
Committee on the Undergraduate Program in Mathematics, 307
Complete contrapositive, 233
Computer, 74
Conant, Levi Leonard, 10, 127
Conclusion, 218
Condron, B. F., 310
Conics, 343, 352
Constructive methods, 222
Contrapositive method, 226, 230
Converses, 234-238, 247
Cook, F. S., 78
Coordinate geometry, 290-298, 349-353
Counting, 97
Courant, Richard, 396
Coxeter, H. M. S., 271, 299, 304
Coxford, A., 199
Coyle, F., 43
Creativity, 49-50
Crescimbeni, J., 79
Crist, R. L., 145
Cross-staff, 241-244
Cummins, Kenneth B., 42, 411, 427
Cunningham, G. C., 145
Curves, 326-331
Cylindrical projection, 266

Dadourian, H. M., 43
Dale, Edgar, 239, 271
Dantzing, Tohias, 96
Davis, David R., 43, 199, 310
Davis, E., 310
Dead reckoning, 257
Decomposition methods, 115-116
Deductive teaching approach, 26-30
Deductive thinking, 218-226
Definite integral, 407-410
Denbow, Carl H., 58
Destructive methods, 222
Developmental approach, 58-63
DeWitt, M., 312
Dilations, 336
Discovery, 48-52
Discovery teaching approach, 41-42, 47-79
Distributive law, 162
Divisibility, 125-126
Division, 117-122
  rationals, 136-139
Dogmatic teaching approach, 30-32
Dunnrankin, Peter, 43, 310
Duodecimal system, 130
Durell, Fletcher, 397

Edwards, R., 367
Egyptian level, 239-241
Eicholz, Robert E., 300
Ellipse, 343
Equal additions method, 116-117
Equations, 152-165
  quadratic, 197-198
Equiareal transformation, 280
Equivalent open sentences, 161
Euclid, 210, 269-272, 299
Euclidean transformation, 280
Eudoxus, 210
Euler line, 298
Eves, Howard, 270, 271, 272, 299, 311
Exhibits, 6
Exton, E., 78

Fallacies in argument, 222
Fawcett, Harold P., 2, 42, 151, 311
Fearnside, W. Ward, 225
Fehr, Howard F., 17, 23, 32, 91, 215
Ferguson, W. E., 427
Field approach to equations, 157-165
Fine, Nathan J., 357
Finite arithmetics, 130-132
Fischer, I., 13
Fleckman, B., 79
Fletcher, T. J., 74
Forbes, J. E., 367
Foster, B. L., 199
Francis, R. L., 427
Frazier, A., 79
Fremont, H., 145

Friedland, A., 199

Gager, W. A., 44
Gagne, Robert N., 44
Galileo, 35
Gemignani, M. C., 311
Generalization, 222
Genetic teaching approach, 32-35
Geometry,
absolute, 271
affine, 275
classified by transformations, 278-284
coordinate, 290-298, 349-353
history of, 201-203
in space, 263-269
logical beginnings, 207-218
postulational basis, 284-290
projective, 275
study guide, 71
use of applications, 239-258
vector, 258-263
Geyser, G. W. P., 145
Gibby, W. A., 18
Gillcrist, William A., Jr., 145
Gillings, R. J., 202
Glennon, V. J., 44, 146
Glicksman, A. M., 199, 311
Gnomonic projection, 265
Goedicke, Victor, 58
Gold, S., 79
Graesser, R. F., 428
Graustein, William C., 291
Greater Cleveland Mathematics Group, 88
Griffin, H., 79
Groenendyk, E., 44
Grosch, H. R. J., 13
Grossman, G., 428
Growth, 50
Guess and check method, 169
Guggenbuhl, Laura, 13, 311
Gurau, P. K., 199, 311

Hamilton, E. W., 146
Hamilton, H., 80
Hammer, Preston, 13
Harmeling, Henry, 44
Harris, E. M., 311
Harwood, E. H., 44
Heimer, R. T., 79
Henderson, K. B., 44
Hendrix, Gertrude, 49, 79
Hesser, F. M., 311
Heuristic teaching approach, 32-35
Hewitt, F., 311
Hight, D. W., 428
Hilbert, David, 284-285, 287
Hippocrates, 210
Hirschi, L. Edwin, 79
History, 9
Hohn, Franz F., 47
Holcomb, John D., 44
Holden, Lyman S., 391, 428
Holther, William B., 225
Horton, G. W., 199
Houston, J., 79
Humphrey, J. H., 146
Hutchinson, C. A., 50
Hydrographic Office, 311
Hyperbola, 343

Ideal point, 244
"Improving Mathematics Instruction," 44
Indirect proofs, 226-234
Induction, 353-361
Inductive sets, 357
Inductive teaching approach, 26-30
Integers, 178-189
Intuitive approaches to multiplication, 187-189
Intuitive experimental point of view, 84-85
Invention, 49
Inverses, 234-238
Involution, 301
Isaacs, Nathan, 93
Ivey, John F., Jr., 50

Jackson, Humphrey, 51
Jackson, R., 44
Johnson, Donovan A., 5, 44, 79, 146, 199, 311
Jones, Phillip S., 13

Kadushin, I., 13
Kant, Emmanuel, 41
Kaplan, Jerome D., 80
Karst, O. J., 428
Kattsoff, L. O., 311
Keiber, Mae Howell, 44

Kelly, Paul J., 303
Kemeny, J. G., 44
Kennedy, E. S., 13
Kennedy, Joseph, 146
Kersh, B. Y., 80
Keyser, Cassius Jackson, 67, 92, 211, 270, 291
Kinney, Lucien B., 13
Klein, Felix, 42
Kline, Morris, 13, 270
Klinkerman, G., 311
Kluttz, M., 44
Kovach, L. D., 44
Kucinski, R. A., 428
Kvaraceus, W. C., 78

Laboratory teaching approach, 35-40
Ladd, Norman E., 303
Lambert, 270
Land, F. W., 89
Landon, M. W., 13
Langer, S. K., 199
Languages, 9
Lankford, Francis G., 13, 31, 32, 34, 44
Lansdown, Brenda, 51
Laper, L. M., 428
Latitude, 249-252
Law of contradiction, 227
Law of excluded middle, 226
Law of identity, 226
Laycock, Mary, 44
Lazar, Nathan, 230, 238
Least integer principle, 357
Lecture teaching approach, 32-35
Lee, E. C., 12
Lee, Everett S., 13
Legendre, 270
Leonard, W. A., 199
Lerch, Harold, 51, 80
Levi, Howard, 83
Levin, G. R., 311
Lewis, E., 44
Lichtenberg, Donovan, 9
Line,
  segment, 316-318
  slope of, 319-321
Lloyd, D. B., 199
Lobachevsky, 270, 271
Loeb, A. L., 311
Logarithms, 189-193
Love, M. G., 311
Lowry, W. C., 45
Lysaught, J. J., 80

Macarow, L., 45
MacDonald, I. D., 311
Major, J. E., 312
Major premise, 218
Mallory, C., 199
Manning, H. P., 271
Mark, S. J., 13
Marks, John L., 80, 199, 312
Mathematical induction, 29
Mathematics Staff of the University of Chicago, 13
"Mathematics to the rescue," 63-67
Matrices, 321-325
Mauro, C., 146
Maxwell, E. A., 125, 286
May, K. O., 80
May, L. J., 146, 312
Mayor, J. R., 45
McCreery, L., 199
McDonald, I. A., 199
McKinney, W. M., 13
McNabb, W. K., 302
Meland, Bernard Eugene, 67
Mercator projection, 267
Meserve, Bruce E., 275, 278
Methodology, 10-11
Metzner, S., 146
Midpointness, 277
Miller, E. E., 80
Mills, C. N., 312
Milne, W. P., 291
Miniature arithmetics, 130-132
Minor premise, 218
Mock, G. D., 45
Models, 6-7
Modenov, P. S., 281
Modular arithmetics, 130-132
Modus ponens, 223
Modus tollens, 232
Moise, E. E., 45
Moore, E. H., 41, 209
Moorman, R. H., 13
Morton, R. L., 146
Moser, J. M., 199, 312
Mucci, Joseph F., 14

Multiplication,
binomials, 195-197
integers, 180-181, 185-187
intuitive approaches, 187-189
matrices, 321-325
natural numbers, 105-107
rationals, 135-136
table, 139
Murnaghan, F. D., 369, 371
Murphy, Katherine, 91
Murphy, Michael, 91

Nannini, A., 312
National Council of Teachers of Mathematics, 4, 14, 29, 36, 41, 75, 87, 88, 146, 199, 212, 312, 357, 378, 404
Natural number system, 93-132
Navigation, 252-258
Nemeck, P. M., 45
Neugebauer, O., 202
Neureiter, P. R., 312
New age in mathematics, 74-77
Newman, James R., 122, 149, 202, 223
Newsom, Carroll V., 270, 271, 272
Niswonger, Dan, 121
Non sequitur, 222
Nulton, Luch, 51
Number-line approach to integers, 181-187
Numeration systems, 126-130
Nypaver, Elizabeth, 353

Oda, Linda, 313
O'Donnell, J. R., 146, 200
Olberg, R., 146
"On the Mathematics Curriculum of the High School," 12, 216
Open phrase, 174
Open-sentence approaches to equations, 156
Ordered-pair-of-natural-number approach to integers, 178-181
Organization for Economic Co-operation and Development, 89
Organization for European Economic Cooperation, 370
Organization of materials, 67-68

Pack, E., 14
Pangeometry, 271
Papy, F., 303
Parabola, 343
Parallelism, 343
Parallelogram law, 252-258
Parallelograms, 309
Parkhomenko, A. S., 281
Partial contrapositive, 233
Partitioning, 118
Payne, Joseph N., 45
Peck, Lyman C., 121
Peirce, C. S., 35
Percentage, 141-142
Perham, A., 45
Perry, John, 41
Perspective drawing, 244-249
Peterson, David, 15
Phillips, H. L., 45
Phillips, J. M., 146
Physics, 8
Pictorial beginnings to equations, 152-156
Pieters, R. S., 428
Piwnicki, F., 312
Plato, 210
Playfair postulate, 271, 274
Polya, George, 42, 49, 80, 365, 385, 396
Postulational approach, 165-167, 188
Powers, L. S., 14
Prenowitz, Walter, 291
Projection, 341-345
Projective geometry, 275
Projective transformation, 381
Proofs, 50, 214, 284
Psychological teaching approach, 30-32
Ptolemy, 210
Purdy, C. Richard, 13
Pythagoras, 210

Quadratic equations, 197-198
Quadratic expressions, 365-366
Quinimal forms, 129
Quotitioning, 118

Ranucci, E. R., 45, 367
Rasmussen, Othom, 14
Rate of change, 371-381, 386-394
Rationals, 132-140

Read, C. B., 45, 200
Rectangle, 143-144
Reducto ad absurdum, 222, 226-232
Rees, Mina, 14
Reeve, W. D., 45
Reflection, 302
Report of the Harvard Committee, 216
Resultant, 252
Rettaliata, J. T., 14
Rhind papyrus, 201
Riemann, 270, 271
Rising, Gerald R., 146, 199, 311
Roberts, J. B., 363
Robinson, Donald W., 79
Rosenbaum, R. A., 14
Rosenberg, Herman, 14, 312
Ross, Ramon, 146
Rosskopf, M. F., 14
Ruchlis, H., 14
Ruderman, H. D., 311
Rummell, F. V., 45
Runkel, Philip J., 14

Saccheri, 270, 271, 272-273
Salzen, R. T., 80
Salzer, R. L., 146
Sanford, Vera, 202
Sawyer, W. W., 29
Saxelby, F. M., 351, 371, 396
Schaaf, Oscar, 42
Schaaf, William L., 14
Schaff, M. E., 312
Scheid, Francis, 15
School Mathematics Study Group, 88, 300
Schultze, Arthur, 18, 22
Schuster, Carl N., 312
Schuster, Seymour, 261
'Scratch" method, 108-110
Secants, 308
Series of skills approach to teaching, 86-91
Sganga, F. T., 146
Shanks, Merrill E., 300
Similarity transformation, 281
Simson line, 297
Situation approach, 52-58
Skinner, B. F., 80
Slook, Thomas H., 15
Slope of a line, 319-321
Slope on a curve, 381-386
Smart, James R., 80, 312
Smiley, Charles H., 15
Smith, David Eugene, 1, 4, 20, 223, 286
Smith, L. B., 312
Smith, L. T., 428
Smith, M. D., 80
Smith, W., 80
Snow, John T., 147
Snyder, Henry D., 80, 200
Sobel, Max A., 42
Social science, 9
Speech, 8
Speed of a body, 371-381
Springer, C. F., 15
Springs, 404-406
Stein, H. L., 45
Steinbrenner, A. H., 428
Stover, Donald W., 15
Strehler, Allen F., 147
Stretton, William C., 15, 428
Struthers, Joseph A., 147
Study Guides, 68-74, 141, 195-198, 307-309, 365-367, 423-427
Substitution, 174
Subtraction, 110-113
  integers, 184-185
Successor sets, 357
Sum of two natural numbers, 97-99
Sundara, Rao, 36
Sweet, Raymond, 43, 310, 312
Swineford, Edwin J., 15
Syllogism, 218-226
Synthetic teaching approach, 18-26
System-emphasis point of view, 85-86
Szabe, Steven, 80, 313

Tangents, 308
Tautology, 223
Teaching, 77-79
  points of view, 91-92
Teaching aproaches, 361-365
  analytic, 18-26
  deductive, 26-30
  discovery, 41-42, 47-79
  dogmatic, 30-32
  genetic, 32-35
  heuristic, 32-35
  inductive, 26-30

laboratory, 35-40
lecture, 32-35
psychological, 30-32
synthetic, 18-26
Teller, Edward, 15
Thales, 209
Theorems, 409-413
Thompson, R. A., 15
Thorndike, Robert L., 67
Time factor, 68-74
Todd, J., 45
Todd, R. M., 313
Topology, 275
Townsend, Myrtle, 147
Transformations,
formulae, 331-336
interpretations, 336-341
Translation, 333
Triangle, 307
Trigg, C. W., 313
Trimble, A. C., 15
Tulock, M. K., 45
Two-point converses, 247

United States Navy Department, 9
University of Illinois Committee on School Mathematics, 87
University of Maryland Mathematics Project, 87

Vacca, G., 353
Vance, E. P., 428
Van Engen, Henry, 45, 147, 367
Varberg, Dale E., 15
Vaughan, Herbert F., 58
Vector, 309
defined, 258
Vector geometry, 258-263
Viertel, W. K., 428
Vigilante, Nicholas J., 288

Wahl, M. S., 313
Wallin, Don, 15
Walter, M., 313
Wehrman, K., 80, 147
Well-ordering axiom, 357
Welsing, W. C., 147
Welson, J. W., 147
Wentworth, George, 223, 286
Wernik, W., 313
Whitman, N. C., 313
Whitemore, Edward, 274
Wick, J. W., 15
Wilansky, Albert, 15
Wildermuth, Karl P., 15
Wilks, S. S., 15
Willerding, M. F., 15
Williams, H. E., 428
Williams, J. D., 147
Wilson, Raymond H., Jr., 16
Winthrop, H., 16, 147
Wiseman, John D., Jr., 238
Wolf, A., 225
Wolff, Peter, 286
Word problems, 167-177
Wren, F. Lynwood, 12, 16, 43, 145, 198, 238, 310
Wright, A. L., 313
Wright, Frank, 80, 200

Young, J. W. A., 18, 63, 212, 385

Zero, 122-125, 345-349
Zweng, Marilyn, 9